THE COLLECTORS OF LOST SOULS

The Collectors of

Lost Souls

TURNING KURU SCIENTISTS INTO WHITEMEN

Warwick Anderson

The Johns Hopkins University Press

BALTIMORE

© 2008 The Johns Hopkins University Press
All rights reserved. Published 2008
Printed in the United States of America on acid-free paper
9 8 7 6 5 4 3 2 1

The Johns Hopkins University Press
2715 North Charles Street
Baltimore, Maryland 21218-4363
www.press.jhu.edu

Library of Congress Cataloging-in-Publication Data

Anderson, Warwick, 1958–
The collectors of lost souls : turning kuru scientists into whitemen /
Warwick Anderson.
p. ; cm.
Includes bibliographical references and index.
ISBN-13: 978-0-8018-9040-6 (hardcover : alk. paper)
ISBN-10: 0-8018-9040-3 (hardcover : alk. paper)
1. Kuru—Papua New Guinea—History. 2. Gajdusek,
D. Carleton (Daniel Carleton), 1923– 3. Virologists—United
States—Biography. 4. Neurologists—United States—Biography.
5. Fore (Papua New Guinean people) —Diseases. I. Title.
[DNLM: 1. Gajdusek, D. Carleton (Daniel Carleton), 1923–
2. Kuru—history—Papua New Guinea. 3. Biomedical
Research—history—Papua New Guinea. 4. Cannibalism—
Papua New Guinea. 5. Funeral Rites—Papua New Guinea.
6. History, 20th Century—Papua New Guinea. WC 540 A552C
2008]
RC394.K8A53 2008
616.80092—dc22   2008007840

A catalog record for this book is available from the British Library.

*Special discounts are available for bulk purchases of this book. For
more information, please contact Special Sales at 410-516-6936 or
specialsales@press.jhu.edu.*

The Johns Hopkins University Press uses environmentally friendly
book materials, including recycled text paper that is composed of
at least 30 percent post-consumer waste, whenever possible. All
of our book papers are acid-free, and our jackets and covers are
printed on paper with recycled content.

For Adele E. Clarke

CONTENTS

THE COLLECTORS OF LOST SOULS

THE DISEASE EUROPEANS
CATCH FROM KURU

IN 1965 WHILE STUCK IN PORT MORESBY,
a sweltering colonial outpost in the South Pacific, D. Carleton Gajdusek sat
brooding on the drama of research into the disease called kuru. The American scientist recognized he was acting in a "vast human comedy of modern
man's frailties . . . in the fantastic setting of a stone-age culture under a horrible plague . . . and the investigation plagued by sweeping personal involve-
ments which girdle the earth!"[1] Yet Gajdusek was also aware he was the principal character in an astounding story of medical research. Since 1957 he had
led efforts to understand the fatal brain disease that afflicted the Fore people
of the eastern highlands of New Guinea, threatening their extinction. Another
eleven years after his Moresby reflections, Gajdusek would receive the Nobel
Prize in Physiology or Medicine for his discovery of the cause of this plague:
a transmissible agent that could take years to become manifest, the first human "slow virus."

Anthropologists and epidemiologists suggested that Fore cannibalism, the
ritual consumption of loved ones after death, might spread this mysterious slow
virus. All the same, most Fore remained skeptical: they continued to believe
kuru was a particularly malign form of sorcery. Scientists like Gajdusek insisted
this was superstition, yet their methods of diagnosis proved no more sensitive than local techniques, and all treatments and responses were equally futile. While the scientist could sometimes present his involvement in kuru research as a picaresque adventure, he was always fully aware of the tragic
dimensions of the epidemic.

This book tells the story of the Fore (pronounced for'-ay), their suffering

and resilience through plague, their transformation into "modern" people, and their compelling attraction for a throng of eccentric and adventurous scientists and anthropologists. Fore people drew these white men and women to the highlands and held them there with bonds of sympathy and sociability, developing intense and often intimate relationships with the outsiders. The study of kuru would open up a completely new field of medical investigation, challenging conventional notions of disease causation, but this is more than a tantalizing case study of scientific research in the twentieth century. It is a story of how a previously isolated people made contact with the world through engaging with its science, rendering the boundary between "primitive" and "modern" completely permeable. It tells us about the complex and often baffling interactions of researchers and their interlocutors on the colonial frontier, tracing their ambivalent exchanges, passionate entanglements, confused estimates of value, and moral ambiguities. Above all, it reveals the "primitive" foundations of modern science.

Throughout the first decade of research into kuru, medical investigators desperately sought to acquire Fore blood and body parts for study in laboratories in the United States and Australia. They exchanged valuables with the Fore people, pleading with them, threatening them, cajoling them into transacting brains for blankets, urine for knives, blood for tinned fish. "Naturally, everyone would like to get their hands on kuru brains," Gajdusek observed in 1957, writing from his hut in the New Guinea bush.[2] When exchange went awry, some scientists thought wishfully of easy appropriation of Fore goods or even "medical cannibalism": that is, they dreamed of consumption without reserve. But relations in the field could never be so simple. Fore kept making claims on the investigators as persons, entangling them in local communities and sometimes managing to transform the white men in the process.

The New Guinea highlanders persisted in collecting these outsiders even when the new affiliates disappointed them and displayed ignorance, stubbornness, and insufficient generosity. Meanwhile, scientists contended among themselves for possession of the Fore, fighting for research access and credit, making their own quixotic colonial claims on scientific territory. Many became obsessed with the Fore, so deeply involved that they never got over it. Kuru research broke the reputations of some of them; for others it led to acclaim, even a Nobel Prize. Some found the Fore enthralling; others retreated into laboratory work. Kuru took over their lives. This is what Gajdusek meant when he referred to "the disease Europeans catch from kuru."[3]

THE BASIC CHRONOLOGY of kuru investigation is clear enough. When whites first encountered the twelve thousand or so Fore people in colonial New Guinea during the 1940s and 1950s, they found that many of them, mostly women and children, suffered from strange tremors, shaking, shivering, muscle weakness, and a general lack of control of the limbs, which led eventually to death. Ronald and Catherine Berndt, social anthropologists from Sydney, Australia, who entered the region in 1953, explained kuru as an extreme psychosomatic reaction to the stress of culture contact, and they related some features of it to earlier expressions of emotional insecurity attendant on interaction with whites. But as soon as Gajdusek and Vin Zigas, the district medical officer, entered Fore territory, they began to translate kuru sorcery into a medical vocabulary. The field practices of social anthropologists and medical scientists would give quite different meanings to kuru.

Competing with scientific teams from Australia, the colonial power, Gajdusek attempted to standardize the clinical and pathological features of what was now the *disease* of kuru, and through traffic of specimens to distant laboratories he sought the biological cause of the ailment. Was it genetic? A reaction to toxins? An infection? With the assistance of Joe Gibbs and Michael Alpers, Gajdusek in the early 1960s began the crucial transmission experiments at the National Institutes of Health in Bethesda, Maryland. After a couple of years, chimpanzees inoculated with the brain tissue of kuru victims developed a neurological degeneration resembling kuru. Thus in 1966 Gajdusek could announce the discovery of the first human "slow virus" infection. Meanwhile, the evolving medical understanding of kuru required a renewed attention to ethnography—not the social anthropology of the Berndts but the more applied medical anthropology emerging in the 1960s. Through yet another set of field practices, the virology of kuru was linked to Fore mourning rites, particularly cannibalism.[4]

The concept of the slow virus later became a model for understanding Creutzfeldt-Jakob disease (CJD) and familial fatal insomnia—and to a lesser extent, Alzheimer's disease, AIDS, and some other human ailments. But the slow virus proved remarkably elusive—no one could isolate the agent and grow it. Later, the discovery of a pathogenic protein called a *prion* supplanted the idea of a virus as cause of kuru and earned Stanley Prusiner (who visited the Fore in the late 1970s) his Nobel Prize. Since 1996, John Collinge of the Medical Research Council Prion Unit in London has led a new kuru investigation, this time studying the genetics of susceptibility and resistance to prions in order to elucidate patterns in the variant CJD outbreak in Britain, the re-

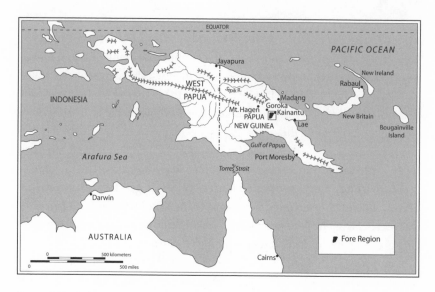

Papua New Guinea

sult of consumption of cattle with bovine spongiform encephalopathy, or "mad cow" disease. Thus an obscure disease of the isolated Fore became crucial for understanding human and other animal pathology across the globe.

When Gajdusek ventured into the region, kuru was killing hundreds of Fore each year and the people faced annihilation. Into the twenty-first century, one or two people die each year from the disease—indicating not recent contraction, since cannibalism ended around 1960, but the occasional long incubation period of the agent. Even so, most Fore still argue the cause is sorcery, which, they correctly point out, also has diminished since 1960.

SINCE THE EARLY 1960S, epidemics have often been regarded as "social sampling devices," revealing structural inequalities, hidden patterns of behavior, and dormant cultural assumptions in the societies they afflict.[5] Fore responses to kuru, an especially devastating epidemic, thus disclosed their most firmly held convictions about causation and responsibility, threat and security, blame and trust. The epidemic stripped life bare, showing the most visceral conflicts in Fore society. The emerging disease disrupted and rechannelled once-stable modes of social reproduction. Yet it also caused sufferers to explore and mine traditional sources of resolution, palliation, and consolation. It precipitated accusation and warfare, sorcery and divination, as well as mourning and reconciliation.[6]

Gajdusek and other scientists among the Fore watched as all this was happening. They desperately wanted to help, but soon they realized there was little they could do—except investigate. For them, kuru was not simply the cause of suffering or a social sampling device: it also represented a sort of biological and cultural sentinel—alarming, puzzling, and monitory at the same time. Kuru was a danger signal to the rest of the world, full of pathological foreboding, if only scientists could properly decipher its true configuration. A potent stimulus to scientific activity, the epidemic therefore becomes for us a sensitive sampling device—or perhaps telephoto lens—capturing the culture of biomedical investigation in the second half of the twentieth century.

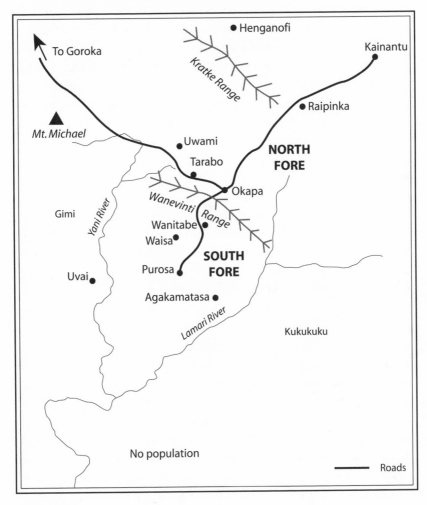

The Fore Region

As I focus here on the fieldwork of Gajdusek and others among the Fore people, I want to probe the material cultures of modern biomedical science.[7] I take kuru brains, with related objects, and use them to think more generally about the circulation of goods, the creation of value, the making of relationships, and the fashioning of identities in "global" science. I look at how the mobilization of organs and tissues can cement alliances, conjure up communities and kinship, fracture respect, create animosity and resentment, produce knowledge, and extend markets. What does the framing of material transactions between Gajdusek and the Fore tell us about their particular relationships and social identities, and how they were transformed? How did the scientist manage to draw a boundary between exchange regimes, and what happened when the boundary proved dangerously permeable? What were the operations whereby Gajdusek attempted to distance himself from the Fore, fashioning himself as a consumer, in order to extract a kuru brain? What did it mean for Gajdusek to imagine himself (only to dismiss the thought) as a "medical cannibal" in his transactions with the Fore? How did he come to circulate the same brain—now inalienably "Gajdusek's kuru brain"—as a gift to colleagues in Bethesda and elsewhere? Put simply, how did the Fore get turned into scientific things and these scientific things take on the qualities of other persons, coming to stand for their scientist transactors? In addressing these questions I track the transformation, or incorporation, of things into different social worlds and describe the people they make visible.

Aspects of this story, generally subtle and unmarked, can show us what practical ethical engagement looks like in an out-of-the-way place. The history of the interactions of scientists and Fore vividly demonstrates the mundane complexity of relationships forged in this contact zone, the recurrent potential for insight and misunderstanding, the intricacies of negotiation and ingratiation, the fragility of trust, and the durability of suspicion. It shows us the ordinary paradoxes and ambiguities of cross-cultural research at a place on the indistinct margins of the map. It quietly suggests we should locate ethics in these messy, confusing circumstances—not in the realm of abstract universals. Kuru research began just before the widespread acceptance of ethics protocols and institutional review boards, on the eve of intensive surveillance and codification of the behavior of scientists toward research subjects. Careful scrutiny of the actual material cultures of cross-cultural research can make such bureaucratic notions of informed consent and ethics audits look facile, dysfunctional, or redundant. Regulation of medical investigation is no doubt necessary, but it must be based on a realistic understanding of scientific practice and social contingency. Rather than appeal to our contemporary ethics adjudications, I therefore seek to retrieve the complex moral sensibilities and structures of feeling of the participants in kuru research, their sensuous apprehension of the relationships they formed and awareness of the expectations and obligations entailed. I want also to show their fallible value judgment, confusion over the character of exchanges, and fears of entering into moral peril when communication failed and intentions were misinterpreted. Without this knowledge, one can hardly touch on the ethics of kuru research.[8]

We might also learn from the Fore how to understand the social dynamics of global science. Kuru research occurred in the shadow of World War II and on the edges of the cold war. It took place as scientific institutions flourished in advanced settler societies such as the United States and Australia, extending their reach into "primitive" colonies like Papua and New Guinea. The traffic in specimens, equipment, reagents, and texts linked laboratories in large metropolitan centers with bush huts where autopsies were performed and tissues prepared. Scientists came and went between these diverse sites, becoming cosmopolitan as they made their careers. Parts of the Fore circulated, too, turned into globally available specimens. Fore thus became medicalized even as they first were colonized. They found themselves caught up at this striking conjunction, though never completely subsumed in it. Rehearsed at a multitude of local sites, yet performed as though on a global stage, kuru research dramatized claims and contests over territory, bodies, and persons. It shows us how science travels in the modern world and what it does to people when it arrives. But we need to know what to look for: to understand how science makes these global claims, how it becomes worldly, we must attend to dislocation, transformation, and resistance; we need to recognize partially purified and hybrid forms and identities, making visible the modern primitives and sorcerer scientists. It requires us to retrace the work that allows standardization, shifts scales, and constructs boundaries. It means we have to observe ostensibly global science as though it were assembled out of a set of rather peculiar local achievements.[9]

BACK IN PORT MORESBY, Gajdusek eventually realized his triumph had been to make science the accomplice of his restlessness. He was fond of quoting from Herman Melville's *Moby Dick:* "I am tormented with an everlasting itch for things remote. I love to sail forbidden seas, and land on barbarous coasts."[10] It was part of Gajdusek's genius that he found the science that matched his desires.

Chapter 1	**STRANGER RELATIONS**

IN AUGUST 1949, GERRY TOOGOOD conducted a rough traverse of uncontrolled country far south of Kainantu in the eastern highlands of New Guinea. It was not long after the Australian colonial authorities had returned to the high valleys following the defeat of the Japanese in the Pacific war. Accompanied by an Australian medical aide, six police and forty carriers, the assistant district officer hiked over the ridges into unmapped territory toward the border with Papua. Toogood was surprised to encounter so many curious and cooperative people. He gathered it was the first time they had met a white man.

Situated between five thousand and seven thousand feet, the country was cool and misty during the day, and usually cold and wet at night. The patrol trudged through thick forests and past terraced gardens containing sweet potato, taro, and yam, carefully cultivated and fenced. Clusters of hamlets, often dirty and neglected, occupied commanding positions on spurs and ridges. Toogood observed stockades surrounding the circular grass houses, which had been reinforced to protect against arrows. While friendly enough toward the patrol, the inhabitants of the region—whom some called the Fore—lived in perpetual fear of attack, alternately suspecting and threatening neighboring settlements.[1] Sorcery accusations and consequent fighting were common. Calling together the warring villagers, Toogood tried to communicate with them, using forceful gestures and the help of a Kamano man who claimed to know some Fore words. The patrol officer hoped to persuade the men to surrender their bows and arrows, promising them that a police constable would soon be stationed in the area.[2]

Australia had acquired New Guinea from Germany in World War I, effec-

tively adding the territory to the colony of Papua in the south but governing it separately under a League of Nations mandate.[3] Although the coastal areas of New Guinea were well charted, the whole of the interior remained a mystery well into the 1920s. Then, in advance of Australian authorities, and along the far horizon of the colonial archive, missionaries and prospectors passed through parts of the highlands, mostly skirting the Fore region. In 1926–27, a Lutheran missionary reconnoitered the headwaters of the Purari River west of Kainantu and found plenty of opportunities to evangelize. Early in the 1930s, Seventh Day Adventists began preaching in this area. Heartened by gold discoveries in the foothills, prospectors ventured west from the coast up the Markham Valley and into the mountains. Mick Leahy and Michael Dwyer explored some of the eastern highlands in 1930, and within a few years the Leahy brothers discovered huge populations in the high valleys farther to the west.[4] Soon the government's "outside men" were leading expeditions across the highlands and attempting to control and pacify the people they found on patrol. "Who are you?" the highlanders asked them. "Who are *you*?" was the white men's response.[5] In 1934, Jim Taylor, a pioneering assistant district officer, introduced the first police posts to the New Guinea highlands, assigning constables from the coast to Kainantu and Bena Bena, where they mediated between patrol officers, or *kiaps,* and villagers. By 1936, with further intrusion westward, fifteen police posts were strung across the highlands.[6] Still, the Fore people remained largely unknown to the colonizers.

During the 1930s, the area within two days' walk of Kainantu was thoroughly explored, if not fully pacified.[7] Most highlanders welcomed the missionaries, prospectors, and *kiaps,* observing them closely and giving them food from the gardens and pigs in exchange for trade goods like salt, tobacco, cowrie shells, steel axes, and knives.[8] Soon many of the people around Kainantu were donning *laplaps,* wearing their hair differently, questioning the value of separate men's houses, and learning a smattering of Pidgin.[9] Sometimes, though, these encounters went awry. The nearby Tairori became known as "difficult," and the Kukukuku (later Anga), in the far south straddling the border with Papua, were notoriously fierce and "treacherous." It was reported they killed a man fossicking for gold on the Upper Watut. In 1932, J. K. McCarthy, leading an expedition into the area to investigate, found the bodies of other prospectors. His party was attacked and he was carried out with arrows in his thigh and abdomen.[10] Ian Mack, a tough, wiry patrol officer posted from New Britain, relieved McCarthy, but a series of infractions and misunderstandings at a large Agarabi settlement northwest of Kainantu led to Mack's death in 1933.[11] On these early patrols resort to the rifle was frequent, and scores of highlanders were injured and killed. In the minds of young, adventurous Australians in New Guinea such violent encounters confirmed the need for vigilance, self-reliance, and control in the bush. Colonial authorities demanded a monopoly on force.

The *kiaps* tended to look skeptically at the few ethnographers who sought in this period to live among the highland "natives." Sometimes they may have had good reason to do so. In 1935, Reo Fortune arrived in Kainantu, brooding and angry over the breakup of his marriage with Margaret Mead during their Sepik fieldwork, and looking for conflict. Hiking into Kamano territory with just a backpack, trade goods, notebook, and pipe, he found plenty of it. Although the authorities compelled him to stay at an empty police post, the thirty-two-year-old New Zealander often felt threatened and used his rifle at least once, having later to pay compensation for shooting a man in the leg. The government ordered him out after six months of desultory fieldwork. Despite fantasies of running a trade store in the highlands, Fortune returned to live among anthropologists in the Northern Hemisphere, where he remained defeated, querulous, unproductive, and afflicted with facial tics. The Kamano had never been sure why he was there, or what sort of white man he was. Fortune may have shared their puzzlement.[12]

The police worried even more about the sole other anthropologist to venture into the region: tiny, intrepid Beatrice Blackwood from Oxford who spent nine hard months among the Kukukuku of the Upper Watut in 1936 and 1937, studying their stone technology while she carried around her kitten Sally in a string bag. She was restricted to a few "safe" villages but found the people dour and suspicious, and as violence became more intense, the authorities refused to allow her to return after her health failed. "There can be few people so difficult and so secretive," she reflected on her time with the Kukukuku, "but their eyes would light up at the sight of an iron blade."[13]

Although the Fore people must have heard of the predations of these missionaries, prospectors, government officers, and even anthropologists in adjacent parts of the highlands, they had little direct experience of intruders until the 1950s. A few prospectors wandered through their land in the 1930s, and at least one plane crashed there during the war. New diseases began to penetrate the high valleys in the vanguard of white contact.[14] After Toogood's efforts in 1949, government patrols passed further into Fore territory.[15] In the 1950s, patrol officers would line up each hamlet and try to conduct a census and seek out local "big men" who might represent the colonial authorities as *luluais* (headmen) or *tultuls* (deputies). Usually they lectured the villagers on the importance of hygiene and road construction and the need to give up warfare, sorcery, and cannibalism. A police post was set up at Moke (later Okapa) among the North Fore in 1951, with a *kiap,* John R. McArthur, stationed there from 1954 when the rough track from Kainantu opened. By 1957 it was possible to take a Land Rover or motorbike down to Purosa in the South Fore. It became clear to the colonial authorities that there were at least ten thousand Fore to know and control. Ronald and Catherine Berndt, Australian anthropologists, spent some time on the northern margins of the district in 1953. Missionaries and trade stores began to penetrate further south during this period. The smell of trade tobacco and kerosene mixed with the older highland odors of wood fires and pig fat, or "grease." Within a few years some northern villages were planting and harvesting coffee, and the economy had begun to monetize. But only at the end of the decade did the scientific developments that so altered the lives of the Fore start to exert their influence.

In these postwar years, New Guinea became a magnet for young Australian men hankering for adventure and anthropologists eager to make first contact with "primitive" societies. By the end of the 1950s, even the previously isolated Fore people had come to know these types. Their interactions with government officers and anthropologists—as with missionaries, traders, and even dilatory medical doctors—became increasingly adroit and rewarding. What are we to make of the early encounters, before they turned routine and banal? When the first patrol officers and the first anthropologists ventured among the Fore, the interactions between all parties were tentative and exploratory. It seems the highlanders were often more pragmatic and adaptable than the newcomers, less prone to make assumptions or to prejudge. The intruders commonly showed overconfidence in dealing with "primitive" people, their imagination impoverished through too much reading about "similar" situations or facile generalization from past experience. But all parties to these encounters demonstrated a combination of blindness and insight, and they could feel both related and distant at the same time, confused about their status. Slowly, though, new ideas crystallized, unprecedented views from different perspectives emerged, and some features of the strangers came into focus, still colored by previous experiences and modes of perception. The Fore noticed above all the intruders' rich cargo; and the newcomers fixed on kuru, though it took them some time to do so.

IN HIS REPORT ON THE PATROL of the Fore region in May 1951, Gordon T. Linsley wrote unusually extensive "anthropological notes" on the

Fore hamlet, 1957. Photograph courtesy Peabody-Essex Museum, Gajdusek Collection

people he saw. Above all, the pace of social change astonished him. Although the patrol post at Moke had only recently opened, the surrounding area was already well controlled and village warfare had abated. Young Fore men seemed glad to find an excuse to give up fighting. Rest houses for patrol officers appeared in the larger villages, and some hamlets were moving down from the ridges and closer to the gardens. Separate men's houses persisted in only the more isolated settlements. Broad paths now linked communities, and many people regularly used the rough track to Kainantu, which allowed them to view new fashions and hear Pidgin. Beyond Moke, people no longer appeared nervous when the patrol approached. Close to Purosa, however, sorcery accusations abounded and fighting between districts continued. Two weeks before the patrol, the fresh constable at Moke found himself in the middle of a battle down south and shot a man. When Linsley heard this he sent the constable back to Kainantu. But the victim's family "line" still welcomed the patrol and sought trade.

Like other "red men," Linsley found the country picturesque, though he deplored the constant drizzle, mud, and leeches. His patrol followed the steep wooded ridges that divided open valleys falling away to the Lamari River. In the more heavily populated northern valleys, he admired neat gardens and stretches of *kunai* grassland with isolated casuarina groves and stands of sugar cane. Conventionally, the patrol officer regarded the inhabitants as simple and childish. Worse than children, they lacked "any spirit of enquiry and do not appear to care if anything is explained to them or not." Revealingly, Linsley noted that as he lectured them they looked on him "with the polite inattention civilized people are in the habit of giving to a distinguished bore."[16] He found the same "lack of initiative" in their dress and adornment, which was plain and scanty. The males seemed obsessed with tobacco and rarely stopped smoking their bamboo pipes. As only agile, healthy adults greeted the patrol, Linsley found it difficult to assess the incidence of disease among the Fore, but they appeared happy and well "by native standards."[17]

Linsley anticipated other interpreters of the Fore in focusing on sorcery and cannibalism. The belief that illness and death were consequences of sorcery was "rampant," prompting most of the distrust and fighting among villages in the region. Linsley despaired at ever convincing Fore that sorcery was mere superstition.[18] Although undoubtedly troubling, sorcery evidently offered them a coherent and convincing explanation of unexpected illness and death, and *kiaps* struggled to find a satisfactory substitute. The patrol officers' grasp of contemporary microbiology and theories of disease causation was rudimentary at best, and they lacked a ready means of translating the little scientific knowledge they possessed. Simple disparagement of sorcery was futile. Visiting the village of Yagusa in June 1952, patrol officer W. J. Kelly saw small leaf-wrapped packages containing flesh of some kind, apparently used to detect the origins of sorcery.[19] Later the same year, a cadet patrol officer heard that sorcerers took something intimately connected with the intended victim, such as hair, discarded food, or feces, then wrapped it with a stone in leaves and placed the bundle on swampy ground. As the coverings decayed, the victim weakened until death occurred. After divination of the source of the enchantment, the offending settlement would be challenged and attacked.[20] On patrols in 1953 and 1954, John McArthur identified this form of sorcery as *kuru,* a word that described the characteristic shivering and trembling of the victim and distinguished it from other magical operations. "Certainly it is difficult to reason out 'kuru,'" he wrote. "It must surely be psychological." He called together the villages around Moke and made a great bonfire of all materials that might be used in kuru. After that he heard few sorcery accusations.[21] But the following year, John Colman observed that sorcery was still the chief "impeding factor" in development of the Fore. In one hamlet the patrol officer watched the wife of a *tultul* "sitting dejectedly, shaking violently all over," a victim of kuru sorcery. "All this to the outsider is unbelievable and difficult to reason, but to the native it is a fearful shadow."[22]

The Fore practice of consuming their loved ones after death also fascinated and disturbed many early patrol officers. Linsley reported that the people around Moke had begun to conceal their cannibal feasts since learning that

Sorcery bundles for making kuru. Photograph courtesy Peabody-Essex Museum, Gajdusek Collection

white men disapproved. But farther south he received detailed accounts of the practice. Evidently, no part of the body was spurned, with even the bones pounded into meal and consumed. It proved hard for him to determine how Fore distributed the bodies, whether to specific relatives or the community as a whole. "These people rationalize the custom," he wrote, "by saying that by consuming their own dead they incorporate them into themselves and so lessen not only the sorrow, but even the idea, of loss."[23] The following year, R. R. Havilland felt sure that cannibalism still prevailed among the South Fore. "Everywhere the patrol halted," he noted, "the European members, and the well-fed police escort, were greeted by native men who rubbed their hands over arms and legs, making enthusiastic noises as they did so." At first he thought this was a normal manifestation of curiosity, but then he observed that it "accompanied a motion of conveying the flesh of the arm and leg to the mouth and chewing eagerly." Aware that the party could scarcely constitute eligible loved ones, the *kiap* assumed this was simply a gesture of amity and compatibility. Indeed he soon realized that "the highest degree of commendation bestowed by these natives is the remark: 'I like you so much I could eat you.'"[24] Not Havilland nor any other outsider claimed to have seen or engaged in a cannibal feast, but there was little denying the plausibility of the plain, and sometimes pedantic, Fore descriptions of the event. Even so, as the years passed, fewer patrols reported cannibalism, and by the end of the 1950s the combined influence of mission and government seems to have suppressed the practice among the Fore, except perhaps in the most isolated hamlets. Thus cannibalism disappeared while sorcery accusations persisted. Patterns of consumption proved far more flexible, or more readily substitutable, than moral convictions about the cause of illness and death.

Colonial officers in New Guinea were seeking to couple social anthropology to administrative goals. Even before the war, government anthropologists had operated in both Papua and New Guinea, and the Department of Anthropology at the University of Sydney offered training for recruits to the colonial civil service.[25] The leaders of the Sydney department convinced Australian authorities that it was necessary to understand native languages, customs, and habits of thought in order to change the lives of Aborigines and Pacific Islanders.[26] The administrators of Papua and New Guinea therefore permitted and sometimes even sponsored anthropological investigations in their territories, though often they regretted that formal results turned out irritating or irrelevant. The training of patrol and district officers in applied anthropology after the war at the Australian School of Pacific Administration seemed to offer greater practical benefits.[27] Some sensitive and observant prospective *kiaps* like John Colman appreciated this exposure to basic anthropology, while others could never work out why it mattered. In any case, their training was relatively brief and narrow, emphasizing the recognition of salient cultural forms and movements and especially identification of those cultural sentinels, such as cargo cults, sorcery, and cannibalism, that might issue challenges to social control and development efforts. *Kiaps* generally demonstrated little interest in conventional aspects of anthropological inquiry such as kinship and child development, though knowledge of land tenure could be helpful. Basic training provided the more receptive patrol officers with an anthropological alert mechanism which they might add to armed force, bush skills, character, and orderliness to enhance their repertoire of social and political control in out-of-the-way places.[28]

INCLINED TO STUDY ABORIGINAL AUSTRALIANS, Ronald and Catherine Berndt went reluctantly to the eastern highlands of New Guinea. A. P. Elkin, professor of anthropology at the University of Sydney, advised the ambitious young graduate students to gain experience of a field site different from the places to which they had become accustomed over the previous decade. Moreover, Elkin was eager to extend his influence on Australian policies of native administration to Papua and New Guinea. For most of his career, problems of culture contact, mutual adaptation, and social change in continental Australia had preoccupied him. He abjured simple functionalist accounts of "primitive" societies, in which every cultural form and event was integrated into an unchanging whole; such studies were "free from, or abstracted from, factors of contact and change." He urged his acolytes to remedy the omissions of functionalists like Bronislaw Malinowski and to conduct a "searching inquiry into the history and effect of contact on the social structure, culture and population." In July 1950 he toured New Guinea and passed through Kainantu, where he recognized at once a "great and pressing opportunity for research . . . amongst almost unknown peoples." At such a site his anthropologists might provide the administration with "an understanding not only of the social and political organization, means of livelihood and ritual, but also of the system of values and attitudes to life's problems, old and new, which are part of a people's cultural heritage, and which become part of the individual's 'make-up' during the process of growing up and 'education.'" Above all, Elkin demanded that his sort of anthropologist should appreciate the Melanesian meaning of life, the values, attitudes, and reactions of the people.[29]

Elkin's rivalry with S. F. Nadel, the new professor of anthropology at the Australian National University (ANU) in Canberra, hastened his efforts to colonize New Guinea with his own confederates and disciples. Nadel, an Austrian musicologist who had studied anthropology with Malinowski at the London School of Economics, found Elkin facile, pushy, and deplorably provincial. Field experiences in the Sudan made Nadel a firm believer in "cooperation between anthropologist and administrator," and he too saw opportunities for applied research when he toured New Guinea in 1951.[30] A research fellow at the ANU, Kenneth E. "Mick" Read, was already living among the Gahuku-Gama in the Asaro valley outside Goroka, studying their social life and rituals.[31] Although it was sometimes claimed that Nadel's students tended to focus on social structure and deploy African models of kinship, while the people Elkin sponsored looked mostly at social change and carried with them assumptions derived from Aboriginal Australian societies, the work done in the field in New Guinea was never so neatly differentiated.[32] The exigencies of life in colonial New Guinea did not permit a simple transposition of theoretical models and institutional rivalries. Read, for example, was always prepared to proffer advice to the Berndts, and his later reflections on the Gahuku centered on social change, not fixed social structure.[33]

The Berndts were among the few anthropologists Elkin trusted: they were tigers for work, lived on next to nothing, and showed their aloof professor deep respect. It helped that their dislike of Nadel, whom they found unbearably arrogant and abrasive, was obvious.[34] Since their marriage in 1941, the Berndts had studied Aboriginal communities across Australia, initially in association with the South Australian Museum and then Sydney University. Granted a military exemption during World War II, Ronald became preoccupied with problems of conflict, violence, and social control. Catherine focused on the lives of Aboriginal women, child development, and the social role of folklore and mythology.[35] She had grown up in New Zealand and knew Reo Fortune, a close friend of her uncle. The Berndts may have modeled themselves on Fortune and Mead: they replicated the older couple's division of labor in the field; and they too concentrated on eliciting native temperament and personality. Catherine regarded Mead as an exemplar for women anthropologists, and although there is no evidence Ronald admired Fortune in the same way, the men shared a strong interest in studying violence and sex. Certainly the Berndts knew that Fortune had failed in 1935 to complete his field work at the headwaters of the Kamamentina River, and this was one of the sites they would choose. Of course, other factors influenced their decision on where to conduct research. They spoke with Elkin and corresponded with Read, as well as examining patrol reports and accounts of the expeditions of the 1930s, such as Leahy's *The Land That Time Forgot* and Jack Hides's *Through Wildest Papua*.[36] And after arriving in Kainantu in late 1951, they just set off and explored for themselves. But Fortune would haunt them.

When they left Kainantu seeking their field site, Catherine, who had sprained an ankle in Sydney, rode one of the few horses in the highlands, ahead of a long line of porters:

> An excited mob accompanied us and our carriers, grabbing at us and pulling us back, making sucking and hissing sounds, shouting and calling to us, greeting us with their welcoming words, "I eat you."[37]

Still later, Ronald recalled:

> Our progress consisted of scenes reminiscent in some ways from those of *King Solomon's Mines*. The terrain was rough and at times very steep; the track, often edged with jungle, was slippery and narrow. A long line of carriers (more than we wanted), with our boxes lashed to poles, stretched out further than I could see. At the head was Catherine, mounted on her horse

Catherine Berndt in the highlands, c. 1953. Photograph courtesy Berndt Museum of Anthropology

and surrounded by plumed and decorated men with their bows and arrows, singing as they walked and danced along.[38]

Unlike some of the earlier missionaries and patrol officers, Ronald did not record any efforts to sprinkle pig's blood on the couple in order bring them into the orbit of the human, or to hold his genitals to confirm he was a man. Yet their stories of contact contain more drama and romance than the dry accounts in all but a few patrol reports.

Arriving at Maira in Kogu, two days walk beyond the Lutheran mission at Raipinka, they saw a house had been built for them so they stopped. Later they realized to their dismay that the Kamano, Usurufa and Yate (Jate) language groups overlapped around them. It was mountainous, heavily wooded country, and unexpectedly cold. The site that in effect chose them had been "opened" only for three years. "A day's walk south is still uncontrolled, with inter-village fighting," Ronald wrote. "Only a few European men have been through, and Catherine is the first white woman to visit these parts."[39] Their arrival occasioned two big *singsings*. "Garden produce was heaped before us, pigs were killed, and dancing and singing went on until well after dark. We were viewed as returning spirits of the dead who had forgotten the tongue of our fathers and wanted to relearn it."[40] The house had been made ready for the cargo the couple were carrying from the spirit world.

During the first period of field work, from November 1951 until May 1952, the Berndts focused on culture contact, violence, and issues of personal responsibility and social control. From their own observations and from what the local inhabitants told them, they attempted to put together a coherent account of the local society, its kinship patterns, feelings of insecurity, recourse to warfare, and the character of economic relations. As usual, Ronald focused on understanding men's business, while Catherine concentrated on women, mythology, and child development. Their need to deal with four or more different language groups greatly complicated these tasks. Initially they engaged two interpreters at Raipinka, but the mission soon demanded the return of one, and the other had a "singular habit of starting quarrels among the people." Catherine complained he was "a little irresponsible and quite unreliable."[41] Eventually she gained some competence in Kamano and Usurufa, while Ronald learned Yate and, later, a little Fore. But it was proving a troubling field experience. "At times we did not like these people, but just as frequently we did; and this fluctuation is mirrored to some extent in their own response to life—aggressive and violent excitement contrasted with extreme and sometimes tearful sentimentality."[42] The Berndts found it an "exhausting and difficult region—much more so than we expected"—and they felt highland communities lacked the "subtleties" of Aboriginal societies.[43]

Compounding their problems, Fortune had met them at Raipinka and announced his intention of working again among the Kamano. In fact, he may just have wanted to compensate further the man he shot in 1935, but the Berndts feared that he was poaching on their field site.[44] When Ronald first saw Fortune, he "appeared in need of a good interpreter and seemed rather lonely and depressed."[45] The Berndts wrote to Elkin and Mead seeking their advice. Elkin suggested that they go farther south to the Fore on their next visit, while Mead's response was mainly solicitous and protective toward her former husband. Fortune, the ghost of ethnographies past, soon disappeared, but in 1958 he would again materialize, this time among the Fore, attaching himself to the medical investigators.

Accustomed to gauging distinct and durable features of temperament and personality in Aboriginal and European communities, the Berndts struggled to comprehend the provisional and relational character of personhood or self among the highlanders.[46] Exchanges of goods established and confirmed identities and relationships, whether at the formal ceremonies ensuring social reproduction that marked weddings and deaths, or in everyday encounters with friends, family, and strangers. Assessment of the sort of person you are, your rank and character, could not readily be disentangled from how well you give. Thus in transacting with the highlanders, the Berndts, like other strangers were becoming human and acquiring a certain status and relationship. But they lacked a strategic understanding of prestation and reciprocity: they neither gave nor received well. Ronald recalled:

> We were assumed to be capricious and undependable, possessed of "power" such as ghosts or malignant spirits have, to do them harm if we felt so disposed—beings who had to be propitiated by lengthy recordings and descriptions and explanations. This led to a certain strain in interpersonal relations and served as a basis for some misunderstandings.[47]

Mick Read had suggested they stock up on "gold-lip" shell, tobacco, matches, salt, *laplaps,* steel implements, mirrors, and beads, in order to barter for pigs and garden produce and to provide gifts in return for information and attentiveness.[48] But the Berndts soon tired of importunate locals. Ronald complained there were "so many people and all of them want not an anthropologist so much as a trader. They have large gardens and produce great quantities of vegetable foods which they try to sell to us."[49] To Elkin he wrote that they "all require 'payments' for stories: salt, tobacco, newspapers, wool strands, matches, razors, and so on."[50] The highlanders' dissatisfaction and bewilderment seems to have increased along with the strangers' resentment and irritation. A story began to circulate among the locals that the anthropologists planned "to round them up and take them to jail, first cutting off their hands and even their heads!"[51]

After withdrawing from the field for a few months, the Berndts dutifully returned in November 1952 and hiked even farther south, this time to Busarasa and Moke in Fore territory. They were now "on the fringe of the restricted region, where inter-district feuding and cannibalism are still prevalent."[52] The people appeared even more timid than before, and the Berndts found conditions far more frustrating. Mist and rain swept across the mountains, and they spent most of the next few months shivering in the cold and attempting to dry themselves. Catherine admitted to finding the Fore "rather trying at times." She lamented they viewed the outsiders as a "source of supply and constantly expect us to restore to them as rightful owners the 'cargo' of which they had been cheated."[53] She was disappointed the society was so materialistic and dominated by men. There was "plenty of cannibalism too," she reported. "It's reasonably safe to visit some of the nearer villages, but the others are only for strongly armed patrols with plenty of native police. We haven't any illusions about these people."[54] Ronald wrote to Raymond Firth, his Ph.D. supervisor at the London School of Economics:

> Even where open fighting is no longer practised, there is always the alternative of sorcery (very popular in these parts, at least by repute). It is certainly an interesting area, despite the existence of some features which we, personally, find somewhat uncongenial.[55]

He told Elkin that the Fore were "difficult people to deal with, and we have had to be stricter here than we were in Aboriginal Australia." Ronald slept with a loaded pistol under his pillow, and one day resorted to firing it in the air to disperse some villagers who pestered him.[56] The Berndts gladly left the highlands in March 1953. They never returned to New Guinea.[57]

RONALD BERNDT LATER DESCRIBED a society organized in villages, or clusters of hamlets, each containing one or more patrilineage or "line." But the sense of kinship was surprisingly vague and diffuse, and claims of common descent were often hard to justify genealogically. Most hamlets contained some matrilineal kin, affines, refugees, and casual visitors, any of whom might participate in exchange and feasting, or gardening, or building, or raiding. Some of the affines and cognates might be adopted into the line of agnatic descent, through a process that John A. Barnes later called "cumulative patrifiliation."[58] That is, affiliation was less bounded and biological than in many other societies. In ceremonial transactions individuals could reach out beyond agnates and recruit more widely. Thus network cohesion was far more important to people than group solidarity. Leadership was not hereditary, but rather it depended on the development and maintenance of personal exchange relationships, on the structuring of mutual ties of sentiment and reciprocal obligation. Through the careful management of such networks, a village leader or "strong man" would emerge for a time.[59]

When the Berndts encountered the Fore, men shared a communal house, while women and children occupied smaller houses. Already among the Kamano this pattern was breaking down, as it soon would farther south. Indeed, Catherine recalled that when they first visited in 1951, people looked askance at a husband and wife living together, but by 1953 many highland couples co-

Ronald Berndt in the highlands, c. 1953. Photograph courtesy Berndt Museum of Anthropology

habited, hoping this would enhance their prospects of acquiring cargo.[60] The last initiations of young men were still taking place, and the sacred flutes could be heard. Ronald described obsessively the display of male strength and aggression in initiation rituals, sexual activity, and warfare. *Excess and Restraint,* not published until 1962, was notoriously salacious and imprudent—some called it an "ethnographic extravaganza."[61] If the advent of the Berndts had once brought to mind Rider Haggard, Ronald's reflections now conjured up the Marquis de Sade. Unfailingly he dwelt on excess, whether sexual or violent. In a society rife with interpersonal and intergroup conflict, it seemed to him that actions exhibiting excess were the only restraint on other sociopathic conduct. If he had considered exchange relations more seriously, he might have found another explanation for social control and cohesion. Instead, his work is a threnody on incontinence and dissipation. Its reception was poor, and both he and Catherine came to regard their New Guinea investigations as a blot on their careers.[62]

According to Ronald, sorcery and the payback for alleged sorcery represented the main arenas for male aggression. It was, the anthropologist argued, both an instrument of crime and a means of enforcing conformity to certain social norms. At the same time that it might resolve conflicts, it also served to exacerbate them. "The use or threat of sorcery provokes fear," Ronald wrote, "and an accepted local response is aggressive action."[63] The Fore considered sorcery the real or underlying cause of many serious illnesses and most deaths. As anything that had been in contact with a person might become an ingredient in sorcery, great care was taken in secreting any "leavings" or discards. Ronald learned about a variety of local types of sorcery, including kuru. This primarily afflicted women and girls, making them shake uncontrollably and leading to partial paralysis and loss of muscular coordination.[64] Although the Berndts never observed the final stages, they heard that death was inevitable. At the time, such clinical manifestations appeared compatible with a hysterical reaction, a psychosomatic response to stress. As late as 1959, Ronald still maintained that social or cultural events might have such "far-reaching effects on the human organism itself, even to the extent of interfering so drastically with it that it ceases to function."[65] All the same, Catherine recalled talking about kuru with Mead, who suggested they should get the doctors involved; but nothing came of this advice.[66] Ronald in particular was determined not to be diverted from his theme. He regarded sorcery accusations chiefly as a means of promoting social discipline and group cohesion, and therefore classed sorcery with interdistrict warfare as a failed or failing mechanism of control. "Here we have fear and anxiety, with revenge and counter-revenge, the release and building up of tensions in a repetitive fashion, halted from time to time only to recommence within the same or another constellation."[67]

The Berndts imagined cannibalism as a manifestation of conflict and outlet for aggression, a sign of lack of control. Unfortunately Ronald did not detect any patterns in Fore cannibalism, and asserted they ate any human flesh except the victims of dysentery and kuru sorcery. Later analysts insisted the Fore ate only loved ones after death, and matrilineal kin prepared the body and determined its distribution. "Dead human flesh," Ronald expansively claimed, "to these people is food, or potential food." A human was "not eaten to absorb the 'power' or strength of the deceased." Rather, the Fore supposed that the dead would prefer to be eaten and their wishes should be respected.[68] Certain parts of the body might be kept and hung around a child's neck to promote growth. But Ronald wanted to focus on what he called the "orgiastic feast," lingering over gross fantasies of copulation with corpses and other forms of sexual violence toward the dead. Many of his statements perhaps better reveal his own preoccupations than any true pattern of Fore behavior. He did not like these people. "In many ways this is a most unpleasant culture," he wrote to Elkin early in 1952. "Actually these people are 'bestial' in many ways—perhaps an unanthropological statement, but nevertheless one we can't help making."[69] The Berndts' inclination to believe the worst about highlanders could vitiate and distort their social analysis. William Arens later condemned Ronald's "lengthy, titillating descriptions of often-combined cannibalistic and sexual acts," and suggested that *Excess and Restraint* was "aptly titled only in the sense that on intellectual grounds it displays too much of the former and too little of the latter."[70]

During their period in the field, the Berndts watched vigilantly for "cargo cults" or "adjustment movements." They expected to find such phenomena in the New Guinea highlands. For the Berndts, a cargo cult would help to crystallize the diffuse examples of suspicion and social disruption they were witnessing and link such cultural turmoil to contact with Europeans. They defined the cargo cult as a social movement that "provided adjustment to peculiar conditions, or rather attempted to do so at a psychological level in the first instance, and secondly at an economic level."[71] Many of the highlanders they first encountered seemed to believe white people were spirits of ancestors who had returned bearing valuable and novel goods, but these white specters, like most ancestors, behaved capriciously or unpredictably and refused to recognize their obligation to distribute cargo. As news spread about mysterious and impetuous agents who possessed incalculable wealth, some locals resorted transiently to cargo cults, engaging in magical performances in the hope that planes would thereby bring goods directly to the people. When they identified a cargo cult, the Berndts were then able to expatiate on the emotional insecurity of highlanders, their awe and fear of white people, and their expectations of material benefit.[72]

Ronald reported on one such cargo movement among the Kamano, fed by resentment and frustration. When the *zona* or ghost wind blew across the land, it brought with it the spirits of the dead. Many of the people who felt it became possessed by these spirits and began to tremble or shake. The Berndts used the Pidgin term *guria* to describe this collective manifestation and distinguished it from kuru, an individualized "shaking sickness caused by sorcery."[73] The "contagion" of involuntary reeling, staggering, and violent shaking resembled other adjustment movements. Once the spirit of the *zona* possessed them, people set about building a large house and filling it with leaves, stones, and wood, in the expectation that these items would be transformed magically into paper, knives, and rifles. After killing a pig, the villagers anointed the objects and the house with blood and awaited their transformation. The

Berndts appreciated the symbolism of blood better perhaps than most of the later medical investigators who sought to collect it:

> Blood (whether of pigs or men) is a "human" element, and is thus a desirable substance from the spirit's point of view. Blood is, in essence, "life," so that in presenting blood gifts to the spirit, the inference is that it will come into the human orbit. Moreover, blood being a symbol (more than that, a necessary component) of life or reality, the sprinkling of blood over leaves, sand and stones which are placed in the special house means that their reality is ensured: they are bound to turn into real objects.[74]

But the Berndts, unlike the later medical investigators, did not realize that the blood that saved and connected might also be sprinkled in reproach.

It seemed to the Berndts that they were witnessing the end of an era. Cargo movements sprung up sporadically while the anthropologists were in the region, but *kiaps* and missionaries rapidly suppressed them. Ronald observed two missionaries entering a village and telling the Fore "to bury all human bones and skulls, which had been placed in the village clearing as a result of the cold wind and shivering accompanying the initial manifestation of a cargo movement."[75] Soon cargo cults, warfare, and cannibalism would be just memories. The couple counted themselves fortunate that they had been able to document so many traditional practices and behold the beginnings of social transformation in what they regarded as blighted New Guinea societies. Fifty years on, it appears that Fore adjusted more readily to these strangers than the Berndts ever adjusted to the Fore.

EARLY PATROL REPORTS implied that the health of the Fore was generally good, apart from some tropical ulcers, scabies, and the yaws that disfigured so many of those greeting the outsiders. But *kiaps* were unlikely to see many invalids on a brief visit. In June 1952, however, W. J. Kelly reported that the health of South Fore was "deplorable," and people were "literally rotten with yaws and infected injuries." He tried to send fifteen of the worst cases to Kainantu for treatment, but they refused to travel through hostile territory. The only solution was to station a medical assistant in the area.[76] The following year a native medical orderly was posted to Moke, and it was planned to send another further south to Purosa, but a medical patrol advised caution as "the natives . . . are just beginning to come under control."[77] In the meantime many did go willingly to Moke for antibiotic injections against yaws. The *kiap* won-

Kiaps *John Colman and Harry West on patrol in Kukukuku Region, c. 1956. Photograph courtesy Peabody-Essex Museum, Gajdusek Collection*

dered if the availability of an effective cure had dispelled assumptions that the disease resulted from sorcery.[78]

The Berndts and a few patrol officers had noted the occurrence of kuru sorcery, but at first it did not appear to represent a major problem for the Fore or the government. The condition was rarer in the north around Moke, and in any case, most sufferers were hidden from strangers. Some *kiaps,* though, began to suspect that something was seriously wrong. In October 1953, W. J. Hibberd wondered why he saw so few young women in Fore villages, but he did not press the issue.[79] John Colman was sufficiently concerned and puzzled by kuru to send a sufferer to Kainantu late in 1955, where the district medical officer, Vin (Vaclovas) Zigas, observed her. Zigas made a provisional diagnosis of "acute hysteria in an otherwise perfectly healthy woman."[80] In the middle of 1956, Frank Earl, an Australian medical assistant, accompanied Colman on patrol to the South Fore to investigate a suspected gonorrhea outbreak. They discovered little venereal disease, but kuru was rife. At Yagusa they found five early "cases" of the "disease," then at Amusi they heard that thirteen women and eight children had recently died of kuru, and they saw three comatose victims. The absence of young girls and women in the region was marked. "These people," Earl wryly observed, "appear very prone to sorcery." Farther on, "a small female child aged eight years was carried to patrol suffering from 'kuru.' This child died before examination could be carried out," he reported. In early August at Kamira, Earl finally expressed his growing skepticism and irritation toward psychogenic explanations for the condition: "Ten cases of 'kuru' were brought to patrol today, all in the final stages of the disease, and it seems almost certainly to be encephalitis. Previous reports claim this disease to be psychological, but personally I am unable to see such effects on children under the age of two, if this were so, as I have seen today." The impact of yaws treatment, and better knowledge of these strangers, may have encouraged Fore to disclose more kuru victims. But in this case the medical officers could offer no injection, or "shoot," to help sufferers. On August 4, Earl was called back suddenly to Kainantu to care for his wife. He was "advised that Dr. Zigas will be proceeding to investigate further."[81]

The colonial health services in New Guinea were already thinly stretched and focused on combating the major diseases, especially malaria, tuberculosis, yaws, and leprosy. When John T. Gunther had been appointed director of public health in 1946—at the age of thirty-five—he could deploy five doctors, five nurses, and twenty-three "European medical assistants" to control disease among more than 1.5 million people. A brusque and domineering but amazingly effective bureaucrat, Gunther immediately began to recruit and train "native medical assistants"—later called aid-post orderlies (also *dokta bois* or *liklik doktas* or simply *doktas*)—to run small clinics in their home villages across the territory. Often illiterate, they knew how to give injections and provide first aid and basic treatment as well as improve village hygiene and mosquito control. Gunther also recruited more "European medical assistants," many of them from the army. Some may have done a few years of a medical course, others claimed no more than a first-aid certificate: the best took on the same responsibilities as medical doctors and patrolled with a native affairs officer or alone. The major problem remained recruitment of physicians. In 1950, the director of public health managed to get permission for fifty refugee doctors mostly from Eastern and Central Europe to practice in the territory.[82] Belying the "curious xenophobic outlook" of most Australians, these recruits became the "backbone of the service." In his usual blunt fashion, Gunther recalled that "some were first rate, a few were misfits, the majority were good industrious public servants."[83] Having conducted research on malaria and scrub typhus during the war, Gunther was also committed to expanding Australian scientific investigation in the territory, though he

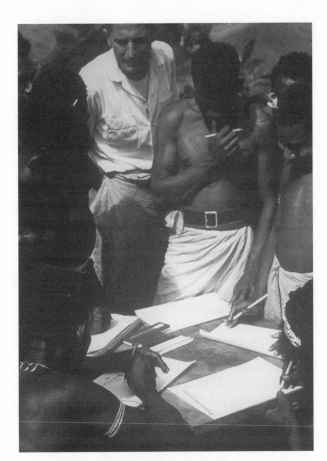

Vin Zigas oversees census. Photograph courtesy Peabody-Essex Museum, Gajdusek Collection

struggled to find the resources to do so. Still, he was a forceful advocate who benefited from the colonial government's shift in emphasis from patrols and pacification efforts to social and economic development projects. By 1956–57, health expenditures made up almost a quarter of the Papua and New Guinea budget—and yet, considering the health problems of the territory that was hardly enough.[84]

Born in Estonia and educated in Finland, Lithuania, and Germany, Zigas was one of the more volatile of the refugee medical doctors admitted in 1950. Tall and handsome, with curly brown hair, blue eyes, and an impenetrable accent, the bush medico often brought to mind the actor Danny Kaye. His moods fluctuated between exuberance and depression, but he proved himself in the highlands to be a skillful and dedicated physician. In his unreliable memoir, *Laughing Death,* Zigas claims that John McArthur, the diffident, easygoing *kiap* at Okapa, told him about kuru in September 1955 during a heavy drinking session in Goroka and invited him to see some sufferers.[85] But there is no record of any visit to the Fore before October 1956, when he went to investigate Earl's suspicion of an encephalitis outbreak (which Zigas would never later attribute to him). Accompanied by "Liklik," a "very shrewd medical orderly," Zigas managed over the following month to identify and study some twenty-seven cases at Okapa and a further eleven in neighboring hamlets. Colman, whom he regarded as a "proud young man" and an unusually thoughtful government officer, assisted him and even, according to Zigas, took him to the Kukukuku—though this was not recorded in the patrol report.[86] Returning from Okapa with twenty-two samples of blood and a brain, he wondered what he should do next. He discussed the problem with Clarissa Andrea de Derka, the librarian at the Public Health Department, a stunning Hungarian blonde who presided over Port Moresby's only intellectual salon.[87] She suggested that he send the materials to S. Gray Anderson, a virologist at the Walter and Eliza Hall Institute in Melbourne, Australia, who had recently investigated Murray Valley encephalitis near Moresby.[88] She had heard too that F. Macfarlane Burnet, the director of the Hall Institute, was interested in conducting further research in Papua and New Guinea. Zigas, imagining a new career as a clinical investigator, bypassed the government pathology laboratory in Moresby and posted the specimens directly to Melbourne.

Not until late December did Zigas write to Gunther—a figure both respected and feared, whom he later described evocatively as having a face "hewn of rough granite, with two front teeth missing, and the expression of a bald eagle."[89] Zigas reported that a number of people in the Okapa region, mostly women and children, were suffering from a new form of encephalitis, called kuru and attributed to sorcery. The disease began with lethargy, headache, vertigo, and vomiting, progressing to tremors, unsteadiness, erratic eye movements, and then to loss of control of sphincters, inability to swallow, and finally to death, usually within nine months of onset. The condition seemed limited to the Fore people. Zigas suspected that a virus caused the brain inflammation, and he speculated on the contribution of insects or birds to its transmission. A "large number of local influential natives" approached him, begging him to "rid them of this killer," just as medical treatment had recently eliminated yaws, which they "always thought was brought about by Hoodoo." He concluded ambiguously: "They, in their primitive minds, now think that the Doctor has the power to beat their sorcerers."[90]

Vin Zigas resting on patrol in South Fore, Sinoko standing above him. Photograph courtesy Peabody-Essex Museum, Gajdusek Collection

Gunther immediately wrote a letter to Anderson and Burnet, formally inviting them to become involved, and then asked Charles Julius, the shy, methodical anthropologist in the Department of District Services and Native Affairs, to visit the region. In early 1957, Julius went to the area around Wanitabe and Kamira, further south than the Berndts had ventured four years previously. He was able to differentiate for the first time local groupings and to describe variations in participation in ritual cannibalism. Significantly, he noted that in most areas "only females and uninitiated males had the right and duty of eating dead relatives." Julius detected only six different types of sorcery among the Fore, but the intensity of their obsession with kuru—"for which lists of remembered dead were invariably long"—more than made up for such an impoverished repertoire. He recalled one of the kuru sufferers:

She was trembling spasmodically, and was not able to walk without assistance, her leg movements being jerky and uncontrolled. Her face seemed to have a rather fixed or "rigid" appearance, and in speaking her words were slurred and indistinct. In discussion with her husband, he told me that she had been suffering from kuru for about two months, and that her condition was steadily becoming worse, although she was still being helped down to her garden every day.[91]

The woman's relatives had quickly settled on a known enemy as the culprit. They told the anthropologist how the sorcerer had obtained a piece of clothing and performed magical operations on it. In this case, divination rites had not been necessary, but Julius learned how they detected more elusive sorcerers too. In Wanitabe, people remembered eighty-five kuru deaths; in Amora, seventy-two deaths; and in Okasa, seventy-one. Most sufferers were women and children, and few if any seemed to recover. Fore remained convinced that kuru, unlike some other forms of sorcery, was beyond the scope of European treatment. Julius advised Gunther that until a cure is discovered, "it seems improbable that any amount of explanation or propaganda will achieve much in removing what is the main cause of suspicion and insecurity in an otherwise unusually harmonious group."[92] Neither Julius nor Gunther made the connection between the gendered and generational patterns of human consumption and the apparent epidemiology of kuru.

The evident pragmatism and adaptability of the Fore during the 1950s made it hard for some outsiders to believe that they were dealing during this time with a devastating epidemic. Under the circumstances, their dignity and fortitude, and their generosity to strangers, were remarkable. Soon, though, the Fore region became for many the kuru region. The bodies of the Fore and their social lives were reframed in relation to kuru; the census of the Fore became a kuru census; and the map of the Fore delineated a kuru map. Thus within a decade the term *Fore* went from colonial linguistic fabrication to signifying medical stigma: henceforth the people would be medicalized as they were colonized.

MASASA SHRUGGED WHEN I ASKED HIM at the Okapa market in 2003 about the first time he saw a white man, as though it was not a question that interested him any longer—white men perhaps having become rather ordinary. Fifty years ago, as a teenager, he had seen Harry West on patrol. "When the white man arrived," he told me, "we saw it as a dead man come back to life, we saw that and were afraid. We hid all over. But then they put away their knives, and took the salt and said, 'Come, come, try, try,' with salt in their hands. The food was good. 'Come, come.' So we gathered and looked." Masasa was curious and followed the patrol back to Kainantu, where he stayed with the *dokta boi* and learned Pidgin. When he left the hospital and returned to Yagusa, he said, "'I've brought salt, tobacco, soap, newspaper, and such things,' and they saw it, my parents, and said: 'Good things! We must stay in the open, listen to them, follow them, and bring up the village.' They said that."[93] The exchange of goods was the most common and most powerful means of overcoming mutual fear and suspicion. Fore thus recognized similar needs and tastes in the strangers: these outsiders also knew the sort of things that moved people, they shared interests in consumption.[94] There were other, less conclusive means of deciding whether these white or red newcomers were human, or what sort of human they were: Fore studied their physiognomy and color, touched their genitals, and observed bodily functions and sexual activity. They watched the strangers closely and generally without prejudice. These highlanders were, above all, pragmatic and creative, receptive to new ideas and social change, and talented at manipulation. It was not long before they noticed differences between government officers, missionaries, anthropologists, and doctors, partly because each group expected and offered different goods—each group carved out a distinctive pattern of transaction. Exchange proved the most sensitive and efficient mechanism for working out who these people were, what they wanted, and what use they might be.

The strangers often used much blunter tools, and less rigorous analysis, to identify the Fore, largely because their encounters with "primitive" people were quickly routinized. Dogmatic assumptions about native childishness, ignorance, sorcery, and cannibalism sometimes impeded understanding the local society. To many government officers on patrol, the Fore primarily were uncontrolled natives; to missionaries, heathen ready for salvation; and to the first anthropologists in the region, they were endangered (and dangerous) primitives, conveniently filling for a short time the "savage slot."[95] While the Fore sought relatedness with the newcomers, some way in which to *interest* them, the Australians often resorted to asserting a binary opposition to distinguish themselves from natives—as though they could not afford to become "human" and related. Both Fore and outsiders were prone to try to fit others into their various preexisting categories, whether of relation or incommen-

Crossing the Lamari River near Agakamatasa. Photograph courtesy Peabody-Essex Museum, Gajdusek Collection

surability, but the Fore seem more readily and adeptly to have revised and adjusted their first impressions. They often did better at "accommodating what is given to the circumstances of interpretation," while the outsiders appeared distracted, thinking of something—or someone—else, as though they had been there, or done that, before.[96] That is, until the novelty of kuru focused their attention.

Chapter 2

PORTRAIT OF THE SCIENTIST
AS A YOUNG MAN

GROWING UP OBSESSED WITH SCIENCE IN
Yonkers, New York, during the 1930s, Daniel Carleton Gajdusek eagerly read
René Vallery-Rodot's *Life of Pasteur* (1913) and Paul de Kruif's *Microbe
Hunters* (1926), a set of racy stories about medical detectives. A bookish youth,
he painted the names of de Kruif's scientific heroes on the stairs to the attic
of his childhood home. Later he immersed himself in Sinclair Lewis's *Arrowsmith* (1925), fascinated by the scientist's tragic tropical quest, and spent
weeks leafing through H. G. Wells, Julian S. Huxley, and G. P. Wells's *Science of Life* (1931), a vast compendium of contemporary biological knowledge.[1]
He always knew he would be a scientist, like his mother's sister Irene Dobroscky, who worked as an entomologist at the nearby Boyce Thompson Institute for Plant Research and was a close friend of the geneticist Barbara McClintock.[2] Tante Irene took him on expeditions along the Hudson River
collecting insects and bugs. She introduced him to William J. Youden, a mathematician and physical chemist at the institute, who let him play with his
calculator and molecular models and perform simple experiments in the laboratory. One summer, young Carleton helped some other scientists to synthesize 2,4-dichlorophenoxyacetic acid, one of the first hormone-based
weed killers, and his boyhood notes helped to validate the institute's later
patent application.

Ottilia (Mahtil) Dobroczki Gajdusek wanted her son Carlti to do well in
America, to work hard and become a famous scientist and physician. Mahtil,
or "Mimsi," whose Hungarian parents had migrated in the nineteenth century, wanted him to do better than her husband, Karl, a Slovak-born butcher
who lacked Mimsi and Carlti's dedication to books, learning, and the cele-

Carlti (right) *as a pirate, age eight. Photograph courtesy Peabody-Essex Museum, Gajdusek Collection*

bration of European cultural achievement. She hoped that Carlti and his brother Robert (Robin) would distinguish themselves from other boys in Yonkers, then a polyglot, largely Eastern European town. She fostered the sense they were an uncommon family, set apart.[3] As a mother, she intensely wanted them to stand out and justify her pride in them. The boys rejected their father's commitment to commerce, though they appreciated his humor and enthusiasm, the joy he found in work, dance, music, and conversation. His industry allowed them to live in a large, solid house high on a hill above Yonkers, surrounded by fields. Trained as a social worker, Mimsi expressed more somber academic and esthetic aspirations, an attraction to eugenics and studies of human development, and interest in the classics and culture. She read the boys Homer, Hesiod, Sophocles, Plutarch, and Virgil, and played Bach and Beethoven on the gramophone. They grew up talking about Kant and Darwin.[4]

Lanky, awkward, and ambitious, Carlti studied hard. He was immensely talkative and eager to learn, and only frequent migraines checked his energy and enthusiasm. At Roosevelt High School he excelled in science, edited the student newspaper, and became president of the French and German clubs. On his way home, burdened with books, he tried to avoid the rough Italian boys—"the guineas"—who taunted him. After school he took classes in Egyptology at the Metropolitan Museum of Art, and in entomology, geology, and botany at the American Museum of Natural History. Lacking any aptitude for team sports, Carlti attached himself to the local scout troop, which gave him opportunities for outdoor exercise and companionship. In 1937 he was the senior patrol leader of the Yonkers delegation to the National Jamboree in Washington, D.C., and in 1939 he led the delegation to the World's Fair at Flushing Meadows, New York. When he was sixteen, he guided camping trips to the Adirondacks and the wilder reaches of upstate New York, discovering a love of untamed nature.

No ordinary boyhood, it was more an intensive period of training. Emerging from the Depression in New York City, a vessel for immigrant aspirations, D. Carleton Gajdusek possessed a sense of destiny along with an abiding need to prove himself worthy. A gloss of self-confidence and boundless self-assertion covered over intellectual neediness and desire for recognition and approval. He was restless, burning for achievement.

CARL GAJDUSEK ENTERED the University of Rochester at the age of sixteen, determined to study physics. He wrote weekly to his mother to discuss what he was learning and to ask for more money; she responded with sentimental exhortation and occasional admonition and cash. Although it was the eve of the war, they corresponded partly in German, extolling German authors and composers. Carl, however, became increasingly humanistic and opposed to racism, even chiding his father for residual anti-Semitism. He refused to join a fraternity and mixed little with his classmates. He slept only a few hours, preferring to spend most of the night reading or walking. He thought a lot about the emotional and educational development of children, slowly coming to admire Soviet methods of human advancement. "Complete communism," he told his mother, "where loyalty to the state (not to a few leaders, but the entire world community) replaces parental and other family loyalties, is what I look to."[5] She scolded him for his youthful idealism and naiveté, then sent him another check.

It was not long before Carl realized that he wanted to conduct research into the biochemical and biophysical aspects of medicine. For a while, after taking a history course, he wondered about becoming a "medical administrator in occupied territory."[6] But mostly he found he was interested in "the application of quantum theory and other means of physical analysis to the problems of biophysical chemistry, the very basic problems of biology."[7] A few years later, he read Erwin Schrödinger's *What Is Life?* (1945) and saw this manifesto as affirming his enthusiasm for applying the laws of physics to biology. While at Rochester, Carl introduced himself to Edwin B. Wilson, the Harvard public health statistician, who encouraged him to gain his medical qualification and begin laboratory research. In 1943 he entered Harvard Medical School, and in his spare time studied protein chemistry—which he called "thrilling"—with John Edsall.[8] "It is wonderful to study the human body and its adjustments to disorders," Carl, still a teenager, wrote to his mother. "The study of its pathological conditions is exciting and bewildering."[9] But in his journal he wrote, "Pure or theoretical science shall be my vocation."[10]

After completing his medical degree at Harvard, Carl endured an internship at Columbia Presbyterian Hospital and a residency in pediatrics in Cincinnati. Although easily bored with clinical routine, he found encounters with children always refreshed him. Other entanglements, however, might disturb and unsettle him. In 1947, while in the Midwest, Carl spent an awkward evening with Qin Zhenting, whom he had met when she was a pediatrics resident in New York, which resulted in the birth of a boy in Beijing nine months later. Surviving the turmoil of the Chinese revolution, Qin separated from her husband, devoting herself to her child and her work at the Beijing Medical College. Carl never saw the boy, Chen Wuguang, though he did write to him suggesting a new name, Tadzio-kai, as homage to Thomas Mann. Many years later, in 1973, Qin would tell him his son had died from "prostate cancer," harassed by Red Guards, his medical care neglected during the Cultural Revolution.[11]

"I still cannot even consider practicing medicine," Carl wrote in 1947, "for I have NO interest in private practice at all. My only hope is that . . . I shall have enough uninterrupted opportunity for research work to produce some completed investigation."[12] He knew already that he was heading west to the California Institute of Technology, to undertake biological and biochemical research. But it remained unclear whether he would be able to combine laboratory studies and clinical work in the future, or just what sort of investigation might satisfy his curiosity.

Caltech emerged after World War II as a center for the application of methods and insights from chemistry and physics to the understanding of life. In the chemistry division, Linus Pauling employed x-ray crystallography to elucidate protein structure and describe molecular mechanisms of disease, and John G. Kirkwood studied physical chemistry using electrophoresis and the centrifuge.[13] In biology, Max Delbrück, who had trained in physics with Niels Bohr, was taking the bacterial virus, or bacteriophage, as a conceptual model of gene action, an example of infectious heredity. The potential of phage for explaining microbial genetics had been recognized in the 1930s, when F. Macfarlane Burnet discovered a mutant virus whose character in the bacterial cell changed permanently and hereditarily.[14] Phage soon joined the fruit fly, *Drosophila*, as an experimental model for studies of gene activity.[15] Delbrück, an acerbic eccentric and iconoclast, encouraged physicists and mathematicians to join him in studying the biological problems of phage replication. Although Schrödinger had argued that Delbrück was seeking new laws of physics through the investigation of biological events, the phage scientist was skeptical of simple reductionism and mechanistic thinking, urging instead a deeper appreciation of the complexity of life forms, even the humble bacterial virus.[16] He expressed a playful dislike of biochemists who favored a mechanistic approach to the study of life, employed cumbersome techniques, and became deadened by medical inanity. A "scientific cult," as Lily E. Kay put it, developed around Delbrück in the late 1940s: he recruited scientists at the annual phage course at Cold Spring Harbor, and with charm and authority, leavened by the occasional practical joke, he forged a cohesive and admiring community in Pasadena.[17] A coterie of young scientists, including Gunther S. Stent, James D. Watson, Elie Wollman, Benoit Mandelbrot, Carl Gajdusek, and others, partied at the Delbrücks' house, ate the elaborate meals that Manny Delbrück prepared, and spent weekends on camping trips together.[18] Through this "blend of merrymaking, rarefied intellectual atmosphere, and romance of the wild," Delbrück sought to implant Bohr's "Copenhagen spirit" in the foothills of the San Gabriel Mountains.[19]

Carl found himself on the margins of the Californian phage cult. Pauling, on Edsall's recommendation, had drawn him to Caltech, but he soon swayed toward Delbrück, though he never became fully initiated into the phage group. He liked working nights in the laboratory, mostly using electrophoresis to characterize "nucleoprotein," or putative genetic material, sometimes studying influenza virus hemagglutination. But bench work resembled "menial labor," and he preferred reading novels and anthropological treatises and learning Russian. "The self-control and application, the constancy and the unity of purpose of a good research worker I lack, I fear," he wrote in the winter of 1949.[20] Increasingly, he inhabited instead the imagined worlds of Franz Kafka, Thomas Mann, André Gide, and Albert Camus. He might pass the whole day on his bed at the Athenaeum reading James Joyce's *Ulysses,* and venture into the laboratory only when it was dark outside. He recognized his behavior as "hypomanic," but did not seem to mind. In California he also learned that he was undoubtedly a "hedonist," despite the "apparent converse on first inspection."[21] Each weekend he enjoyed guiding the phage group on hikes into the Sierra or high desert. Jim Watson recalled: "Innumerable camping trips occupying two to four days, long weekends often led by Carleton Gajdusek whose need for only two or three hours sleep a night allowed him to spend five or six days each week in the wilderness while maintaining the pretense that he was interested in the world between John Kirkwood and Delbrück."[22] Eventually they would return to Pasadena, past the endless rows of bungalows, along the sunny, empty streets. There Carl would look again for some definitive experiment he might do; awash with ideas, he was searching for a solid landing.

Like others at Caltech, Carl was drawn into speculation on the function and structure of genes, especially virus genes. Wendell Stanley, working up north at Berkeley, recently had shared the Nobel Prize in Chemistry for his crystallization in 1935 of tobacco mosaic virus, which he claimed then was effectively an autocatalytic (or self-forming) protein. That is, the virus seemed to reproduce through protein replication. Distrusting any attempt to reduce a biological agent to a chemical object, Delbrück refused to get excited about virus structure.[23] Some other scientists regarded phage as a sort of enzyme, as a protein native to bacteria, but for Delbrück and his associates it was most importantly an exogenous infective agent with fascinating biological properties.[24]

In any case, nucleic acid was challenging polypeptide hegemony during this period, even at Pasadena where protein chemistry had long been ascendant. Controversially, O. T. Avery at the Rockefeller Institute in New York, in 1944 had identified deoxyribose nucleic acid (DNA) as the "transforming principle" of the pneumococcus, making attenuated variants of the bacterium pathogenic. In his recessive, quiet style, Avery argued that nucleic acid, not protein, was the substance of the gene—thus passage or mutation or reassortment of nucleic acid potentially changed the observed features of microorganisms and other life forms.[25] Nucleic acid had once appeared such a boring, repetitive molecule, lacking the variation and complexity of protein,

Carl hiking in California, c. 1948. Photograph courtesy Peabody-Essex Museum, Gajdusek Collection

yet Avery's studies slowly undermined the complacency of protein chemists. By the early 1950s, members of the phage group were able to prove that nucleic acid alone was responsible for the genetic effects of viruses on bacteria.[26] Already Carl had abandoned his studies of nucleoprotein. He became resistant for the remainder of his career to explanations based solely on the active, especially autocatalytic, properties of protein molecules. Like his friend Watson, who a few years later with Francis Crick would discover the double-helical structure of DNA, he became a convert to nucleic acid as an operational unit and prone to mock mere proteins.

Still in his early twenties, Carl was drifting along, unsettled, chafing against bourgeois conformity and the "family system." "Oh, that we might be Peter Pans and live always in Never-Never Land," he wrote to his young friend Corydon B. Dunham. "What could be worse than 'middle-aged respectability' and all its vanishing idealism, bourgeois complacency, dormant romanticism, sacrificed loyalties and ambitions!" He wanted to remain a free agent, an unburdened adolescent. "I then am dreadfully neurotic and find nothing more terrifying than the respectable, conservative, realistic values that I find creeping up on me."[27] He wrote to his mother telling her that he found "nothing which holds me to the last shreds of convention." Now he planned to "roam the beaches and the deserts of the world, sit in public libraries, ponder in hermitages, where pride, conformity, and convention become meaningless nothings."[28] But the subject of his pondering was still obscure even to him.

At the end of his time at Caltech, Carl knew that he wanted to work on viruses either with Burnet in Melbourne, Australia, or with John F. Enders back at Harvard. He had heard Burnet deliver the 1944 Edward K. Dunham Lecture at the Harvard Medical School, and the Australian's advocacy of the "natural history" of disease appealed to him. Like Delbrück, Burnet was an ambitious thinker: as the physicist sought to use quantum theory to transform biology, the antipodean microbiologist was exploring Darwinian approaches to disease, hoping in particular to add immunology to the new evolutionary synthesis. Both scientists despised mere biochemical reductionism and clinical routinism. But until he completed military service or its equivalent, Carl could not work outside the United States, so in 1950 he returned to Boston. As a medical student, Carl regarded Enders—"the chief"—as reserved and a little distant, and he had preferred to work with Edsall, who was far more relaxed. But in the 1950s he came to appreciate Enders's dedication to developing tissue culture techniques for polio and other viruses, and he found the laboratory work unusually stimulating.[29] The "chief" came to like and even admire young Carl, telling Robert S. Morison, the director of medical sciences at the Rockefeller Foundation, to watch out for this "somewhat flamboyant but extremely original Central European."[30] Toward the end of Carl's Boston sojourn, Enders made sure he found a research position with Joseph E. Smadel at the Army Medical Service Graduate School, part of the Walter Reed Army Medical Center in Washington, D.C.

For Carl it was an opportunity to fulfill his selective service requirement while conducting some creditable laboratory research. He was "drafted" into Smadel's department of virus and rickettsial diseases, where he investigated infectious hepatitis and hemorrhagic fevers, which were major problems for American troops in Korea. Soon he developed another filial relationship, this time with the profane, ribald Smadel, who treated his odd protégé with indulgence. Smadel admired Carl's inexhaustible energy and broad enthusiasms. He believed his prize recruit was "one of the unique individuals in medicine who combines the intelligence of a near genius with the adventurous spirit of a privateer."[31] Carl found himself in the "uncomfortable position of being a rather independent and somewhat favored flunky" in Smadel's laboratory.[32] They shared a commitment to widening the scope of U.S. biomedical research, to supporting international investigations and trials. Smadel had studied tropical medical problems during the war and conducted a trial of chloramphenicol in the treatment of typhus in colonial Malaya in the early 1950s.[33] He encouraged Carl to travel to Korea, the Levant, and the Pacific, asking him to send back reports on disease outbreaks, samples of blood from isolated communities, and gossip from outlying research institutes. It soon became clear that the young scientist had discovered his vocation.

A few years earlier in Boston, Carl had been stuck in an existential morass. "I have lost all my incentive, all my ambition, all my former zeal and energy," he wrote to his mother after reading Søren Kierkegaard in the spring of 1950. "I have lost confidence, lost desire for all practical forms of endeavor, and lost or am rapidly losing all opportunities which might be grasped."[34] Hiking in Maine over the summer cheered him up a little, but his reading on that trip— Joyce's *Portrait of the Artist as a Young Man*—did not help. "In the last few years I look at fame as hollow, at position as ridiculous, at pecuniary goals as beneath me (or of little interest), and scientific discovery as trifling," he wrote soon after his twenty-seventh birthday. "The skepticism and nihilism with which I have become infected has brought to an end—a final death—all possibilities of my worldly advance, title and position-striving, climbing—I scorn all this and cannot help it."[35] But a few years later, in 1953, Carl found an answer. "I am planning a wild, moving and restless next few years which will bring me much further from 'settling down' than I have ever been," he warned his mother. "I am well aware of how busy I must keep myself."[36] More and more, he felt drawn toward the equator and to the exotic. "My childhood made me one to haunt Valéry's arctic seas, ever inquisitive about the tropical seas," he wrote to Barry Adels, a teenage friend. "I know the fate of icebergs in the tropics. In their search for the pleasant warmth and comfort they lose themselves. The narcissus must be a hardy flower!"[37]

AS HE TRAVELED ACROSS the United States, Carl continued to participate in scouting activities and lead groups of young men on hiking and climbing trips. He felt a sympathy with troubled adolescents, especially those boys who would attach themselves to him as though he were their elder brother. Some of these fraternal bonds proved ephemeral but many lasted a lifetime. A few younger associates, including Barry Adels and Richard "Dick" Sorenson, later came to work with him at the National Institutes of Health, though relationships often soured in the laboratory. Still, Carl always liked to be surrounded by a group of admiring younger men, whether fellow researchers in the field like Michael Alpers, or the sons of senior colleagues, or a gang of cheeky and occasionally insubordinate *dokta bois,* or the many Melanesian children he acquired in the 1960s. Like the Fore, he possessed a talent for attracting and adopting outliers and affines, for sponsoring a sort of cumulative patrifiliation, for making young males kin—a band of brothers, so it seemed.

After visiting Germany and observing its devastation in the summer of 1948, Carl had arrived in Pasadena obsessed with the thought of bringing a deserving boy to the United States for education. "It was a thrilling thing to receive all the letters from German children," he later wrote to his mother. "It remains a pathetic situation, that a foreign doctor can in so short a time exert such an emotional effect on so many German children. They are certainly looking for attention and grasp at it when they find it available."[38] He hoped that Mahtil might soon accept another boy into the house at Yonkers. In late 1950, after months of persuasion, he was able to get a German family to part with their teenage son Wolfgang. He assured Mahtil that the lad was from good stock and would cause no trouble. Carl came down from Boston to enroll him in the local high school, drilled him in English, and set him examinations. But Wolfgang was unhappy and failed to thrive in Yonkers. Disappointed, Carl was determined not to give up. "I have, in lieu of marriage and family, accepted this much responsibility," he wrote in 1953. "I have accepted it knowingly, willingly, and gladly, and plan to continue to do so."[39] But when he traveled abroad, Carl left Wolf increasingly in his mother's hands; she tried hard to support and discipline her adopted son. In time, the young man would complete graduate studies at the University of Pennsylvania and become an expert on foreign relations, before eventually drifting away into the American hinterland.

Carl soon realized how much he liked being on the road. He traveled well and often got on better with the people he met along the way than with those back at home. As he moved around he rarely stopped investigating and collecting: he acquired measurements, charts, blood and other specimens, and sometimes even boys like Wolfgang. The unpredictability and complexity of such exchanges and transactions in foreign places never failed to excite him. A road trip from Pasadena down through Mexico with Stent, Jack Dunitz, and Reinhart Ruge had proven particularly rewarding.[40] In their beat-up car, the young scientists followed rough tracks to isolated villages and sometimes encountered Tarahumara Indians along the roads. Besides hiking and climbing, Carl made sure to talk with the more "primitive" groups about the education and development of their children and to take measurements of the people and some of their blood.[41] He soon realized that the most effective means of making contact with hesitant strangers was through playful engagement with the more curious and brave of their children.

Smadel encouraged him to travel farther. In 1954 Carl went to the Middle East to track some mysterious hemorrhagic fevers that might be related to the Korean diseases he was studying. Thinking of compiling a "medical geography" of the hemorrhagic fevers, Carl based himself in Tehran, occupying an elegant apartment on the grounds of the Institut Pasteur, with access to its modern laboratories.[42] The British and United States governments had overthrown Mohammad Mossadegh the year before, and anti-American feeling and animosity toward the reinstalled Shah still ran strong in the capital. The city was virtually under martial law. Carl preferred to spend most of his stay in Iran farther north in high isolated valleys, looking for hemorrhagic fevers, collecting ticks and fleas, taking blood specimens from villagers, and observing child development.

He gained confidence from his "special project of village medical snooping and blood collecting" and became accustomed to dealing with resistant research subjects.[43] In May 1954 he visited a small community in an "enchanted valley" where he "collected some forty serum specimens after much difficulty in persuading the people to part with some blood."[44] The community was intensely religious, so he tried to convince the people that medical research was one of the dictates of Islam. He charmed the children, who willingly gave blood, but most adults remained reluctant. In August, closer to the Russian border: "The villagers resisted our every effort to collect blood specimens and absolutely refused to let us bleed them. They exerted their proud right as individuals—said their bodies were their own and that they wanted to live and die as Allah decreed; they cared nothing for our medical research and refused to have anything to do with it."[45] A few days further on, in Azerbaijan, "several known communists . . . went about decrying us and claiming we were taking the blood to use in soldiers in Korea . . . The political sensitivity and suspiciousness of the region is so great that we cannot hope to get cooperation."[46] Later, having finally obtained a Jeep, Carl was able to venture even farther away from Tehran, into the northern valleys of Afghanistan near the Soviet border. He continued to try to collect serum specimens and to ship them out with the courier back to Smadel's lab in Washington, D.C. He was "no longer the cowering youngster" avoiding the "guineas" on his way home from school.[47]

"To be engaged in research work," Carl wrote to Smadel, "wherein so much is centered far-removed on this other side of the world, and so much depends upon receipt of shipments and letters about which I always worry, is a trying experience."[48] The risks in transferring materials, and the tenuous links between field and laboratory, became a common refrain over the next few years. The logistical demands were daunting. Each specimen required proper handling and preservation as well as a secure means of tracing its provenance and context. Often the global chain broke, and segments of the network dropped out. Specimens might be lost, or deteriorate, or forfeit their connection to a person and disease. Sometimes lab workers threw out imperfect samples without understanding the difficulties in acquiring them. Carl therefore wanted the laboratory in Washington to appreciate the effort that went into obtaining and transporting specimens from distant parts of the world. He needed Smadel to realize the value of this exotic material. Writing of the Afghanistan bloods, Carl informed his patron that "these are of greater importance than the previous IRANIAN sera, for they come from one of the most sensitive Soviet border areas in the world, and they have caused much difficulty, and perhaps more of an 'incident' than we now suspect, to get them. It will be no easy job to get duplicates of these!"[49] The drama of extraction, the aura of danger, the allure of the exotic, all might add value to the samples Carl was presenting to Smadel and others. He was becoming an expert in obtaining, declaring, and dispensing rare and fascinating medical things. As Robert Traub, a recipient entomologist put it, "Only those of us who have been at the ends of capillaries allegedly serving as lines of communication can really appreciate how much you have accomplished."[50]

During a prolonged bout of infectious hepatitis, Carl lay around the Institut Pasteur, listlessly reading Gide, J. D. Salinger, and William Faulkner. After immersing himself in the poetry of Arthur Rimbaud, he briefly succumbed to despair. "In my first two decades," he lamented, "having had naught of the genius, naught of the profundity, naught of the creation and none of the life of this fine youth and boy, I go on to mimic his second two decades rather farcically." He wondered if he was producing only "meaningless scraps of bleeding ego."[51] But as he recovered, Carl regained his optimism and enthusiasm, realizing again just how rewarding his Middle East jaunt had been. Above all, he was developing an effective *modus operandi* for investigating isolated communities: engage first with children and adolescents, ask about prevalent diseases, examine the sick, and take as much blood as possible. Specimens could be studied at leisure in a local laboratory or back in the United States. "Although most of what I came to do I cannot do and most of what I seek I cannot find," the young scientist wrote to Adels from Tehran, "I am encountering so many other, unexpected medical problems for study that I shall have my hands full with the Middle Eastern work for years, IF I wish."[52] On leaving Iran, he slowly wandered back to Washington to analyze the sera in the labs at Walter Reed.

In the United States he still yearned for life beyond the beaten track, and soon set off for Bolivia and Amazonia, collecting data and specimens wherever he went and sending the material back to a grateful Smadel. "Mother, I love the tropics," he wrote from Peru in March 1955. "The filth, the lassitude, the indifference of the jungle-living Latins and Indians is distressing, yet to sacrifice their humanity, their love of life and their uninhibited expressions of their desires and loves for a bit of Nordic cleanliness and German efficiency I would be unwilling to do."[53] But it was only a brief tropical interlude, and soon Carl was planning new adventures in the South Pacific.

WRITING FROM THE INSTITUT PASTEUR in Tehran in 1954, Carleton told Joe Smadel, "The gene-cancer-virus problem at the basic level of dynamics of gene action and the replicative process, and the physico-chemical basis thereof, have always remained my primary interest and concern." He believed that a research fellowship with Burnet at the Walter and Eliza Hall Institute in Melbourne, Australia, might provide him with time and resources to investigate these matters. It promised a "non-high-pressure liberal research opportunity in which I can putter at the seemingly ridiculous to the best of my ability."[54] Carleton eventually arrived in Melbourne in November 1955, after the detour to South America and a visit to Turkey (where he collected 220 serum specimens) and Iran again *en route*. Initially it seemed like "just another city," Calvinist and dull, "provincial, self-consciously isolated from European cultural trends and all the more snobbishly seeking them out."[55]

Under Burnet's direction, the Hall Institute had become a major international center for the study of virus diseases, especially influenza, attracting researchers from Europe and the United States.[56] An assiduous laboratory worker, Burnet, in addition to his influenza virus studies, had contributed to the elucidation of the bacterial cause of Q fever, a nonspecific febrile illness of humans spread through contact with livestock. Furthermore, the phage group had belatedly recognized his pioneering studies of bacterial virus replication, Enders admired his brilliant technique of growing influenza virus in chick embryos, and Smadel closely followed his investigation of rickettsial or arbovirus diseases. Burnet's broader reputation, however, rested primarily on his wide-

F. Macfarlane Burnet. AP via AAP. © 1960 The Associated Press

ranging biological analysis of the causes and patterns of disease. His book *Biological Aspects of Infectious Disease* (1940)—in later editions *The Natural History of Infectious Disease*—was the key text in disease ecology and perhaps the most important monograph in twentieth-century biomedicine. Attending the 1944 Dunham Lecture, Carleton had heard Burnet distinguish his work from those disciplines "which deal essentially with the organism in isolation and which are chiefly concerned with its structure, chemical and morphological, and with its functioning as a single unit." Rather, Burnet associated his own research interests with "the sciences whose object is to interpret and control the phenomena of living things as they are, in such environments as the presence of nature and the activities of mankind have allotted them." He wanted to understand how organisms are "distributed in space and time," and the "long-term historical aspects of the interaction between organism and environment."[57] The influence of Delbrück and Burnet would help Carleton to view biology and pathology as complex intellectual problems, not just the objects of technical, biochemical tinkering.

By the late 1950s, interest in disease ecology led Burnet to focus increasingly on immunology, as he tried to invoke fundamental biological principles to explain the body's mechanisms of resistance. His "biological" concept of antibody formation was predicated on an internalized cellular version of natural selection, an "application of the concept of population genetics to the clones of mesenchymal cells [i.e., those capable of differentiation] within the body." Burnet was unhappy with Pauling's earlier "instructive" hypothesis which suggested that the antigen (or foreign substance) acted directly as a template for antibody production. The idea was far too chemical for Burnet to countenance it, and it failed to explain why the body did not produce antibodies to "self" constituents and how it continued to make antibodies in the absence of the antigen. Nor was he comfortable with Niels Jerne's recent "selective" hypothesis, which stated that an antigen would combine with the best fit among the organism's diverse natural globulins (proteins) and transport it to antibody-producing cells, ready to incorporate and make multiple copies of the presented globulin. Burnet reconfigured this idea, arguing that it would make sense if cells had produced a pattern of globulin for genetic reasons, and the arrival of the corresponding antigenic determinant caused the proliferation of the appropriate clone of antibody-producing cell, or lymphocyte. That is, the antigen was not bearing instructions for antibody specificity, rather it was selecting from diverse cell lines or clones.[58] The clonal selection theory would come to represent the culmination of Burnet's "ecological" thinking.

Carleton found Burnet authoritarian and rather prudish yet a very shrewd and impressive intellectual. In the laboratory the Australian was conscientious and careful and usually willing to share his findings and enthusiasm. He combined a "deeply analytic approach to problems of virology and an astounding command of all biology and basic life science." But Carleton was also aware of Burnet's "propensity to 'steal the show'" at crucial moments.[59] Later, Carleton noted in his journal, "above and over everything in the Institute hovers Sir Mac's envy of any fame in biology other than his own and his inability to tolerate originality and independent excellence in his staff."[60] Certainly the institute director made it clear to everyone that nothing should happen in the small community of Australian biomedical science without his backing. For his part, Burnet regarded Carleton's personality as "quite extraordinary." Although the young scientist was obviously very bright, "you never knew when he would leave off work for a week to study Hegel or a month to go off to work with Hopi Indians." Burnet informed John Gunther in New Guinea that the American interloper seemed "completely self-centered, thick-skinned and inconsiderate," but equally he did not let "danger, physical difficulty, or other people's feelings interfere in the least with what he wants to do." Smadel told Burnet that the best way to handle this legendary figure was "to kick him in the tail—hard."[61] On the whole, the Australian scientist suspected his new colleague was not sufficiently pliant and respectable to be considered completely trustworthy.

Burnet arranged for Carleton to stay at Ormond, his old college at the adjacent University of Melbourne, but the Presbyterian foundation was not to the American's taste. The atmosphere was "inhibitory to my spirit," he told his mother.[62] Frequently he spent time in the nearby neighborhood of Carlton—which he insisted on spelling as "Carleton"—a rundown, largely Italian section of the seedy Victorian metropolis. Migrants had brought "some life to the corpse-like city," and he discovered a few good cafés, theaters, and repertory cinemas. At the edge of the former British Empire, in the middle of the cold war, he was surprised to hear "Communists rant and rave in the parks."[63] During the oppressively hot summer, he passed the time wandering through the city, watching a new Australian film *Jedda* (1955), attending a performance of Ray Lawlor's *Summer of the Seventeenth Doll* (set in Carlton), and reading Robert Musil's *The Man without Qualities* (1951). Carleton pondered this Nietzschean anatomy of nihilism, in which Musil represents scientific rationality as essentially dehumanizing, as dispensing with the claims of experience and human sympathy. Early in the new year, he returned to studying Russian, astounded to find a thriving language program at Melbourne University. Nina Mikhailovna Christesen, his lecturer, introduced him to the circle of liberal intellectuals around *Meanjin,* a literary magazine edited by her husband Clem Christesen.[64] She urged him to read Gogol, Tolstoy, and Dostoevsky in the original Russian. The city was beginning to feel more lively.

Soon Carleton, with a group of American Fulbright scholars, began renting a large house in Mont Albert, a distant, leafy suburb. "It will be rather ostentatious and rather elegant," he boasted to his mother. "We shall live in rakish splendor."[65] The Americans held vast parties in their "mansion," inviting staff from the Hall Institute and other research centers, the *Meanjin* crowd, and many of the Central European athletes attending the 1956 Olympic Games in Melbourne. Carleton spent a lot of time at the Olympic Village getting to know in particular the Hungarians. He also frequently "went bush," taking some Australian boys hiking in Barmah Forest, the Grampians, and the Otway Mountains. "The bush and the Aborigines, the flora and the fauna and geology," he wrote to Adels, "have more to offer the visitor than the suburban sensibility of the conservative population."[66] Early in his stay, he caught a slow bus to Canberra, which to him resembled a heavily wooded Princeton, to consult with a Hungarian virologist at the Australian National University. The "weird and wild temperament" of Stephen N. Fazekas de St. Groth was a relief from the British stuffiness he encountered so often in Melbourne.[67] While in Canberra, Carleton also managed to talk his way into S. F. Nadel's group of anthropologists at the ANU. There he deftly and irresistibly attached himself to Charles A. "Val" Valentine, an anthropologist visiting from the University of Pennsylvania, who was on his way to a field site on the island of New Britain, north of New Guinea. Not long after Val and his wife Edie settled into Rapuri village they wrote to Carleton: "We are more eager than ever to have you here to work with us for a while if you can possibly manage it."[68] He could, of course.

Carleton heard of a measles outbreak among previously unaffected Aboriginal communities in far north Queensland, which he thought might interest Enders or Smadel, so he decided to drop in on tropical Australia before traveling to New Britain. Through duress of personality he persuaded Dr. Tim O'Leary to let him join the Flying Doctor Service, based in Charters Towers, Queensland, on their "runs" in order to reach otherwise inaccessible Aboriginal communities in the north.[69] "I cannot hope to describe all that we did and saw," Carleton wrote to his mother in June 1956. He learned about Aboriginal child development, especially the secondary sexual changes in boys and the age of menarche in girls. He also obtained more than five hundred samples of blood, which he claimed was "the most extensive serum collection ever made on Australian Aboriginals."[70] Smadel had sent him venules to supplement the meager supplies at the Hall Institute, but on this occasion Carleton sent the specimens not directly to the United States but first to new friends in Melbourne, Roy Simmons at the Commonwealth Serum Laboratories and Cyril Curtain at the Baker Institute, thus usefully extending his network of collaborators.[71] Carleton liked dealing with Aboriginal people in the outback. "I loved these people as I seem to love all primitive dark-skinned people," he wrote to Mahtil. "They are really beautiful people and people I admire and respect." As usual, he also appreciated the drama of bush research. "Sharks, crocodiles, taipans, red-backed spiders, giant ant hills, kangaroos on our airstrips and emu eggs . . . all seemed to add excitement to the trip."[72]

From Townsville in Queensland Carleton took a flight to Papua, touching down at Jackson's Field in Port Moresby, where he found five thousand or so whites living among thirty thousand "natives" in a "truly British colonial pattern."[73] Still, he was relieved to meet some whites beyond the pale. "Have no fear," he quickly wrote to his mother. "Hungarians abound here."[74] Soon, though, he flew on to Rabaul, the provincial center of New Britain, and then west along the coast on a Catalina to meet the Valentines, who were working among the Nakanai. It was a rugged part of the island, mostly a series of narrow ridges divided by deep gorges. The Nakanai had acquired a bad reputation in the 1920s after murdering four gold prospectors at Silanga, though it seems there was a history of provocation in the area. In response, a punitive expedition sent from Rabaul in 1926 killed twenty-three Nakanai.[75] R. W. Cilento, the director of public health in New Guinea, participated enthusiastically in the expedition, setting up a medical post at Malutu when not shooting at the inhabitants. "We turned a machine gun on them, wounding them," he reported to the Australian press. "We established a medical station on the site of the village, and some days afterwards . . . the natives returned and eventually handed over those concerned in the massacre [sic]."[76] Having thus tainted medical involvement among the Nakanai for a generation, Cilento went on to a distinguished career in public health in Queensland. It seems Carleton had little knowledge of this history when he embarked on his Nakanai adventure.

Initially the Valentines were not quite sure what to do with their new assistant. They put him to work assessing child health and development, as he planned to do in any case. Eventually Carleton also managed to collect more than four hundred serum samples from the Nakanai, which he distributed to S. Gray Anderson and Eric French for virus studies at the Hall Institute, and to Enders and Smadel. But he sometimes irritated the anthropologists. When he accidentally destroyed some photographs, Val angrily wrote that the "fool is so overconfident of what he can do that he makes a hash of most everything . . . We're furious with him." The Valentines decided to work in the more heavily populated areas "rather than trekking off into the mountains where the scenery is lovely, but where there ain't many people."[77] So they packed Carleton off to investigate the bush people—"with evil reputations which include cannibalism"[78]—thus pleasing him greatly.

At the time, the anthropologists were more generally frustrated by their poor relations with the local missionaries and government officers, which disturbed their rapport with the local people. "Open conflict has been avoided and it's at a much lower and more insidious level," Val wrote. "In a way it will be a relief to leave the field this time as field work hasn't been the duck soup it was last time we were here."[79] He had hoped to study the local cargo movements springing up along the coast, but the missionaries vigorously suppressed them.

According to Val, the priests went on to exploit the Nakanai and conduct a "terror campaign" against some of the dissidents.[80] His critical report on missionary activities received coverage in the New Guinea press and particularly enraged Gunther, who would recall Carleton's association with the troublesome anthropologists. Val heard that his report later became "an issue in the [Papua and New Guinea] Legislative Council, their poor white supremacist excuse for a legislature."[81] But little was done to moderate missionary activity and the Valentines left the field embittered and disappointed. "Our affection for the Nakanai has been somewhat sobered and saddened by the intrigues of the last six months," Val wrote to his parents.[82]

Carleton was developing definite views on Melanesian exchange relations during the couple of months he spent in New Britain. He certainly appreciated the hunger for cargo and the pressure to participate in material transactions. After leaving Rapuri in August 1956, he wrote to Val, telling him he had asked the *kiap* to take "the maximum I can afford for those lads who taught me most while I was there. I would like to send more to others, but cannot, and, being highly individually oriented, I feel responsibility ONLY to individuals of the community and NOT TO ANY COMMUNITY AS A WHOLE." He sent sixteen *laplaps* with instructions that each was to be donated to a specific person "with the clear explanation that it is a gift to HIM alone in gratitude for confidence and friendship shown me."[83] To Vuaroa, one of his favorites, he wrote that "although I am your close friend and hope to remain so, I am no unlimited source of cargo." Rather poignantly, he signed off: "I am sorry that at various times you got frightened of me . . . or at least afraid to 'talk straight' . . . and hope that is all gone now."[84] But the transactions appear to have been satisfactory in the end. When Carleton returned to Rapuri in February 1960, "the kids swarmed about me, moved into the house I was given, and cared for me wonderfully."[85]

"Whatever the time I have given to this New Guinea–New Britain project may have cost me professionally," Carleton typed into his journal, "I have gained in the enrichment of my life beyond my fondest boyhood dreams."[86] In a letter to Adels, he claimed his connections with the Nakanai were "some of the closest and warmest interpersonal ties" he had ever formed. Their acceptance of him into "every aspect of their lives won me completely and permanently to their soul." The thirty-three-year-old scientist continued rhapsodically: "As an intellectual experience my last three months were overwhelming and awe inspiring. As an emotional experience, they were richer and more intense than I had expected life would still have in store for me."[87]

The letters to his mother were hardly less restrained. He had lived "so closely with the Nakanai children that it is all as traumatic separating from them now as it was from Germany in 1948." He knew he must return to New Guinea soon. "My heart is with these people with whom I have slept and eaten and shared a life unknown to us and one into which they were willing to take me with confidence and kindness," he told his wary mother, who a few weeks earlier had been accusing Burnet of letting her son get lost in the wilderness.[88]

Once back in Melbourne, Gajdusek immured himself in the Hall Institute, working now with mounting frenzy. "My work has taken a most exciting and promising course," he wrote to Adels, "and for a full two months I have hardly left the bench and the clinic for at last I have a really 'hot' discovery."[89] To his mother he declared: "I am really on a 'hot' trail and the discoveries of the past month have been exciting and important."[90] He had stumbled on to a means of detecting antibodies to one's own tissues, based on the complement fixation test that Enders taught him, which indicated that many old "collagenous" or rheumatoid conditions were in fact autoimmune diseases. "I believe that my discovery will open up a fully new approach and a completely new field of human pathological physiology," he disclosed to Adels.[91] "It is the first really original and important thing I have turned up," Smadel was told.[92]

Initially, Burnet compelled Gajdusek to settle down and finish writing any incomplete scientific papers, but soon the young man became distracted. Gajdusek thought he could devise a complement fixation test for viral hepatitis, the disease he suffered in Iran. The test would reveal if complement, a protein involved in the immune response, was binding to the product formed from the union of a known antigen and antibody in the patient's serum, thus indicating an immune response to the antigen. In this case, it might distinguish viral hepatitis from other causes of jaundice. Gajdusek obtained some liver from the autopsy of a patient who had died from acute infectious hepatitis to serve as the antigen, and tested the blood of patients suspected of having the disease, but the results were disappointing. Then Ian R. Mackay, a physician-scientist in the Hall Institute's Clinical Research Unit, mentioned a group of women patients with chronic active hepatitis and an unusual blood picture which included very high levels of serum globulin. When their blood was tested, it showed extremely elevated complement fixation titers—that is, when their sera reacted with liver antigen, it led to fixation of complement in a manner indistinguishable from standard antigen-antibody complement fixation. Gajdusek and Mackay, collaborating with Lois Larkin, a brilliant research assistant, set about refining what they came to call the autoimmune complement

D. Carleton Gajdusek at the Hall Institute, 1956. Photograph courtesy Ian R. Mackay

fixation test, since it appeared this group of patients was producing antibodies against a constituent of the body, not a virus. Other patients with suspected autoimmune conditions, such as systemic lupus erythematosis, also tested positive against human tissue antigens. The researchers had discovered a general test for autoimmunity.[93]

In 2005, when I spoke with Mackay on a hot summer afternoon at his house near the beach in Melbourne, he recalled that everyone had been "somewhat in awe" of Gajdusek. They "gazed in wonder at this prodigy," impressed with his intellect and drive, and sometimes on guard and circumspect too. Mackay had taught me medicine some twenty-five years earlier, and he seemed unchanged: even in his eighties he was still fit, relentlessly inquiring, and occasionally sardonic. He described Gajdusek's engaging personality, conversational excess, and extraordinary capacity for work. But he felt that Gajdusek had been too self-absorbed in this case to recognize fully the clinical contribution to the laboratory research, and rationed the credit that should have been distributed more generously. In the end, "nobody felt they could enjoy a collaboration with that person."[94]

In the corridors of the Hall Institute, Gajdusek must have overheard some brief conversations about research problems in New Guinea. Burnet had long considered the birds of Papua and New Guinea as a potential reservoir for Murray Valley encephalitis, which burst out as an epidemic every time inland Australia flooded severely. In 1952 he sent Gray Anderson, a biddable virologist at the Institute, and other scientists to look into the waterfowl and mosquitoes, and any alternative hosts and vectors, of the Eremea River region of southern Papua. Anderson told me he was one of the first white men in the area, and he found the "natives" diffident and evasive. Initially, they did not want him to bleed them, or even to know their names. He felt uncomfortable there.[95] But Burnet, whose son Ian was a *kiap* in the highlands, gradually became more enthusiastic about this new territory for Australian biomedical research, coming to regard it as his own vast colonial laboratory for the investigation of disease ecology. In 1955, Frank Fenner, the professor of microbiology at the ANU, told Burnet that he thought the Rockefeller Foundation might be prepared to fund a "field station in the Australian tropics, and New Guinea is more likely to be a good source of material than Queensland."[96] The following year, Burnet wrote to Gunther explaining that while there were no current plans for collecting specimens, he was however "concentrating on getting the feel of the place, discussing possibilities with those on the spot and where necessary, making preliminary arrangements for a subsequent visit by Anderson to collect what is necessary."[97] Burnet aimed for a systematic investigation of mosquito-borne viruses, including Murray Valley encephalitis and dengue, and the establishment of a small "staging laboratory" at Port Moresby, "mainly concerned in preparing material for transport to Melbourne."[98] He instructed Anderson to get ready to scout out some promising sites for the study of viral diseases.

Burnet's only impediment was that the Rockefeller Foundation was already supporting James Hale, professor of bacteriology at the University of Malaya, in his studies of the prevalence of Japanese B encephalitis in New Guinea. In May 1956, Robert Morison from the Rockefeller Foundation wrote to Burnet: "I certainly don't think the Foundation ought to enter in any way into the decision as to who is to do what where in this interesting region and can only hope that you and Hale and Gunther can work out some arrangement."[99] As usual, Gunther was most emphatic, stating that his "present approach is the work here should be performed by *Australians,* if that is at all possible."[100] The director of public health therefore suggested to Hale that he might get out of the way, which he promptly did. Although Gajdusek was in New Britain during most of these negotiations, when he returned he detected heightened sensitivity to Melanesia at the Hall Institute and unexpected interest in his experiences there.

At morning tea and lunch at the Institute, Gajdusek occasionally spoke with Anderson about New Guinea. "We had a number of conversations to which I didn't contribute very much," Anderson told me cautiously. "But he was interesting." It was 2005, and we were seated at his house in Mill Hill, north of London, looking out onto the garden. Mild-mannered and modest, an exemplary member of the Melbourne middle class, Anderson had felt overwhelmed by Gajdusek's enthusiasm and informality. Nor did he share the American's rapture in the tropics. Regardless of Burnet's insistence, Anderson was happy to leave New Guinea field work to others.[101]

At the beginning of 1957, Gajdusek was preparing to depart Melbourne to return to the United States, intending to stop a few months on the way in New Guinea for further study of child development in primitive communities. He expected to resume work in Enders's laboratory in Boston, but Smadel urged him to consider moving to the National Institutes of Health (NIH) at Bethesda, Maryland, an organization that was rapidly becoming the powerhouse of U.S. biomedical research. Recently appointed associate director for intramural programs, Smadel was presiding over a massive expansion of resources at the NIH for laboratory investigation, and moreover, he wanted to extend the global reach of American science.[102] In July 1956, Smadel wrote to his prodigal son:

If you have finished wandering and are ready to settle down, it might be possible to work you into a base laboratory with the option of expeditions limited to several months a year. On the other hand, if you are still anxious to spend several years in the jungle, living with those wild Indians, it might be possible to support you on the staff while you are doing this.[103]

Gajdusek considered the offer and told Smadel that he might accept a role in his empire, but only "if the NIH is really interested in further tropical virus study . . . and if I can get some leeway in pursuing what I believe are equally important interests . . . comparative child development in different races and environments." He demanded freedom to roam and work hard on whatever topic took his fancy. "All I ask," he told Smadel, "is not to be restricted to weekend beach parties year in year out."[104]

IN APRIL 1957, D. Carleton Gajdusek typed a letter to his brother Bobby from a bush hut in Moke in the eastern highlands of New Guinea, where he had stayed for over a month. "I am stuck with one of the most interesting problems of my life," he tapped out. "A new disease to modern medicine spotted by Vin Zigas, the doctor who is intensively studying the illness with me." Gajdusek was working in Australian territory, digging in Burnet's own backyard. But already he was "enchanted" by the Fore, especially the two curious boys, Anua and Tiu, who attached themselves to him and called him Kaoten. "Melanesia has captivated me, seduced me," he wrote, "and I am in no hurry to leave."[105]

The passage from Carlti to Kaoten, through the avatars of Carl and Carleton, was almost complete. Similar transformations of names and identities and places were occurring rapidly around him, drawing him into the future. His mother Ottilia had become Mahtil, then Mimsi, and finally Mom. Robert became Bobby, then Robin, and was now a poet. The Pamusakina, Keiakina, Yanarisakina, or whatever they once called themselves, became the Fore, and now sometimes the "kurus." Moke, or perhaps it was Pintagori, was turning into Okapa. The scientist at the edge of the phage cult was about to imagine himself a "medical cannibal"—in a sense, going from bacteriophagy to scientific anthropophagy. Soon sorcery poison would be refigured and condensed into putative genes, toxins, and infectious agents. Nothing was stable anymore; everything was shifting form and density, altering its meaning. Like everyone else, Gajdusek was changing, but some fairly consistent patterns were emerging too. "Primitives" captivated him; his appetite for new experience was voracious; he worked ceaselessly. Everywhere he went, he found new fathers and brothers, made affines into agnates, and achieved intimacy through reinventing his family. He was ready to join the Fore in their high valleys.

Before leaving Melbourne he took Nina Christesen's advice and read Nicolay Gogol's *Dead Souls* in Russian. For a while I liked to think of Gajdusek as that novel's Chichikov, an ambiguous figure who scoured the country purchasing the names of dead serfs, or "dead souls," then using them as collateral on loans and mortgages, hoping to become a sort of Czarist big man. But that was too simple. Then I thought of him as Kaoten, and he was transformed instead into one of the lost souls the Fore collected.

Chapter 3

A CONTEMPTUOUS TENDERNESS

ON MARCH 15, 1957, A FORE MAN TOOK HIS bright eleven-year-old girl, Aoga, to see Carleton Gajdusek in a makeshift clinic at Moke. Aoga's mother and sister had died from kuru a few years earlier. In January, after attending a *singsing,* the girl had begun to feel hot and sick. On the way home she walked unsteadily and soon developed tremors and slurred speech. "Patient claims she has been made sick by sorcery of an unknown adult man," Gajdusek recorded in the case notes. "She has heard her father say this! Others about her state clearly that she is doomed, and that progress is inevitable to complete invalidism within a few months." As he examined her, the young pediatrician observed her head trembling and occasionally jerking uncontrollably. She was rational and cooperative, but her speech was indistinct, and she staggered. The fluid from a spinal tap proved clear and normal. Gajdusek and his fellow investigator, Vin Zigas, began to ply her with some new antibiotics and corticosteroids, but she would continue to deteriorate over the following months.[1]

A week before he first saw Aoga, Gajdusek had been in Port Moresby, enjoying the company of Clarissa de Derka and her salon of displaced Hungarian intellectuals. He was still planning eventually to travel around Papua and New Guinea to chart child development in "primitive" communities. He also intended to call on Ian Burnet—the son of his former boss Macfarlane Burnet—who worked as a patrol officer at Lufa nearby in the highlands. But Gajdusek felt obliged first to introduce himself to Roy F. R. Scragg, the acting director of public health now that John Gunther was promoted to the position of assistant administrator of the territories. From Scragg, Gajdusek

learned about kuru. The discovery of a mysterious illness among a primitive, isolated people immediately fascinated him.

On that fateful day in Moresby, Scragg showed him the extensive correspondence between his predecessor Gunther and Burnet and Gray Anderson at the Walter and Eliza Hall Institute. "I was rather furious to think of how secretive [Anderson] and Burnet had been," Gajdusek wrote in his journal:

> I arrived in Melbourne filled with enthusiasm for the Territories, dug up encephalitis reports from here, which Anderson had filed away, and pressed my point of the need of studying arthropod-borne virus diseases here! I urged and discussed my every find to disinterested ears and started planning my own trip here. Six months later I was presented with the fact of Sir Mac's and Anderson's extensive correspondence with the Territory and plans for such studies here. All had been quietly done without even bothering to tell me . . . I went on with my plans and my trip, never being offered the chance to take part in theirs and treated as an intruder into "their territory."[2]

He resolved immediately to go and look at the Fore. "It is not these purloiners of ideas and ambitious politicians who count," he wrote. "I shall simply get to work and *produce* while they are busy trying to steal the field from under me."[3]

Evidently Gajdusek was bitter and angry, but it is not clear whether he realized yet that Burnet and Gunther were allocating the rights to future kuru research solely to Anderson. At the Hall Institute, Anderson had already processed the brain and sera that Zigas collected and sent to him late in 1956. After inoculating brain emulsion into fertile eggs and suckling mice, he detected no virus. The pathologist at the Clinical Research Unit was still examining brain sections.[4] But Anderson had discounted any infectious cause of kuru, and as he did so, his interest waned. Burnet, however, continued to insist he travel to New Guinea to investigate the disease.[5] In February 1957, Anderson wrote to Gunther expressing concern at the "possible danger . . . from hostile natives" who might not submit to medical investigation.[6] He feared that he might risk contracting the disease if he were to venture into the region.[7] He delayed further, dithering, seeking insurance coverage, looking for a way to avoid exposure to the Fore. Meanwhile, Burnet was becoming more frustrated and irritable.

Soon after his conversation with Scragg, Gajdusek took a plane from Moresby to Goroka in the eastern highlands of New Guinea. Even if kuru proved disappointing, he expected that the Fore were sufficiently isolated and untouched to make good subjects for child development studies. In Charles Julius's anthropological report he read how "adults and children have until recently been eating their dead relatives in ceremonial cannibalism. They use counter sorcery to prevent or cure the sorcery-induced disease, kuru, which caused the most deaths. Women and children, particularly, partake of the human flesh."[8] Such exotic practices greatly appealed to him.

After landing in Goroka, Gajdusek quickly arranged a ride over to Kainantu, "a beautiful garden community" set in a "wide, kunai-grass filled valley," with "mountains rising in the distance to great heights on all sides." Zigas was there waiting, ready to show him two Fore women at the hospital, both kuru victims, scarcely alive. The Eastern European medico seemed to Gajdusek "a most vigorous, enthusiastic, intellectually curious and inquiring person whose energy and zest and joy in work and life immediately attracted me and infected me."[9] He detected something of himself in Zigas and knew they could work well together. For his part, Zigas immediately felt drawn to the scruffy, crewcut American striding around in shorts and dirty T-shirt—a skinny, almost frenzied figure with piercing blue eyes and protruding ears. Both of them talked constantly for the next few days, scarcely stopping to draw breath, until they reached Moke (or Okapa, as they came to call it) on March 13, 1957.

On reaching the Fore, Gajdusek described kuru in a letter to his mother. "I received your letter about that dread disease," she responded. "Please do not fool around with the native's taboos or the sorcerers."[10] She must have known her warning was futile.

GAJDUSEK WROTE TO JOE SMADEL in Bethesda a few hours after examining Aoga: "I am in one of the most remote, recently opened regions of New Guinea, in the center of tribal groups of cannibals, only contacted in the last ten years and controlled for five years—still spearing each other as of a few days ago, and cooking and feeding the children the body of a kuru case, the disease I am studying—only a few weeks ago." Kuru was such an astonishing illness that Gajdusek had doubted descriptions of it until he saw it himself:

Classical advancing "Parkinsonism" involving every age, overwhelming in females although many boys and a few men also have had it, is a mighty strange syndrome. To see whole groups of well nourished healthy adults dancing about, with athetoid tremors which look far more hysterical than organic, is a real sight. And to see them, however, regularly progress to neurological degeneration within three to six months . . . and to death is another matter and cannot be shrugged off.

Gajdusek still had no idea whether the disease was psychological, genetic, infectious or toxic, but he was sure that "all of medicine should be interested." In a coy aside, he suggested that Smadel and others at the National Institutes of Health (NIH) might like to acquire some specimens, some tissue and blood for pathology testing—so far, everything had been sent to Melbourne. Also, he asked his old patron if he could secure an extension of funding from the National Foundation for Infantile Paralysis, which had expired in Melbourne. "All I need," he wrote, "is enough to ensure my keeping alive once I get out."[11]

As it rained steadily on March 15, Gajdusek typed another letter in his shelter. This time he wrote to Burnet and Anderson, with Aoga on his mind. He described the girl as "ambulatory, cooperative and a good patient who could still travel easily." A complete clinical and pathological study in Melbourne might elucidate her condition. "Faced with a fatal disease and a young, cooperative patient on hand, with about as early a case as one can have with kuru, this is the case for you, if ever you want one." Gajdusek expected that transfer of his patient would help to mend bridges with Burnet as well as contribute to her diagnosis. He also shared with the Australian scientists information about more advanced cases he had examined over the previous two days. He described a seven-year-old boy who was carried to him, incontinent and incapable of speech. Only three months earlier he was a boy of normal intelligence and physical development. Though still a nice, cooperative lad, he was now too far gone for transport to Melbourne. Death was expected within days.[12]

In his letter to Burnet and Anderson, Gajdusek sketched out a distinct and uniform clinical picture. It was a novel disease syndrome, probably involving damage to the cerebellum or basal ganglia at the back of the brain, the parts responsible for motor coordination, posture, and gait. "Native diagnosis of kuru is as reliable as any modern medical appraisal would be," he wrote. "They know what they are talking about, and once you have seen a few cases, you know also."[13]

Already the two investigators on the spot were prepared to doubt any infectious causation. These cases were not feverish, and any evidence for previous encephalitis was dubious at best. The cerebrospinal fluid always was normal in composition. "Nothing in the course suggests microbiological studies," the pediatrician assured the Melbourne virologists. The behavior of sufferers was so bizarre that Gajdusek and Zigas kept thinking of hysteria, but

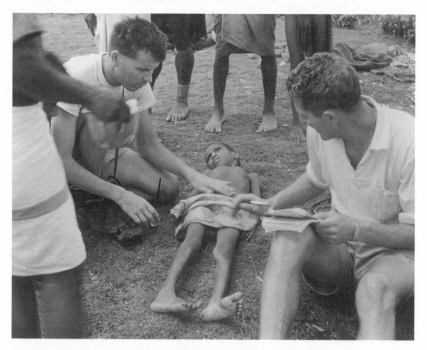

Gajdusek (left) *and Zigas examining child kuru victim at Okapa, 1957. Photograph courtesy Peabody-Essex Museum, Gajdusek Collection*

this did not seem to account for the age range or severity of the condition. They wondered too about a possible "heredo-familial neurological condition," and promised to conduct further genealogical and genetic studies.[14]

Over the following few months, Gajdusek tried to develop a clearer impression of the clinical course of kuru, assembling dozens of sufferers in Okapa at his improvised hospital. Within a week of his arrival he had identified twenty-seven cases; after a month he could locate more than sixty victims. With Zigas sometimes in attendance, Gajdusek attempted to take detailed case histories, eliciting rough genealogies according to a Western biological model. Some curious adolescent males who boasted rudimentary Pidgin soon announced themselves and acted as interpreters, assistants, and playmates. Gajdusek rapidly learned some basic Fore too. He found "the natives as accurate and as good informants and observers as any parents I ever encountered in civilized pediatric practice, if one takes into account the two to three hour discussion a history requires."[15] It appeared the Fore had recognized kuru for at least two decades, and they could identify focal hamlets. The disease evidently ran in families. Having laboriously obtained a personal and family history, Gajdusek carefully examined the patient, looking in particular for signs of neurological trouble. Before long, he could differentiate "real" kuru from rare "hysterical" kuru, and he was able to stage the real disease, recognizing early and advanced forms. At Okapa he followed the rapid deterioration of sufferers and thus inferred the typical clinical course of kuru, which inevitably ended in death, usually (so he was told) within a year of its first manifestation. The disease resisted all medical intervention. Gajdusek and Zigas conducted trials of antibiotics, vitamins, anticonvulsants, tranquillizers, anabolic steroids, corticosteroids, and virtually any other medication they could find, but they detected no effect on the course of kuru.

The early case records, which soon overflowed Gajdusek's workbench, are revealing and moving. Take that of Taranto, a woman from Awande in her early forties, whose sister and young daughter had died of kuru. When Zigas saw her on his first visit in 1956, her gait was unsteady and weaving, and he noticed a marked tremor. She was unable to walk, stiff and swaying, with a shivering tremor when Gajdusek examined her in April 1957. "This must be accepted as a case of kuru," he noted, "at least what the natives mean by kuru." In June he found her "sullen and quiet," and in November she died. In 1956 Zigas had also seen Yo'iea, a forty-five-year-old woman from Busarasa with tremors and unsteady gait. By March 1957 she was incapacitated. Zigas took her to the hospital in Kainantu and treated her with penicillin, vitamin B, and copper binding agents. When Gajdusek examined her after she returned home, Yo'iea was "apprehensive and hyperactive," with a coarse tremor. Soon she was incontinent, lying on the ground, groaning. Late in May, Gajdusek wrote: "Yo'iea is terminal. She lies with teeth clenched in an apparent semi-coma . . . which is questionable . . . and intense rigidity." After her death he performed an autopsy, sending her brain to the NIH. As the cases multiplied, Gajdusek began to concentrate on the children who suffered. Patients like Tasiko, a five-year-old girl from Yagusa, affected him most. He first saw her in April 1957, soon after she developed mild tremors and a tottering gait. Her speech was not yet slurred, and she seemed intelligent and sweet. At Moke he gave her liver extract and metal binding agents, but she continued to deteriorate. "She remains thin and frail, frightened in appearance, but is easily stimulated to smile," he wrote in May. "She talks very little." The following month she was "a wasted, sick-looking child" with severe nystagmus, or wandering eye movements, and coarse jerking of her limbs. She died in August, alert to the end, and Gajdusek conducted the autopsy.[16]

The dozen or so Fore male adolescents gathering around Gajdusek—or

Kaoten as they often called him—watched attentively. Tarubi had come up from Ai, near Purosa in the South Fore district, learned some Pidgin, and trained as an aid-post orderly. When he first saw white men, just a few years earlier, he heard "they were our ghosts, they went and changed bodies and came back." But the patrol officers had "called us and said for us to come. And when we killed pigs to give them they gave us axes, bush knives, that sort of thing." In 2003, as we sat in the dust on the floor of the aid post at Purosa, Tarubi told me how he had eagerly attached himself to the kuru investigators. Pastel postimpressionist murals surrounded us, creating a bush Bloomsbury effect. Tarubi imparted that Kaoten and others would talk with people, looking for cases, then they "observed men and women with kuru, they took their blood and took small samples of feces or urine."[17] Another of Kaoten's retinue, Tiu recalled the young white man approaching him to carry a camera and then asking him to come to Okapa to join the *dokta bois.* "We were always there where the patient was with soap, *laplap,* blankets and other clothing. The first thing we did was give them a full bath because the kuru patients smelled horribly of human feces and other smells." They soon understood "we had to stand close to the people we drew blood from because a lot fainted easily."[18] Paudamba remembered *doktas* or *kuru mastas* "inspecting" sick people and taking blood too. He told me that Kaoten would "put paper in the machine and he hit the machine . . . the little machine, and he'd hit it and hit it. Paper work." If people were sick "he gave them medicine, shots." Paudamba was talking nervously at the coffee plantation south of Purosa. "We sat around him and watched . . . We didn't understand what it was, we weren't clear about it."[19]

Some of the *dokta bois* observed and then rehearsed the intricate performance of neurological examination. Gajdusek was not particularly adept at eliciting signs of neurological disease: he could perform a basic examination, but it was not his specialty. Simple observation and watchfulness gave him the most information. On inspection, rigidity, staggering, tremor, choreic (or dance-like) movements, nystagmus, wasting, and many other signs, soon became obvious. He used a plessor to demonstrate increased reflexes; he tested muscle strength; and he ran through the standard means of assessing coordination, asking each patient to touch her nose, move her index finger between her nose and his outstretched hand, walk heel to toe, stand upright with her eyes closed, and so on. These tests, which largely focus on cerebellar function, seem strange in any circumstances, and they fascinated many of the Fore. In 2003, when I asked Pako, who lives in the hamlet of Waisa, what the doctors would do, he

Gajdusek typing surrounded by villagers. Photograph courtesy Peabody-Essex Museum, Gajdusek Collection

hauled himself to his feet and moved his right index finger between his nose and left hand. He slapped his right hand back and forth on his left hand. He stood on one leg with both arms elevated, as though he might fly off. Finally he closed his eyes and stood swaying. Forty years after he first witnessed it, Pako was still performing the dance of the neurological examination. A week later in Purosa, Tarubi did it too, with minor variations.[20]

The eastern highlands were soon on the medical tourism map. Both Scragg and Gunther came to Kainantu to observe the advanced cases Zigas had brought out—Yo'eia among them. Afterwards, Scragg let Anderson know that kuru "is the most interesting condition I have ever encountered in the Territory." He wondered whether it was a post-encephalitis syndrome, a mineral deficiency, or the effect of a plant toxin.[21] "Clinically they are fascinating," Gunther wrote to Gajdusek, relating his examination of the patients. "I cannot blame you for wishing to partake in any research." The assistant administrator regarded kuru as perhaps "the most interesting epidemiological problem of my genera-tion."[22] Gunther persuaded Sydney Sunderland, the neuroanatomist and dean of the Melbourne Medical School, to visit Okapa. Kuru was a "baffling and fascinating disease," Sunderland later told Scragg, and he "looked forward with great anticipation to examining sections of the brain when they come through."[23] From Canberra, John Eccles, a Nobel laureate recognized for his discoveries in neurophysiology, also journeyed to the Fore to look at kuru. Like Sunderland, he was impressed but puzzled.

Perhaps most helpfully, Donald Simpson, a sage and tactful neurosurgeon from Adelaide, spent some time late in 1957 examining twenty-seven kuru sufferers and provided a complete and detailed neurological description of the condition. Simpson detected mostly cerebellar signs, thus localizing the brain lesion more precisely than Gajdusek and Zigas had managed. Unlike many other Australian visitors, the neurosurgeon became friendly with Gajdusek, and they engaged in discussions of history and literature, dominated by the garrulous American. In 2005, during lunch at the faculty club at Adelaide University, Simpson told me that he went to New Guinea because it sounded like an adventure, an interesting diversion. As they drove down the rough track from Kainantu to Okapa, he felt he was entering a "garden of Eden"—until he saw the ravages of kuru. At the time, the Fore still seemed hopeful that a cure would be found, but Simpson was unsure. And yet, Gajdusek seemed to embody the vast strengths and potential of American medical investigation. For an Anglophile member of the Adelaide medical elite, this was the first direct contact with the exciting world of U.S. scientific research, and Simpson wondered, not for the last time, how to gauge its capacities and what limits there might be to its achievements. He realized he shared this apprehensive sense of expectancy with the Fore.[24]

TOWARD THE END OF MARCH 1957, Gajdusek and Zigas joined Jack Baker, the new *kiap* at Okapa, on a six-day patrol to the South Fore. They were trying to chart the dispersion of kuru. To Smadel, Gajdusek reported that the southern part of this territory was a "little visited region, uncontrolled at its extremity and in which tribal wars, bow and arrow murders, raids and cannibalism, are still rather frequent—in fact, all three infringements of the supposed 'control' have occurred since my arrival."[25] They found thirteen cases even in the most distant clan, most of them women. But it appeared that the disease was restricted to the Fore linguistic area, except for a few groups on its margins who intermarried. When Zigas returned to Kainantu, Gajdusek and Baker—sometimes with John Berkin, a European medical assistant—continued patrolling and case hunting. It was hard walking, mostly through heavy rain and strong wind, contending all the while with deep mud, bees, leeches, and stinging plants. By the end of May they determined that more than one hundred of the twelve thousand or so Fore had died from kuru during the previ-

Gajdusek (left) *and Jack Baker on road to Purosa, 1957. Photograph courtesy Peabody-Essex Museum, Gajdusek Collection*

ous six months, and another hundred active cases were present. Each year at least one percent of the population died from the disease, and in some places almost ten percent of inhabitants were sick with it. "Could any more astounding and remarkable picture be found anywhere?" Gajdusek asked Smadel.[26]

Gajdusek and Zigas amused and impressed Baker, a jocular, sporty young man from Thursday Island, a former Victorian school teacher, light-hearted and insouciant. The new Okapa *kiap* had spent the previous seven years at a torpid post along the Fly River in Papua. Kuru was a shock to him, and he was glad these doctors had already begun to investigate it. He decided they deserved his help, whatever Burnet and Gunther might say.[27] On patrol, Baker concentrated on taking the census, which helped the doctors detect kuru sufferers. Sometimes he also assisted in collecting samples and arranging autopsies. In his report on the South Fore patrol, Baker described an initial resistance to the investigators' request for blood, though "this attitude fairly quickly broke down." Everyone they met was hungry for cargo and keen to interact with wealthy strangers. Baker detected nascent cargo movements around Purosa and high above the Lamari River in the isolated hamlet of Agakamatasa, which soon evaporated. The demand for cash was growing too, even though it could not be spent anywhere within a few days' walk. The South Fore seemed committed to economic and social transformation. Baker thought they were generally well nourished and healthy, apart from kuru, the cause of most deaths in the district. Once the locals found out the *kiap* and his attendant investigators were interested in stopping kuru and wanted to trade, they became engaged and cooperative.[28]

In October 1957, Baker took Gajdusek on patrol across the Lamari River, deep into restricted Kukukuku territory. They led ten constables, a native medical officer, two interpreters, and a long line of carriers. Baker also took with him his pup named "Kuru." Across the Lamari, the party immediately became a sensation. Some hundred men visited their camp at Uriba. "All were armed with axes or bush knives, and tended to become boisterous as the af-

Jack Baker leads a patrol in the North Fore above the Lamari Valley. Photograph courtesy Peabody-Essex Museum, Gajdusek Collection

ternoon progressed," Baker reported. "Several attempts were made to pick my pockets. It became necessary to be quite abrupt in breaking this down with the few who carried this sport to extremes." The Kukukuku were pushing the patrol to its limits and testing its firepower. After bartering his pig, the former owner asked Baker to kill it in front of all the villagers. On seeing his pig shot, the man violently dashed his bow to the ground, pointing to it as he shouted to the assembly, "I told you these things are useless." Elsewhere, the people might be more wary and diffident, but they usually displayed a willingness to trade and sometimes unexpected friendliness. As the patrol progressed, it determined that there was no kuru among these independent people, thus establishing the definite western and southern ethnic boundaries of the disease.[29] "We have put a full boundary around kuru," Gajdusek wrote triumphantly to Burnet, "by well over one thousand miles of hard walking, interviewing for weeks on end, and mountain climbing."[30]

The Kukukuku were a revelation for Gajdusek. Despised in colonial New Guinea for their violence and treachery, these rugged individualists captivated the young scientist. In particular, he admired their penchant for trade and fellatio. The exploring party dispensed axes, bush knives, mirrors, shells, *laplaps,* beads, salt, and tobacco, receiving in return pigs, chickens, vegetables, and safe passage. Gajdusek noted in his journal: "As traders they are keen, reminiscent of Latin Americans. Rather than accept what we offer, they bargain and haggle."[31] He treated yaws sufferers with injections of penicillin and dressed wounds. Sometimes he took blood from local inhabitants, regardless of Baker's intermittent protests that people would blame any subsequent illness on the procedure. "The fact is that I have done so much bleeding of primitive people that I am, in all probability, a bit over-confident," Gajdusek reflected. "However, I think I am a better judge than most of the clime of acceptance, and I am certainly not going to bleed for the research study unless the clime looks good."[32] He found the "Kuks" at first "deceptively quiet, retiring and docile," but they soon demonstrated "guile, avarice, and dogmatic rejection of any coercion."[33] Such duplicity both reassured and attracted him. "Their independence, their acquisitiveness, their curiosity, their arrogance, all tend to make them more fascinating people." Moreover, the determination of the males to fondle the strangers' genitals, and their supposedly insistent suggestions of fellatio, excited him. On this occasion and on later visits, the youths all seemed to be "actively homosexual practicing, ardent fellators, and aggressively seeking homosexual contacts." He loved their odor of pig grease, sweat, smoke, and semen. Gajdusek speculated on the "association of active

Kukukuku hamlets along ridge. Photograph courtesy Peabody-Essex Museum, Gajdusek Collection

homosexuality with ardent individualism, wanderlust, and 'instability.'"[34] "Their vanity and aesthetic sensitivity, and their complex and prudish lasciviousness make them an intriguing people to study," he noted.[35]

As he gained confidence in his bush skills and became more adroit in dealing with highlanders, Gajdusek often patrolled without Baker, with only his band of adolescents and a few additional carriers for company and assistance. Modeled on his experience of government patrols and scouting expeditions, these field trips inevitably acquired a distinctive personal cast. His attachment to his *mankis,* the attendant adolescents, was intense. "I tried to win their favor with friendship and payment," he wrote in his journal, "payment being part thereof hereabouts, I find." The boys' presence gave his investigations an erotic charge. Agurio, one of the "Kuks" in his coterie, would sit beside him "putting his cheek against my shoulder, passing his arms around my waist or shoulders . . . fondling my arm, foot, or thigh. He loves to make a game of such dalliance."[36] Gajdusek felt comfortable with the boys, especially on patrol, and did not miss contact with the outside world. "To me they are, as were my friends of New Britain, among the warmest and closest friends I have had. I respect, admire and love them."[37] Some became particular favorites. Waiajeke, always called Haus Kapa ("tin house"), was an ugly, charming lad who took nothing seriously and played the buffoon.[38] Tiu seemed especially lively and intelligent, while Anua was self-possessed, sincere, and sensitive. Masasa efficiently supervised medical supplies and dressed wounds, but he was "nobody's fool and does no needless work." Irresistibly, Kaoten was drawn to Wanevi, a "cooperative, frisky and fun-loving" boy who stuck beside him on the track, carrying his camera.[39] "He and I responded to each other with understanding and intimate inner knowledge of each other and sensitive understanding of each other," Kaoten wrote after their journey through Kukukuku territory.[40]

Masasa first attracted Kaoten's attention by talking to him in Pidgin when he passed through Yagusa. The fifteen-year-old followed the *dokta* back to Moke, where he acquired a *laplap* and shirt. "Then, wherever kuru was, we went there and brought back people with kuru. We talked to their mothers or fathers, and we carried them with sticks to the *haus sik* over there." Surrounded by his sons, the old man was pointing up the hill at Okapa. "We took their blood and waited. And when they died, we cut them." Unlike some government officers, Kaoten did not get angry and hit the boys—in 2003, Masasa remembered him as an unusually friendly white man. "We were like his children, and we were happy to walk alongside him." Together they searched the region for kuru cases, telling people that "the doctor would give them medicine to revive this person, to heal the sickness." When relatives seemed reluctant to permit any taking of blood, Masasa and the other boys tried to explain what was happening and reassure them. "They didn't understand. They thought he was taking the blood and making medicine."[41]

The boys generally worked happily together, though occasionally they regarded one another as rivals and argued. Sometimes Kaoten caught the boys from Agakamatasa hazing the others. Even if they were paid little at first, they all received food, tobacco, soap, and clothing, and came to learn more about white men. They began to call themselves *dokta bois* or *liklik* (little) *doktas,* and found their status boosted in the community. "We patrolled for sick folks," Tarubi told me one day in Purosa, "and when we found a person laid down by kuru, I would turn him, bathe him, dress his sores. Like I was a doctor."[42] From time to time, their association with the kuru investigators, and their presumed access to cargo, excited envy and resentment in their own communities, but generally they were seen as useful local agents of "development." Among the Fore they traveled unarmed, though some of them became uncomfortable around traditional clan enemies.[43] "We had to sleep at night with people we did not know in order to reach the village with kuru patients,"

Fore dokta bois. From left, *Morieto, Taka, Masasa, Tiu Pakieva, Waiajeke, Anua.*
Photograph courtesy Peabody-Essex Museum, Gajdusek Collection

Koiye told David Ikabala in 2003. "Most of the time we were scared because sorcery still existed and we didn't want any sorcery falling on us, or the villages we slept in accusing us of being sorcerers." He remembered often being "really scared out of our wits."⁴⁴

Farther away, among more belligerent people, the boys might carry bows and arrows but never resorted to using them. On the expedition to the Kukukuku, Masasa recalled using gifts of salt to coax the people from the bush. "We'd give them matches, and salt, and grease, and they'd see it and come out." Sometimes Kaoten would then "hug them and give them good feeling and make jokes."⁴⁵ Tarubi remembered some Kukukuku "held bows and arrows to shoot us with, and Kaoten yelled out, you can't do that, and he gave them salt, or bush knives, everything, to ease their anger, so that we could walk through." A few of the boys sensed that the people beyond the Lamari River feared Fore because they created kuru. "They were afraid," Tarubi whispered to me, "saying that you are the men who make this and you'll kill us."⁴⁶

When Sena, Anua's son, asked Pogasa why the doctors usually got on so much better with the Fore than the government officers did, he offered what I suspect was by then a proleptic and metaphoric response: "Their grandparents went from here not so long ago so they were able to catch up to our way more quickly than other Europeans." Sena was listening to Pogasa talk about the past in Purosa. Doctors, the old man continued, "asked about blood and also took our blood and gave us medicine to take, but did not cure it because I feel it was a man's cause. It would have been done and gone if it was a sickness."[47]

IN THE BUSH, Gajdusek strained to preserve, document, and transport the specimens he extracted from the Fore. He struggled to standardize the conditions in which he obtained and prepared blood, brains, and other tissues so these materials might be studied reliably in distant laboratories. He labored to link inextricably such body fluids and tissues with a particular clinical record and provenance, though frequently the bonds between them proved weak. With time, after the arrival of necessary equipment, some analysis could take place on the spot or in Okapa, but the most important investigations required the scientific resources of Melbourne and Bethesda. The stuff the scientists collected had to travel far. These human materials required careful stabilization, secure identification, and repackaging in the *entrepôt* bush laboratory so as to be legible at their final destination.

Following the logic of his earlier field work, Gajdusek entered the region ready to look for a microbial cause of kuru. "We even delayed our departure [from Kainantu] to obtain buffered glycerin in which to store autopsy tissues for virus studies." They carried "equipment to do further autopsies and to collect further specimens for extensive microbiological studies, especially serological and virological."[48] Soon after arriving in Okapa, Gajdusek realized "our immediate need is a treatment hut . . . with good table for examinations and LPs [lumbar punctures] and a desk-table for writing and for laboratory studies"—for "to study the disease in the home and village is hopeless."[49] Within a few weeks, with Baker's help, he established his "mat-floor hospital . . . in which we have a microscope, hemocytometer, a host of reagents, and all the diagnosis instruments that such a 'bush' hospital would be expected to possess."[50] But once the Fore understood the doctors had no cure for kuru it became harder to entice patients to Okapa. Therefore it was necessary to test, and sometimes later to dissect, most kuru victims in the bush. In such conditions it often proved difficult to prepare tissues as instructed. "We have no appropriate cannulas," Gajdusek warned on one occasion, "nor is the cold wind and rushed excitement of 'bush autopsies' conducive to careful and accurate perfusion."[51]

Keeping track of clinical records, ethnographic notes, and the various body tissues and fluids he collected proved a major challenge for Gajdusek. With-out meticulous inscription and identification, specimens were useless. In the bush, the scientist therefore sought to make the links between family and per-sonal histories, clinical examination and course, and blood and tissue sam-ples as robust and enduring as possible. By August 1957, Gajdusek boasted complete charts on more than 150 kuru sufferers—or "patients," as he was calling them—including their histories, circumstances, genealogies, signs of disease, pattern of deterioration, and results of blood tests and other inves-tigations. Sometimes, if the investigators were lucky, the file closed with an autopsy report. From these documents, he wrote, "we can study all that has been done, all that our laboratory tests have shown, and make all the analy-ses we wish from kuru."[52] In effect, this portable archive was mobilizing many of the Fore, making their diseased bodies, social life and environment avail-able for scientific study in Okapa, Port Moresby, Melbourne, or Bethesda. Gaj-dusek reported to Burnet that the "mere paper work of keeping up with our 150-odd clinical case records, the immense epidemiological project we have undertaken, and the shipping out of specimens in all directions is exhausting every spare moment."[53] To Cyril Curtain, a recipient hematologist at the Baker Institute in Melbourne, he wrote, it is "essential that your specimens be ac-curately and exactly correlated with the phase and stage of illness in kuru sera, and at least with the individual and his ethnic location for the controls . . . We have extensive charts on all patients and complete records on all specimens we send out."[54] Scientists like Curtain tested protein patterns in the blood; others did blood grouping, or looked for trace metals or viruses—seeking any-thing that might hint at causation.

"It is difficult writing and working here in bush isolation," Gajdusek wrote to Smadel in July, "and I sadly feel the lack of colleagues and critical discus-sion."[55] Yet hardly a day went by without him typing letters and clinical records, preparing material for analysis, or framing entreaties for equipment and reagents to Scragg in Moresby, Burnet in Melbourne, and Smadel in Bethesda. "We could use further laboratory supplies, for I am ready to do all the field laboratory work possible," he told Scragg just a few days after his arrival. "I am willing to set up and labor in the field over any laboratory tests that can be done here, when reagents and equipment are on hand."[56] "ANY lab supplies will certainly be used," he persisted the following month, "for we are lab-minded."[57] Specimens, photographs, films, letters, and reports went out by road and then on the small planes, while medications, equipment, and visit-ing experts came in. Brains and other tissues from autopsies, along with containers of blood and urine, were air-freighted to distant laboratories, their destination dependent on Gajdusek's relations at the time with Burnet and Smadel. And if he was unsure how to fix and prepare samples, instructions soon arrived from pathologists and toxicologists in Australia and the United States. "This station is continually inundated with visitors, mainly connected with the kuru research project," Baker reported in November.[58] In the previous months, the *kiap* had entertained the Australian medicos Eccles, Sunderland, and Simpson, as well as H. N. "Norrie" Robson and Harry Lander—both from the Adelaide Medical School, which began to evince interest in kuru research.[59] Lucy Hamilton, a dietician from Melbourne sponsored by the Papua and New Guinea Department of Public Health, spent weeks with the Fore to find out what they were eating and which minerals and plants they used. A public health physician, W. E. Smythe, undertook linguistic and anthropological studies. Although Gajdusek tended to represent himself as the lone scientist in the bush, his distant outpost was becoming crowded.

Efforts to standardize the messy field of investigation drew Gajdusek ineluctably back from medical patrols and apart from his adolescent assistants, toward the outside world and into the orbit once again of the institutions and bureaucratic apparatus of modern biomedical science. He moved between the enchantments of the exotic and the apparent rationality of his own recondite practices. He felt deeply the tension between the romance of the primitive and his commitment to biological investigation. The improvised bush laboratory became a local redoubt for the making or stabilization of scientific facts, the production of novel "epistemic things"—that is, a "local condensation point in the economy of scientific practice."[60] It was a precariously enhanced microenvironment for the staging of scientific work.[61] There, on its benches, items from the field might be displaced, magnified, or shrunk and rendered temporally inert. The laboratory allowed a manipulation of scale and timeframe, an amplification of informational density and extension of its durability. "The crucial matter of the proper handling, preparation, and preliminary treatment of so many different types of specimens," Gajdusek observed, "will determine the value of such specimens to research laboratories."[62] Bodies became disaggregated into tissue and fluid samples, translated into scientific things which were then mobilized as lasting bits of valuable data.

For Gajdusek, entry into the laboratory also implied a shift in emotional register. The laboratory was not simply a place for making and mobilizing scientific facts, it possessed its own affective spatial texture, its special order of sentiment. He could feel the difference between laboratory and society all too

Zigas holds child with kuru. Photograph courtesy Peabody-Essex Museum, Gajdusek Collection

viscerally. The laboratory was a distinctive place for fashioning identities and framing relationships—and sometimes a site where intimacies might be foreclosed or diverted. Sadly, it seemed to favor, in Max Weber's terms, "specialists without spirit, sensualists without heart."[63] Gajdusek experienced an alternation between the libidinal richness of the field and the ego defenses of the lab. For him the laboratory was a necessary node in the global network, but it was also a desiccated place where the native goes to die. "I'm not a sentimentalist; I'm a scientist," Martin Arrowsmith had boasted in his tropical charnel house.[64] Unlike his boyhood hero, Gajdusek was trying to be both.

Gajdusek's ambivalence toward the "ascetic ideal" of modern laboratory science had already found eloquent expression in one of his favorite books, Friedrich Nietzsche's *Genealogy of Morals* (1887). Like Nietzsche, Gajdusek deplored herd instincts and slave ethics, imagining himself sometimes as a "beast let loose in the wilderness."[65] Sadly, most other men, through custom and social constraint, had become calculable and developed a conscience. Even supposedly free thinkers like Gajdusek were not entirely exempt from the morbidity of civilization, represented to him in the form of the laboratory. Repeatedly he asked himself Nietzsche's question:

> These proud solitaries, absolutely intransigent in their insistence on intellectual precision, these hard, strict, continent, heroic minds, all these warm atheists, Antichrists, immoralists, nihilists, skeptics, suspenders of judgment, embodying whatever remains of intellectual conscience today—are they really as free from the ascetic ideal as they imagine themselves to be?

To undertake science, Nietzsche went on, "it is necessary that the emotions be cooled, the tempo slowed down, that dialectic be put in place of instinct, that seriousness set its stamp and gesture—seriousness, which always bespeaks a system working under great physiological strain."[66] Gajdusek later reflected on *The Genealogy of Morals:*

> To see, so long before my own awareness, the clear-cut realization of the "catholicism" of scientists and scholars in their blind faith in "truth" and their blind religious, ascetic espousing of "the scientific method" is ego deflating, but it is good to know what company I have kept and to know that in early adolescence I was a Nietzschean without knowing it![67]

Gajdusek's careful, joyful reading of Nietzsche, as of Robert Musil, confirmed his distaste for the ascetic routines of the laboratory, adding impetus to his desire to venture into the field, unshackled and willful.

Lois Larkin, the research assistant Gajdusek had shared with Burnet, came to Okapa on vacation in August 1957, but she spent most of her time there sorting out records and preparing samples in the bush laboratory. Wry and observant, Larkin soon realized it would not be much of a vacation. In Melbourne there may have been a spark of romance between her and Gajdusek, but they eventually realized it was futile to fan it. "Yes, were I ever to be able to marry, I should have married Lois," he wrote in his journal before leaving Australia. "I am convinced of the injustice of any attempt and never again is there likely to be such a chance."[68] In New Guinea he neglected her, disappearing with his troupe into the bush, while she cheerfully set about organizing the samples and paperwork at Okapa. Before returning to Melbourne, she became friendly with Baker. They were attracted to each other, kept in touch, and married in 1959. Gajdusek had some regrets, but realized he had "a roving eye and roving affection, shifting passion and then uneasiness in any sit-

The band of researchers at Okapa, 1957. Gajdusek drinking beer; Lois Larkin seated.
Photograph courtesy Peabody-Essex Museum, Gajdusek Collection

uation of outwardly, socially imposed duty, restriction or obligation."[69] It would never have worked with Lois. In the early 1960s, the skittish scientist secured funding for both of them at the NIH, where Jack drew regional maps and compiled field notes, and Lois established his American laboratory.

On November 14, 1957, the *New England Journal of Medicine* published an article announcing the discovery among the Fore of a degenerative disease of the central nervous system. Gajdusek and Zigas's report on their study of 114 cases of kuru represented the culmination of their claims to "own" this scientific territory, to occupy the field, thus clinching, or consummating, their labors of diagnosis, case mapping, and laboratory testing over the previous six months in the eastern highlands of New Guinea. They described the typical clinical course of the new disease, the rapid development of incapacitating tremors, disorders of gait, slurred speech, and incoordination, leading inexorably to incontinence, choreiform movements, rigidity, inability to swallow, starvation, and death. In early stages, many patients displayed a "marked emotionalism, with excessive hilarity, uproarious, excessive laughter on slight provocation, and slow relaxation of emotional facial expressions."[70] The researchers provided a sketch of the epidemiology of kuru, its restriction to the Fore people, and its preference for women and children. They noted that laboratory studies so far had been unhelpful. Cerebrospinal fluid and blood serum were basically normal, there was no evidence of excessive amounts of trace metals in the bodies of victims, and no viruses could be found. The

Lois Larkin (left), *Jack Baker* (with guitar, on horse), *Lucy Hamilton. According to Lois, Jack could neither ride nor play guitar. Photograph courtesy Peabody-Essex Museum, Gajdusek Collection*

cause of this illness remained obscure. Gajdusek and Zigas discounted any antecedent epidemic of encephalitis, and failed to find any likely toxins or nutritional problems. They speculated that "strong genetic factors are operating in the pathogenesis, probably in association with as yet undetected ethnic-environmental variables."[71] But the reason for this epidemic was still a mystery.

Gajdusek had been able to extract and preserve eight brains from kuru victims. Although pathologists at Bethesda and in Australia could identify no gross lesions, their preliminary studies suggested widespread neuronal degeneration in and around the cerebellum at the base of the brain. There were no signs of inflammation, of any immune response. At the NIH, Igor Klatzo was the first to describe the characteristic pathological changes in kuru brains, reporting some spherical bodies with radiating filaments in the tissue, later called amyloid plaques. Additionally, he found numerous shrunken nerve cells, some with vacuoles (empty spaces) giving them a moth-eaten appearance.[72] According to Klatzo, it was "definitely a new disease without anything similar described in the literature. The closest condition I can think of is that described by Jakob and Creutzfeldt."[73] At the time, few pathologists could have picked this. Creutzfeldt-Jakob disease is a very rare but cosmopolitan, and inevitably fatal, degenerative brain disease, which occurs sporadically, though it can be inherited too. Generally it begins with fatigues or headaches and progresses rapidly to dementia, muscle spasm, loss of coordination, then to death within three months or so. Most victims are elderly and it shows no preference for men or women. As its cause was also unknown, the similarity to the pathological changes of kuru just added to the puzzle.

Ronald Berndt intensely resented this intrusion of medical scientists into his anthropological domain. When he read the announcement of a disease called kuru, a condition that he had not recognized as a major illness in his own studies of the Fore, the social anthropologist dashed off an article on the devastating effects of sorcery in the region. Berndt now conceded that kuru was undoubtedly an organic ailment, but he argued that hereditary transmission was not established and that the "shallowness of genealogical memory" would make future confirmation difficult.[74] Rather, he offered a psychological explanation of the condition. The anthropologist was convinced that fear of sorcery alone might induce the observed organic pathology—as in the "voodoo death" that Walter B. Cannon had described in the 1940s. The scientific investigators, lacking anthropological training, continued to neglect "intangible" social and cultural evidence of the sort that he and Catherine Berndt had obtained. He castigated medical researchers for ignoring the social and cultural causes of disease and for studying kuru "in the artificial situation of an organized clinic or hospital," not in "relatively traditional conditions." Berndt contended that "the medical evidence is not sufficient to rule out psycho-social factors in contributing to the presence of kuru as a disease."[75] Instead, he asserted the authority of his anthropological expertise and experience in the region. Yet over the next decade he and Catherine would refuse requests to share the information they had acquired on the Fore. The material belonged to them and they guarded it zealously.

As news of "laughing death" began to appear in newspapers around the world, medical investigators, anthropologists, and colonial officers alike were appalled. Gajdusek complained that "we have been plagued by reporters and newsmen on every side and I am furious, helplessly at their mercy, and most disturbed."[76] He wrote to Burnet and Anderson deploring "the horrible publicity which the radio and press have given to kuru."[77] The Australian newspapers were shocking. "'Laughing death,' which is a ludicrous misnomer," Gajdusek told Smadel, "fills most Australian papers and magazines with highly distorted and well-padded accounts." He wished "to hell that kuru were less a thing to fire popular imagination—for it will, I fear, be played for all it is worth by the press."[78] The coverage in the United States was little

Six victims of kuru from one South Fore Village. Photograph courtesy Peabody-Essex Museum, Gajdusek Collection

better. *Time* magazine reported: "Sudden bursts of maniacal laughter shrilled through the walls of many a circular, windowless grass hut, echoing through the surrounding jungle. Sometimes, instead of the roaring laughter, there might be a fit of giggling . . . It was kuru, the laughing death, a creeping horror hitherto unknown to medicine."[79] "The afflicted person suffers convulsions and paroxysms of giggles," noted the *New York Times*, "hence its designation as laughing sickness."[80]

In 1957, listening to the radio at Okapa, trying to tune out stories about "laughing death," Gajdusek heard of the Soviet Union's launch of the Sputnik satellite, and the beginning of an insurrection in Cuba. These were far grander territorial claims than any he might make. On clear nights, lying awake in the New Guinea bush, Gajdusek sometimes wondered if he saw the satellite pass overhead.

MEANWHILE, TROUBLE WAS BREWING in Port Moresby and Melbourne. Already angry because Gajdusek failed to call on him in Moresby, Gunther became concerned that an American scientist squatting on Australian territory might dampen Burnet's interest in New Guinea. Soon after receiving his first letter from Okapa, Burnet had written to Scragg, the acting director of public health, demanding to know "what Gajdusek's status is in regard to the Administration." He questioned the young scientist's "tempestuous enthusiasm" and "unorthodox intrusion," but observed that he may be "well suited for such an investigation." Although wanting to keep New Guinea research an "Australian affair," Burnet recognized "the main thing is to get the investigation done as well as possible." This might mean that the Hall Institute merely continued to test the specimens Gajdusek sent them.[81] Burnet also wrote to Gajdusek, asking him when he planned to leave New Guinea. "Your help . . . has badly disrupted our own plans," he told his former protégé. "I don't like untidy situations."[82]

Anderson did not want to go to Okapa while Gajdusek was still there.[83] He had never really liked the brash, ebullient American and refused to contemplate any collaboration, or competition, with him. After talking with the reluctant virologist, Burnet suggested to Scragg in early April that the Hall Institute might have to withdrew entirely, leaving the field to Gajdusek and Zigas. "I am still considerably irked by Gajdusek's actions," he wrote, "but there is little doubt he has the technical competence to do a first rate job."[84] So long as kuru was not related to any encephalitis outbreak, Burnet would not object to Gajdusek's intervention, particularly if he confined his investigations to clinical studies and collecting pedigrees, which bored the Melbourne scientists anyhow.

Gajdusek protested, perhaps disingenuously, that he was doing no more than undertaking preparatory studies in advance of Anderson's visit. "We are anxiously awaiting ANDERSON," Gajdusek assured Scragg in April 1957.[85] To Burnet he reported that he was "most disappointed to learn that Anderson has interpreted my work here, which I had planned as simply some clinical pediatric observations for my own intellectual curiosity and as material assistance to his intensive microbiological epidemiology, as a field-pinching maneuver." Again he asserted they were "anxiously awaiting" Anderson.[86] But Gunther was skeptical. He felt it was "unethical" for the young scientist to enter the field without approval from Burnet or Scragg and then refuse to withdraw. He deplored the interloper's opportunism and impudence. "Waiting for Dr. Anderson would have given a programmed approach to the problem," he told Gajdusek. "The work done would have been work in which I would have faith."[87] Gunther also wrote to Burnet, telling him how much Gajdusek's rudeness disturbed him. "We will do what we can to prevent Gajdusek having contact with the people. I do hope you will persuade Dr. Anderson to come as soon as possible."[88]

From his bush laboratory, Gajdusek typed a conciliatory letter to Gunther. He apologized to the territories' assistant administrator for not having paid his respects when passing through Moresby. Then he explained that after Scragg told him about kuru he received informal permission to visit Okapa and let Burnet know his plans. Gajdusek was disappointed that Burnet's first meager response "indicated more concern for administrative protocol than scientific interest in elucidating a fascinating problem." Even so, since then he had continued to communicate with the Hall Institute and send specimens there. He regretted that his dedication to investigating this epidemic had landed him in "hot water," but believed it would be irresponsible to desert the Fore now. If this meant he appeared a "boorish American," then he was sorry.[89] Gunther was dissatisfied. "There is no doubt he is erudite," he wrote to Burnet. "I received a most charming letter from him, but without any explanation of what [he was doing] and why he was doing it."[90]

In response to Gunther's urging, Burnet decided after all to send Anderson to New Guinea in May. Gunther was relieved. "We seem to be sorting ourselves out," he told Gajdusek. "I want to see kuru research stabilized." He hoped that Burnet would henceforth direct kuru research, with Anderson as his "field manager." "The humanitarian problem is paramount of course," the shrewd bureaucrat assured Gajdusek.[91] But then Anderson did not enjoy his visit to Okapa and felt no inclination to substitute for the American. A fastidious man, he found the transport irksome and the accommodation "uncivilized." Anderson and Scragg spent a few days at Okapa examining sixteen patients and talking with Gajdusek, who seemed very much at home, confident with the people, and familiar with the clinical features of the disease. According to Anderson, the visitors soon concluded that "the best thing for the propagation of the work was to leave Carleton to carry on, which he was glad to do." When I asked him if he considered collaborating with Gajdusek, Anderson paused a moment and said, "I didn't have any thoughts along those lines. I think I felt it best to leave it to him." He was pleased "to come back to civilization."[92] "Kuru is apparently a new disease which has many fascinating and challenging aspects for the investigator," Anderson wrote in his report on the visit. "Dr. Gajdusek is an investigator of world standing and the administration is fortunate that he is able to work at Okapa."[93] Ian Mackay, in the Clinical Research Unit, remembered Anderson as "a mild-mannered, somewhat self-effacing person" and "one of Burnet's very loyal, very faithful hench-people." The assumption at the Hall Institute was that he "was more or less stared out by Carleton."[94]

Even Gunther came to realize that it was hopeless to try to remove Gajdusek or to achieve more than marginal Australian involvement at this point. Donald Simpson recalled him saying regretfully that the Pacific Islands Regiment was organized along battalion lines and it would take a whole division to get rid of the exasperating American.[95] He therefore conceded that Gajdusek, having established himself in the field, was by default the best man for the research, though he insisted that the Hall Institute take charge of all processing of pathological material and Zigas receive full credit for his participation.[96] Gajdusek was now able to promote his activities as the "KURU RESEARCH PROJECT"—usually in capitals. The Department of Public Health began providing some equipment and funds for research, and it allowed Zigas to work solely on kuru. But this more amicable mode of operating proved evanescent. By September 1957 it had become clear to Burnet that most of the choice pathology specimens, and virtually all brains, were consigned to Bethesda, not Melbourne. Indignantly, he announced that the Hall Institute would have nothing further to do with kuru research. As work on kuru was "channeled more and more by Gajdusek to American centres," Burnet wrote to Scragg, "the time has come for us to withdraw."[97] Thus, until his departure from New Guinea in January 1958, Gajdusek would be fully committed to his relationship with Smadel at the NIH. Gunther was disappointed, but he would not give up.

From one perspective these disputes reflect the tension between the increasingly international orientation of U.S. biomedical science during the cold war and the national ambitions and colonial responsibilities of leading Australian scientists. As in defense and economic policies during this period, Australian protectionism and colonialism, along with some residual loyalty to the British Empire, could impede American expansionism in science. But such a view should not obscure the more personal aspects of the contention over social territory, signified here in moral outrage.[98] As Burnet put it to Gunther in January 1958: "Gajdusek's personality is such that joint work with him is quite impossible for any Australian scientist with a sense of scientific responsibility or the normal ethics of scientific behavior." He was still unhappy that "most of the research material of the Territory will be channeled to American laboratories."[99] In contrast, Robert S. Morison at the Rockefeller Foundation attributed the quarrel to the local scientists' jealousy of Gajdusek, partly "based on the Australian feeling of guilt for not having done anything about the disease before." He noted that Gajdusek in fact "seems to have bent over backwards to provide Australian laboratories with material for study." And yet, Burnet, "who unfortunately has the capacity from time to time of being pretty blindly emotional, has accused G. of an inexcusable act of scientific piracy."[100]

Envy of scientific cargo was an element of the *casus belli*, but there may have been other more disturbing reasons for personal animosity. Morison hinted as much when he observed that opposition to Gajdusek could be overcome "were it not for the fact that it is coupled with another difficulty which is of such a character that RSM prefers not to include it in these notes."[101] There was something about the American that made Burnet and Gunther very uneasy. In particular Burnet was preoccupied with the outsider's peculiar ethnic origins and ambiguous sexuality. Early on, Burnet remarked to Gunther that Gajdusek "has apparently no interest in women, but an almost obsessional interest in children." Later, he complained about his "complete lack of the normal decencies of personal and scientific behavior," and lamented the "continuous intrusion of this completely alien personality." Burnet wished that the research was done by "less flamboyant Australian workers," in effect contrasting straight colonial investigators with the queer scientist he longed to expel.[102] Gunther's annoyance with Gajdusek also gradually deepened into disgust. By 1959, he regarded the intruder as "a most unpleasant person." At first he had seemed merely unethical. But Gunther now believed that he was also "unscrupulous and, in fact, dishonest, both academically and morally. He is not a person that one can encourage to do work for us."[103]

Baker and Zigas were stained by their association with Gajdusek. In particular, Baker seemed too ready to cooperate in unauthorized investigations, to assist this pushy scientist of dubious character. J. K. McCarthy, his boss in Port Moresby, worried that "Mr. Baker has adopted the hobby of amateur surgeon." Such an activity was not part of the patrol officer's normal duties. Indeed, McCarthy ominously regarded it as "a unique recreation."[104] Before long, Baker was transferred from Okapa. Zigas incurred even more odium. Although initially he was assigned to the kuru research project, it was soon obvious that he was too closely aligned with Gajdusek and did not represent the administration's interests. Evidently he was conspiring to send valuable specimens to America. Gunther later decided that Gajdusek's confederate never really gave "any indication of any particular aptitude for research or for detailed clinical observation . . . Dr. Zigas has had greatness thrust upon him." The "Baltic graduate" was "not temperamentally suited to be a research team leader or a liaison officer."[105] Once it became clear to Gunther that Zigas was not as loyal as he should be, he too was transferred out of the region, to a hard post at Mendi in the southern highlands. But as with Baker, Gajdusek countered the move by offering Zigas a research fellowship at the NIH.

In October 1957, while Gajdusek was still in the field, Gunther persuaded Norrie Robson, the medical school dean at Adelaide, to join Sunderland and Eccles in a "small expedition" to Okapa. He was hoping to entice at least one of them to investigate kuru and to substitute for the American interloper. Having examined nine cases and spoken with Gajdusek, they confirmed that kuru was a "unique neurological condition well worthy of continued and intensive investigation." Moreover, they felt that "if the nature of the disease could be discovered and its incidence controlled, it would be a magnificent achievement of medicine."[106] To Scragg, Robson wrote: "The whole trip was, for me, a most thrilling and absorbing experience which I wouldn't have missed for anything." Later that year he took a group of scientists and clinicians from the Adelaide Medical School to Okapa. They examined nineteen patients; began genealogical inquiries; collected blood, urine, and specimens of hair and nails; and conducted three autopsies.[107] Robson thought the disease was probably hereditary, perhaps the result of a gene with highly irregular penetrance, so on his return to Adelaide he sought the advice of J. Henry Bennett, the professor of genetics. Bennett eagerly agreed to collaborate with Robson on genetic studies of kuru. They decided to seek funds from the Rockefeller Foundation for a new kuru research project that would commence as soon as Gajdusek vacated the territory in January 1958.[108]

Through 1958 and into 1959, Gunther dreaded Gajdusek's return from Bethesda to New Guinea. He encouraged the Adelaide investigators to fill the American's place, and he sought to delay or even block his reentry. In November 1958, he heard of a shipment of formaldehyde, salt tablets, and containers bound for Okapa. "These, no doubt, are the opening chords," he remarked sarcastically, "which will precede the maestro formally taking up his baton to conduct the 'kuru symphony.'"[109] Gunther and Scragg managed to delay the *kuru masta*'s advent a few more months, making time for the Adelaide team, but that was all they could do. "Politically," Scragg noted, "Australian activity is more important than Gajdusek's return on the day he desires—USA and USSR notwithstanding."[110] Gajdusek, though, was relentless. He reveled in this "XIVth century tryst of Borgias and Medicis in all its brilliant moves and underhand manipulation and in all its Machiavellian sportsmanship of power and feigned honor."[111] He eventually made it back to the Fore in July 1959 and spent the following month on patrol. Through the 1960s, he kept returning every year or two, staying a month or more each time, much to the irritation of Gunther, Burnet, and the Adelaide group. They put up with him, trying to obstruct him where possible and to make his life there difficult. "Well," Gajdusek observed with satisfaction, "I am a difficult thorn to extract from their skin."[112]

MAKING A DIAGNOSIS, going on medical patrol, establishing a bush laboratory—all were means of staking a territorial claim. This was a claim on the bodies of the Fore and, more generally, a claim on the site of investigation. Like the beginnings of most colonial ventures, this research project was im-provised and chancy, prompted more by the enthusiasm and persistence of the man on the ground than by any administrative design. Indeed, colonial authorities regarded the unplanned intrusion of medical science into the lives of the Fore as rash and provocative, an act of piracy and not a policy of de-velopment. Medical science and colonialism came together into Okapa, but their arrival in the field was messy and contentious, and it was by no means structured or hegemonic. It is possible that neither medical science nor the colonial state ever reached very deeply into the lives and thoughts of most Fore—yet for some they were agents of transformation. These political and intellectual projects etched their marks perhaps most trenchantly into the minds of itinerant scientists and colonial officers, giving them a different sense of them-selves and delineating new careers and relationships.

Medical science and the colonial state excelled at producing—sometimes coproducing—hybrid forms and mixed identities, at bringing people and things into fresh alignment or disassociating and recombining them. Tissues from Fore bodies bonded with reagents and dyes in the laboratory to become new things, mobilized and re-situated in another place and time frame. So too might the scientist experience a flickering of identity between Gajdusek and Kaoten. The colonial setting was suggesting unexpected possibilities for the produc-tion of scientific novelty and conjunction.

The territory and its inhabitants were making claims on Gajdusek. As Kaoten he felt drawn to Agakamatasa, a Fore hamlet high in the mountains above the valley of the Lamari, surrounded by rainforest and cloud, five hours walking along steep and slippery trails from Purosa. It became his home, and the people there treated him well. Although he called many of the boys his children, he addressed no one there as father or mother, and rarely if ever con-tributed to the payment of bride price, which was usually expected of rela-

Kuru victim. Photograph courtesy Peabody-Essex Museum, Gajdusek Collection

tives.[113] But he did help with funeral obligations, giving food and blankets. "We greeted him as our white man," Paudamba recalled. Most of the *dokta bois* came from Agakamatasa. They built him a house—a new men's house, in a sense—and a bridge across the Lamari so he could reach his beloved Kukukuku. Like many others from Agakamatasa, Paudamba disapproved of the site for Kaoten's house. "It was a mountainous place. You had to go down and up, down and up. Not a good place."[114] But Kaoten had insisted on the spot, as he enjoyed the isolation and the rare views through the mist. "The day has been filled with clouds again," he wrote in his journal in 1959. "Sun penetrated through for a short period, but the clouds soon moved in again . . . I have eaten luxuriously, my house has been filled with Agakamatasa inhabitants all day, and I have visited their houses and have worked; nothing could be more pleasant."[115] He liked to bring other scientists from America there,

Fore looking at pictures outside Gajdusek's house. Photograph courtesy Peabody-Essex Museum, Gajdusek Collection

so they could see where he lived and what sort of person he had become. "It is a gentle and beautiful land with a gentle and beautiful people, Mother, and do not accept any other accounts." Kaoten was writing from Agakamatasa to his childhood home in Yonkers. "I love it here, I love the people about me, and I am at home here as nowhere else but at home with you."[116] Later he wrote, "I leave here reluctantly, with misgivings and without enthusiasm, leaving what I love and enjoy and, to a degree 'understand,' for what I fear, distrust and to some extent dislike . . . I shall soon be fully involved again, fully ensnared and entrapped in everything in my office, laboratory, professional and personal and family life."[117]

There were times when Kaoten wondered if he was turning into a character from Joseph Conrad's *Lord Jim*. Like the romantic, morally wounded Jim hiding away in Patusan, he was leaving "his earthly failings behind him and what sort of reputation he had, and there was a totally new set of conditions for his imaginative faculty to work upon. Entirely new, entirely remarkable. And he got hold of them in a remarkable way." As a trader in the tropical outpost, Jim atoned for past disgrace and attained some form of redemption. "He seemed to love the land and the people with a sort of fierce egoism," the narrator Marlow said of Jim, "with a contemptuous tenderness."[118]

Chapter 4

THE SCIENTIST AND HIS MAGIC

"KAOTEN AND ANUA BROUGHT ME GIFTS
like tinned fish, rice, salt, towel, and soap to wash my sister, to keep her clean,"
Andemba Katago told Sena, her and Anua's son, in 2004. Her sister Kaliwani
was dying of kuru in the late 1950s and became one of Gajdusek's first patients
among the South Fore. Andemba believed Anua was obliged to make these
offerings since he was married to her. But she had not expected gifts from
Kaoten. "They gave me blankets and asked me more than five times before I
agreed to let them operate on my sister . . . Anua said the white man wanted
her blood to be warm when he cut her, so they were waiting for her to die."
Andemba and her ailing sister at first did not know what to do. Talking with
Sena, Andemba recalled:

> My sister was worried and asked me not to accept any more gifts from your
> father and the white man because we did not have enough to give back to
> them, and it would be a burden if we kept receiving these things. I told Anua
> and he said they were on our side, they were friends who were trying to
> stop the sickness. It made her feel more comfortable, and also me. They
> dressed my sister up nicely with white bandages around her head, stomach
> and legs before they handed her back to me with a blanket, money, tinned
> fish and rice to cook for the funeral. I buried my sister in a proper way.[1]

Such encounters between scientists and Fore, with *dokta bois* and kin like Anua
as intermediaries, soon became common.

Scientists and Fore speculated on the meaning of their interactions; they
pondered the weight and import of these novel events. It was soon evident to

the Fore that Gajdusek and Zigas did not behave like patrol officers, anthropologists, or missionaries. Of course, like other white men, they possessed cargo, which they might dispense in exchange for food and accommodation. But they also carried some distinctive things they held onto, inalienable possessions like stethoscopes, microscopes, plessors, scalpels, needles, and syringes—a horde of valuables that differentiated them from other strangers. Many of these things played a part in the examination of kuru sufferers. Among their stage props, the *doktas* had something that shone a light into the eye, and this ophthalmoscope caused some Fore consternation. At such times, the medical examination resembled a magical performance, but as conjuring it proved disappointingly perfunctory and ineffective. Somewhat more successful were the cameras and movie equipment many scientists brought with them into the region to document the condition. Taking a picture or making a movie frequently cured "hysterical" kuru, causing patients with the real disease to flock toward these bulky machines for a time—then again, disappointment eventually supervened.

The special desire of the scientists for blood and other specimens also excited curiosity and controversy among the Fore. Bloodletting with small arrows or sharp stones was a common treatment for headaches and muscle pain, but the white men wanted blood from those who were well, too. Fore knew that body discards, including menstrual blood, were potent materials for sorcery, yet they doubted these strangers harbored the social resentments that motivated sorcerers or understood the correct technique to make things like kuru.[2] They were neither magicians nor malefactors nor sorcery adepts. Perhaps, though, they could use these body fluids and tissues for divination of sorcery, or to develop something that counteracted kuru. Perhaps they possessed their own way of finding the people who made this thing.

It was the autopsy that prompted most discussion, in part because the white men seemed so anxious about it. Fore were familiar with examining the corpse for signs of sorcery, so maybe the *doktas* were merely doing that, too. But these peculiar white men seemed obsessed with taking away brains. When they practiced cannibalism, Fore did not esteem this organ. Maternal kin of the victim, who dealt with these matters, tended to favor other parts of the corpse, though they never rejected any human tissue except the bitter gall bladder. Fore had trouble thinking of Gajdusek as a cannibal, even when in early days at Okapa he dissected kuru victims on the "autopsy tea-lab-typewriting-bench-emergency surgery table that must be cleared for meals three times a day."[3] Why would he eat these people? He hardly knew them.

Kuru victim on the way to her garden. Photograph courtesy Peabody-Essex Museum, Gajdusek Collection

Some other highland groups more used to exocannibals—those who eat their enemies—had assumed the first white men they encountered wanted human flesh, but such fears dissipated once the strangers entered into exchange relations.[4] No one ever regarded Gajdusek as an enemy, and he had begun transacting with people as soon as he arrived. Still, moments of crisis in exchange relations sometimes prompted Gajdusek to think metaphorically of cannibalism and to speculate on a potential within himself. To the Fore, however, cannibalism was a ludicrous explanation. As Masasa patiently explained to me at the Okapa market, "They knew he was a white man and that he came to straighten the disease."[5]

Scientists like Gajdusek could never be sure what was happening in these interactions. Tiu, one of his assistants, recalled that after an autopsy, "we gave

A woman with advanced kuru supported by her family. Photograph courtesy Peabody-Essex Museum, Gajdusek Collection

the relatives some blankets and other clothing for proper burial, we even brought some food, such as rice, tinned fish and tinned meat to be eaten during the mourning period."[6] Often it felt like a gift exchange in which Gajdusek was presenting cargo in appreciation and return for brains and other body tissues: that is, he sensed a social debt had been incurred and reciprocity was expected.[7] But on such occasions he was uncertain whether he was giving wisely or to the right parties. Sometimes he tried to barter for brains, but those involved in such haggling always seemed uncomfortable and dissatisfied.[8] Another time, he would find that the brain, or some other organ, was simply not available. Nor was it clear to Gajdusek whether the Fore were associating autopsy exchanges—when they allowed them to take place—with other funeral payments. The possibility that the intermediary *dokta bois* already had some relationship with the victim confused matters further. Indeed, at Agakamatasa

Gajdusek himself might be regarded as kin, though he never knew exactly the obligations this implied. The negotiation of exchange at autopsy meant that he was becoming entangled in new relationships with the victim's relatives, or confirming old ones, but he found it hard to manage the process, to ascertain what might move and satisfy people. Like the Fore, he became concerned with estimating appropriate and inappropriate consumption, and with avoiding mistakes in judgment of value and decorum, mistakes that might put him in moral peril.[9] But he was not always successful at this.

Almost a century ago, Bronislaw Malinowski observed in the Trobriand Islands "the savage striving to satisfy certain aspirations, to attain his type of value, to follow his line of social ambition. We shall see him led on to perilous and difficult enterprises by a tradition of magical and heroical exploits, shall see him following the lure of his own romance."[10] But what might we see now if we substitute "the scientist" for "the savage" in this prospectus? Let us imagine that we are south of Okapa, outside a grass hut, watching him take blood.

AT FIRST, GETTING PEOPLE TO DONATE blood or other specimens, such as hair clippings, urine, or feces, was not too difficult. The Fore generally were willing to make a contribution of blood in order to enter into, or cement, a relationship with the scientist; they sought to make him indebted to them or to reciprocate for his donations of cargo. They were responding to his extractive perspective, just as he was trying to anticipate theirs.[11] In such transactions, the Fore and the scientist sought to identify and make visible the things that possessed value for another. Their exchanges of goods therefore additionally made visible possibilities in someone else, revealed the interests and desires of the other party, and told them what sort of person they had encountered. Thus the gift of body fluids and tissues was doubly personalized. For Gajdusek, at least, the specimens stood intrinsically for the body of a particular Fore person; for all parties to these gift exchanges, the objects conveyed something of the identity of the transactors too, as they made claims on them as persons.[12] To engage in these exchanges was to put oneself on the line, to expose needs and desires, to expect recognition and reciprocity. Accordingly, anxieties about the control of exchange were also concerns about the possible transformation of identities.[13] It is therefore not surprising that a few Fore did from the beginning refuse to give blood, to enter into a relationship with the stranger, to explore his potential as a human. But most took the risk.

It was hard to determine what the Fore made of the scientist's specific

Nine kuru patients at Purosa. Photograph courtesy Peabody-Essex Museum, Gajdusek Collection

interest in blood. Evidently they associated blood with life, and some men expressed fears that a donation would weaken them. In the houses of cargo movements, Fore men were sprinkling their blood on wood and stone objects to turn them into the goods of white men. This suggests that giving blood might attach things to donors and bring them into the orbit of humans, make them alive in a sense. Indeed, the blood of the mother contributed to the growth of the baby—thus women, the majority of kuru victims, were the "natural" givers of creative blood. Yet just as some blood made or extended persons and strengthened ties, other blood could be contaminating and weakening. In a distant hut, far from the cargo house, menstruating women were secluded. Their menstrual blood was deemed polluting and dangerous to men. During initiation ceremonies, the nose bleeding of young males, mimicking menstruation, was expected to purify their bodies and make them into men.[14] Thus it seems that the significance of blood—its gender especially—might depend on the circumstances of its shedding. It is possible even that the blood extracted from non-menstruating females was at times a male substance, while the blood obtained from men was gendered female, as in their initiation. Just how this substance was interpreted on any occasion derived from the timing and structure of its exchange, not from any perception of inherent quality. Among the investigators, the Fore valuation of blood remained ambiguous and tantalizing, but they knew that it was a gift of some special order.

In giving blood and other specimens, Fore became enchained to the scientist. With the transfer of their substances into his custody they were trusting him with part of them. There was never any sense that these substances had been fully alienated from their persons. "They are precious specimens," Gajdusek told Smadel, "and have cost us heavily in time and effort to obtain under these primitive conditions, where even the suspicion of sorcery worked on body parts or excreta is a great hindrance."[15] Fearing that sorcerers might obtain these materials and cause them harm, Fore often seemed more worried about the later security of their specimens than the actual process of donating them. For years they had gone to extremes to limit the circulation of bodily wastes and discards, especially feces. In the villages they willingly constructed deep pit latrines, perhaps the deepest on the planet, and if they needed to defecate in the bush, they did so as secretively as possible, sometimes wrapping the waste in leaves and carrying it back to the latrine in their *bilum* (string bag). They made sure too that no remnants of food or pieces of clothing were left lying around. They hid hair and nail clippings. The kuru investigators had become involved in older anxieties about the improper and dangerous circulation of "specimens," and the need to limit access to inalienable, and hence potent, body substances. Therefore it was necessary to earn the trust of the Fore and to demonstrate the security of specimens, to place these substances ostentatiously in closed containers, lock them in a box or refrigerator, and quickly spirit them out of the region in a Land Rover and then on a small plane. "They trusted the white man," Pako told me in the *dokta*'s house in Waisa. "He didn't leave it in the house for two or three days, he would collect the samples and deliver them to the place he wanted to send them. Thus they trusted him." Pako wanted me to understand this: "If we took the urine or feces sample and gave it to another, they'd be afraid."[16] That no one later attributed illness to sorcery on the specimens that Gajdusek and others had taken attests to their success in assuaging Fore concerns about the illicit circulation and consumption of substances. But each new collection required repeated persuasion and reassurance. Then "we focused our minds and gave blood," Tarubi recalled.[17]

Like the rest of Gajdusek's assistants, Tarubi also collected contributions of blood and other body substances or discards. "We didn't rest in taking blood or feces," he told me in Purosa. "We worked full time sampling—at night and in the day—we worked and worked."[18] At first, according to Pako, Fore

Gajdusek taking blood from Fore woman. Photograph courtesy Peabody-Essex Museum, Gajdusek Collection

"gave lots of samples, feces and urine and such, thinking that now there would be a medicine to cure kuru."[19] The practice of countering sorcery with medicine was familiar. Although detection of the agent of the calamity—divination of the sorcerer—was the principal response to sorcery, some effort was also put into restoring order and balance to the victim's bodily constitution. Usually, sufferers and their families consulted with local *yaba* (stone) experts, who possessed magic stones and dispensed herbs and barks collected in the forest. In the case of kuru, an especially refractory and sinister disorder, Fore began seeking out curers, known as "dream men," in distant parishes or among neighboring language groups. After eating psychotropic plants, these specialists entered into a trance state in which they identified candidate sorcerers. They also supervised a meal of pork, over which they spat ginger and forest barks, medicines presumed to work through inversion or reversal of the influence of sorcery. Then they punctured the skin of the sufferer with miniature arrows shot from a bow, causing extensive bleeding. Such dream men acquired considerable wealth in these exchanges and became local big men for a time.[20] Gajdusek and other *kuru mastas* resembled dream men to a degree. At first they seemed to be using ophthalmoscopes and microscopes as tools of divination. Their choreography of medical examination resembled a magical performance that might unconjure the sorcerer's spell. The *doktas* plied sufferers with medicines, some of them evidently possessing the power of reversing sorceries that induced yaws and respiratory ailments, even though apparently ineffective against kuru. But their exchange relationships were different and anomalous. *Yaba* men and dream men demanded food and other goods from the families of kuru victims; Gajdusek and the *doktas* usually gave cargo and medicines to the victim and her family, hoping to obtain specimens.[21]

Within a year or so, having failed to demonstrate a reversal of kuru sorcery, scientists encountered growing resistance to their requests for specimens. Rare at first, refusal soon became more common. Donation of blood had always been as much an entreaty as a gift, but now, if given at all, it was frequently passed on hesitantly and in reproach. Inamba, who worked with scientists in the early 1960s, recalled people telling him: "'You haven't cured a single sickness, and yet you want to take our blood, and lots of illness is hurting us.' They were angry and crossed us." The scientists did not seem interested in identifying the men who made kuru; nor could they find any medicine to counter the sorcery. "They said, 'You can't have it,'" Inamba told me. "But the sick people, it was easy to get their blood. Just regular people, we gave them a little money."[22] During the 1960s, as they came to appreciate the trade store at Okapa, Fore realized that if they earned money they could buy the cargo that white men otherwise offered so capriciously. Like many

Medical assistants taking specimens. Photograph courtesy Peabody-Essex Museum, Gajdusek Collection

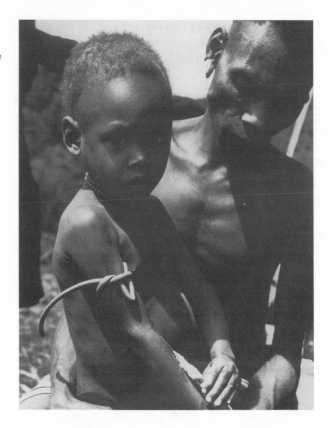

Fore child with needle and syringe. Photograph courtesy Peabody-Essex Museum, Gajdusek Collection

other things, blood was becoming more readily alienated from the Fore: gradually it was turning into a marketable commodity. On rare occasions, blood might now be sold. But the Fore found it difficult to settle on the price of their blood, to determine its value for the white men and the consequences of its loss to themselves. Often it seemed safer not to give it, or simpler not to sell it. After all, it was not helping anyone with kuru.

"I WRITE AT THE MOMENT to let you know that we have had a kuru death and a complete autopsy," Gajdusek informed Smadel in May 1957. He was relieved and triumphant. "I did it at 2 A.M., during a howling storm, in a native hut, by lantern light, and sectioned the brain without a brain knife."[23] The autopsy became the point at which negotiations over transactions of things and persons condensed and hardened. It was around the warm corpse that a tournament of value took place, reshaping relationships and refashioning identities among and between the Fore and the investigators.[24] "We were required to 'fight' for autopsies, as one does in the U.S.—rarely as hard!" Gajdusek wrote. "We never used duress, and our coercion was that of argument, and making pests of ourselves, during the patient's long illness."[25]

As with the transaction of blood and other specimens, barter and monetary payment at first did not have any traction with autopsies. "Kuru brains," Gajdusek observed, "are not a commodity on the open market, nor will they ever be; we are lucky to get any."[26] In seeking to obtain brains, the scientist found himself entangled in a complex and confusing set of gift exchanges. Everyone hoped to give well, but in the circumstances satisfaction with these exchanges, and with the relationships they forged, sometimes proved illusory or evanescent. "I did a complete autopsy in our treatment/laboratory hut by lantern light," Gajdusek wrote to Burnet, "and then at first cockcrow got the body borne homeward with the mourning mother well rewarded with axes and salt and laplap."[27] On other occasions his offerings failed to move the grieving relatives—the gift they most wanted, something that stopped kuru, he did not have. Nonetheless, exchange of gifts in medical research was binding the scientists to the Fore, bringing them into an unstable relationship of mutual obligation and unbalanced reciprocity. As Gajdusek later reflected, it was difficult to "retain equilibrium in the complex plurality of relationships which I have here in this region."[28]

The location of a proposed autopsy influenced Fore responses to the procedure. Gajdusek and later investigators met firm resistance to any attempts to operate on the body away from the home. Initially they tried to conduct autopsies on patients in the hospital at Okapa, but such efforts fed resentment of the investigators. Later, most autopsies took place in the dead person's home or in a makeshift shelter adjacent to it, places that allowed Fore greater control over the flow of specimens and persons. Although Fore were more likely to find such local sites controllable, they presented practical challenges to the scientists.

It was necessary to operate within a few hours of death to avoid decomposition of the brain and other tissues, so a sense of haste and excitement prevailed. Sometimes the *dokta*, staying nearby, heard the shouts and wailing that marked death and rushed to the bedside; at other times, an assistant who was attending the sufferer, ran and fetched him. Within an hour or two, the *dokta bois* constructed a shelter of wood and tarpaulin for the autopsy, hung kerosene lanterns, and assembled a low bed for the body. The *dokta* quickly prepared the operating instruments and found containers and reagents for tissues. A large crowd of mourners jostled the investigator and his assistants as they set up the autopsy. Needing to cut fast, the *dokta* usually asked a senior relative to decide which four or five family members should participate and to send away the remainder. A brother often held the warm head of the corpse while the *dokta* cut; other family members, perhaps more experienced than the white men in dissecting bodies, might offer surgical advice. But white men took the lead in performing these autopsies—with the early exceptions of Liklik, Vin Zigas's capable assistant from outside the region, and Sinoko, who sometimes worked under Jack Baker's supervision. The *dokta bois* stood around holding containers, fixing lights, and on rare occasions assisting with the operation. As soon as the *dokta* removed the brain and some other organs, he and his helpers stuffed the cavities with cotton balls and sewed up the corpse, returning it to keening relatives with gifts of axes, bush knives, blankets, tinned fish, salt, and tobacco. Baker, who supported the scientists, recalled that if the family "permitted an autopsy on the body, there was a present made of a little bundle of trade stuff."[29] Then the scientist or one of his *mankis* sped off with the specimens toward Okapa and on to the airstrip further north at Tarabo.[30]

Negotiations over a proposed autopsy frequently were complicated and trying. As the kuru victim lay dying the scientist would ask repeatedly for permission from her relatives. He attempted to build relationships with them through gifts of blankets and other cargo. But sometimes family members could not be persuaded to donate their loved one's brain to the scientist. On other occasions they would do so reluctantly, after much cajoling, and stipulate just what else might or might not be removed.[31] At the last minute someone might suddenly object to any cutting at all. In the past, the maternal kin or *kandere* had taken charge of mortuary dissections, allocating organs and tissues according to the recipient's relationship to the deceased.[32] But to many white men the victim's husband or father seemed the most senior relative and therefore the chief negotiator in any discussion of dismemberment of the dead, though they respected objections from anyone connected with the victim. Generally, the *kandere* had consigned the brain of a dead woman to her son's wife or her brother's wife, but there is no record of the scientists according any special compensation to these relatives, nor of any particular complaint from them—many were dead or dying from kuru anyhow. Thus the problem of recognizing appropriate transactors could complicate the challenge of giving well in these exchanges.[33] And even when things appeared to go smoothly, the loading up of one family with cargo later might cause resentment among others in the parish who were less adept at exploiting the white man's resources. Many

Fore wanted to know why the scientists ignored their non-kuru dead and re-fused to participate in mortuary exchanges with them. Giving too well to one line often meant damaging the relations of the investigator and beneficiaries with another.

Since most victims were women, the gender of the gift often inflected negotiations over the bodies of the kuru dead. Even if body parts and fluids were not inherently gendered—or more precisely, even if their gender was relational, depending on the character of their transaction—most of these mortuary exchanges with scientists surely brought to mind other settings in which women represented transferable, though inalienable, wealth. Fore men and their kin made claims on living women through contributions to bride wealth; then the woman's maternal kin retrieved her body at death and claimed rights to its consumption. In seeking an autopsy on a female victim, Gajdusek was becoming encumbered in competing claims over the reproductive capacities of women's bodies.

The autopsy fascinated the *dokta bois*, most of them too young to have taken an active part in earlier mortuary rites. It was always an exciting event for adolescent males and sometimes an opportunity for them to make a vicarious and unconventional appropriation of women's bodies. Masasa remembered the first time he helped the *doktas* to cut a corpse. "We washed her and watched her for a period of time, and then when she died, they operated. We gave axes, bush knives, blankets, salt, tobacco, and such, to the family." I told Masasa I wanted to know more about the procedure. "There was a little house, and a saw and hammer. When she died, Kaoten said, 'Carry her and put her in the house, and we'll operate.' He said that and I carried the body and put it there. He took a saw and a razor, and cut. He cut her and I watched . . . We held the body while he cut it. He operated and took the liver and the brain, and we put these in a bottle and sent them off." This time the brother of the girl's mother—representing maternal kin—was the only relative in attendance.[34] A few years later, when Inamba assisted in an autopsy, it was little different. "We sat with them, lit a Coleman [lantern], greased them with the light, and ate together for a week or two. We spent time with the family and asked them if it was alright . . . If one person said no, just one person could stop it. If everyone said it was alright, then we did it." As we spoke at the Ivingoi mission in 2003, Inamba was patting down the smart blue jacket that Kaoten had given him. He proudly told me about his first autopsy. "One of her brothers, one of her fathers, stood and watched us. We said, 'You can't speak, and you can't make noise, and you must cover your nose. Stand by the Coleman.' And we cut the body. To operate we covered our nose and our mouth." A few other relatives stood quietly beyond the glow of the lamp. "The first time, my hands were heavy," Inamba said, looking down at his palms. "And I didn't know what to do. I watched and worked."[35]

Tarubi experienced a lot of autopsies and reflected more deeply on their rationale than many of his mates. "We asked relatives: 'This person here, can we operate on him? Can we operate and get brains and insides from him, his liver, his spinal cord and such?' Then we would send it off. This is what we asked relatives: 'Can we perform an autopsy?' And if they said yes, we did. And if they said no, that was how it was. It was their call." Tarubi explained to the family that the *doktas* "can make a medicine or find the germ and make a medicine. And the people listened to what I was saying and would give permission to operate." In particular, he wanted to make sure I understood his contribution to the procedure. "After they finished operating, we sewed up the body, I sewed it up good, and we'd return it to the relatives to go and bury. It was like that." But along with many other Fore, Tarubi eventually became more skeptical and pessimistic. "We went out over and over and they didn't find the germ. They performed autopsies and sent body parts to the big microscope in America and inspected, they inspected through a big machine. We were asking, 'Did you find it or not?' And the doctors said, 'No.'"[36]

Resistance to autopsies and other medical activities continued to grow among the Fore. In April 1957, Gajdusek warned Gunther that "we have ticklish problems in trying to avoid any trace of coercion of the natives. We have gained their confidence around Moke." For a few months, the interactions of the scientists and the Fore appeared mutually satisfactory. "To humor me and repay my many miles of mountain climbing to track them down," Gajdusek observed, "they haul the litters over miles of cliff-faced and precipitous jungle slopes to bring patients in for another shot at our therapeutic trials and experimental poking . . . I admire and respect them thoroughly."[37] But late in November he was expressing rising frustration to Smadel:

> It looks as though further autopsy materials may be unobtainable. Thus, the natives have given up our medicine . . . they know damn well they do not work . . . and I am fighting (verbal battles in Fore), bribing, cajoling, begging, pleading, and bargaining for every opportunity to see a patient, and strenuously working tongue muscles for hours for every further day we get a patient to stay in hospital, accept therapeutic trials, etc., etc. Vin is sick and tired of the "duress of personality" which is required to pressure every case into our care and I do not like the effort. It means, however, that unless we start curing cases quickly, we cannot expect any clinical material much less any autopsy specimens. I am willing to keep up the push using every ruse short of actual duress by force and authority . . . that we cannot contemplate.[38]

Yet it was not the exertion of any "duress of personality" that troubled Gaj- dusek so much as the more fickle and fraught demands of personification of all his transactions—the exigencies of moving people and being moved through exchange appear to have exhausted him.

The objections of some Fore to the importunate and apparently impotent medical investigators attracted the attention of the colonial authorities. In 1958, Donald M. Cleland, the administrator of Papua and New Guinea, reported to the Department of Territories in Canberra, Australia, that people "have in the past been a little upset by the interference in their lives which has been an un- avoidable accompaniment to the investigation."[39] Roy Scragg, the director of public health, was more blunt: "It is undoubted that previous activity by Dr. Gajdusek in the area did to some extent adversely affect the native feeling. Kuru patients felt inconvenienced in being dragged continually into Okapa and the non-kuru people wondered why the kuru were particularly singled out for bribes and other rewards when they were sick or their relatives died."[40] He was concerned that "itinerant research workers" would "haphazardly enter the area and proceed, at the expense of other research workers, to collect a lot of material in a short time and then leave the area in a state of disorganization and without giving any real benefit to the people suffering from kuru."[41]

IT WAS TEMPTING AT MOMENTS OF CRISIS in exchange relations with the Fore to conjecture on the likeness of medical consumption to native can- nibalism. The Fore were coming to realize Gajdusek was not able to identify the people who made kuru or to counter the sorcery with the substances they gave him. They often lost patience with inept exchanges, ill-timed and poorly targeted, occurring at unsuitable, and perhaps harmful, places. Moreover, they soon found other means of acquiring trade goods and other white men with whom they might build relationships. As Gajdusek despaired of acquiring the specimens he wanted, as the Fore became indifferent or hostile to his needs, he wondered what it would be like to appropriate these substances, to con- sume without restriction or relationship. At the same time, he would have realized that the scientist could never simply become an exocannibal—"*that* we cannot contemplate."[42]

Stories of Fore cannibalism fascinated Gajdusek. In his first letter from the field to Smadel, he had assured his mentor that "although the people are still current warriors and cannibals, they are well 'under control' and very coop- erative."[43] A few months later one of Gajdusek's Fore friends reported that

Zigas, Baker, and Gajdusek at the dining table/desk/autopsy bench at Okapa. Human brains on the plate, specimen bottles on the windowsill, and census books on a folding table. Photograph courtesy Peabody-Essex Museum, Gajdusek Collection

his clansmen had eaten his grandfather "against his advice." "Such recent, nay current, episodes of cannibalism," the scientist wrote, "are not unusual here, but it is highly unlikely that all of our kuru patients have eaten human brain." Yet it was an enticing thought. "It is so unique a concept, and such a romantic one, that I almost wish cannibalism was more prevalent than it is."[44] The practice was dwindling with pacification, but it served to titillate his correspondents. Provocatively, he also flaunted his own medical cannibalism, describing his early dissections of the dead on the table where he ate his meals and wrote his reports.[45] While native cannibalism was forbidden, Gajdusek claimed license to take the bodies of the kuru dead and remove their brains on his dining table.

What does it mean for a scientist to speculate on cannibal tendencies? Among the Fore, endocannibalism, the ritual consumption of loved ones after death, perpetuated the social bonds constituted in bones, tissues, and body fluids and thus helped humans regenerate. It allowed a modulation and amplification of customary exchanges between mourners, enhancing the transfer of social value from one generation to the next.[46] But medical cannibals are more likely to be exocannibals, consuming the bodies not of loved ones but of strangers. In casting himself as an exocannibal, the scientist attempts to simplify and control apparent disorder, to disengage from social debt and demands for reciprocity, as he imagines a means of drawing on the resources of others without becoming other.[47] He indulges in an unsettling fantasy of consumption without reserve, a desire that implies its own impossibility. Fantasies of medical exocannibalism structure the work of colonial science in terms of absolute consumption, while acknowledging that the relations of dominance and submission that might permit such a feast are interdicted.[48] Above all, the emergence of the metaphoric cannibal at this moment marked a crisis in Gajdusek's exchange relations with the Fore.

Gajdusek was not tempted, however, to flirt with head hunting. He suspected the Fore might in the past have resorted to head hunting; and the Berndts had suggested that head hunting rumors circulated widely among the people to the north. But even in flights of fantasy, Gajdusek did not imagine himself as a medical head hunter; nor did he hear he was so accused. It is not surprising that head hunting was less palatable than cannibalism to the scientist. The taking of heads is a violent act, an exercise of force that reveals relations of inequality and exploitation.[49] Thus if exocannibalism implied an unresisted appropriation of the body of the other, an absolute corporeal consumption, then head hunting called attention to its violent expropriation. Gajdusek may have been prepared, imaginatively, to simplify his exchange relations with the Fore, to joke about medical cannibalism, but he could not countenance any violent satisfying of his desire for unreserved consumption.

If medical head hunting was abhorrent and medical cannibalism—despite its romance and queer appeal—remained an impossible fantasy, how else might the scientist modulate his exchanges with the Fore in order to take from them without being able to give back what they want? That is, how might he interrupt the cycle of reciprocity and reduce his social debt and sense of obligation to these people, allowing him to circulate their body parts as his own valuable kuru material? It was a question of how he might gain possession of their inalienable wealth, appropriate their goods and persons, even at the risk of falling into moral peril or showing bad faith. To make a Fore person's brain into one of Gajdusek's kuru brains, it would be necessary to cut the network, to differentiate native and scientific exchange regimes.[50] The principal site at which this transformation took place was the bush laboratory: it thus became both the methodological epitome and the productive antithesis of the house of the cargo cult, where Fore had sprinkled blood on inanimate objects to transform them into valuables. In the laboratory the scientist applied his tools to the transformation and de-animation of Fore body fluids and tissues—they came in as persons and left as things.[51] It was in the laboratory that the scientist performed his magic.[52]

The passage of body parts and fluids through the improvised autopsy theater and bush laboratory to the distant research institute in Bethesda was transforming persons into the scientist's wealth. These sites were stages in the modulation or attenuation of relationships with the Fore, the progressive weakening of bonds formed in the exchanges associated with history taking, clinical examination, and the yielding of specimens. These were places where the relational obligations that limited property claims might be weakened or thwarted; thus they became launching places for Gajdusek's new kuru material. The scientist was not trying to render the field like a laboratory; rather, he sought to use the laboratory and the autopsy to mark boundaries of exchange regimes, to draw attention to his agency as an owner, which amounted to another sort of effort to appropriate others' wealth.[53] In circumventing conventional exchange relations, in attempting illicitly to alienate Fore substances, laboratory transactions most resemble the hidden work of sorcery. Through secret operations, Gajdusek was attempting to transubstantiate bundles of personal material into something else that "alienated" the inalienable stuff of the bundle. Like sorcery, laboratory work seemed to offer a means of interrupting or rechanneling the flow of goods and therefore of upsetting conventional social structure and hierarchy—among the Fore and later among the scientific community. Sorcery represents improper or transgressive consumption, the inverse of "good" exchange and therefore personhood.[54] In this sense, the scientist was caught up in sorcery as much as the Fore became caught up in science. While Gajdusek frequently experienced the laboratory as a place of rationalization and calculation, it was thus also a place of enchantment, of modern magic. Gajdusek's laboratory was in fact haunted by the magic it appeared to exorcise, so that reason and enchantment in practice produced each other.

No matter how hard he tried to use the laboratory to mark the boundaries of exchange, Gajdusek could never fully disentangle himself from his reciprocal relations with the Fore and from his sense of social obligation to them. In the field he was still enthralled with them, and laboratory work was always a necessary chore. He needed the laboratory to make a path for his own transactions with other scientists, but he was never as engaged in its transformative tasks, in its purifications, as he was in his interactions with Fore—indeed, he evinced a romantic aversion to laboratory routine. The laboratory allowed

Gajdusek to mobilize their inalienable substances as his own and circulate them in a global scientific network, yet never were these specimens completely detached from their origins; they retained an aura of the person from whom they came.[55] Even when pulverized and distant from the site of extraction, Gajdusek's kuru brains were also identifiably the brains of Aoga or Tasiko or some other Fore person. Each brain, even in pieces, had a name. These new epistemic things never entirely concealed Gajdusek's relations in the field. In recognizing this provenance, or in failing to misrecognize it, the scientist was drawn back to a particular time and place and entangled again in relationships with local people—relationships that had made him into a certain sort of person. He was unable—for both social and emotional reasons—to close off the exchange system, to cut his Melanesian network, to exit the sphere in which his identity was multiply produced through interaction. Perhaps to other scientists he appeared a singular, possessive individual, but not to himself. [56] He remained as much enchained to the Fore as they were to him.

The Fore also were seeking through the management of exchange to control relationships, to move persons and regulate their flow. As their transactions with Gajdusek became less satisfactory and less hopeful, they became, as we have seen, less inclined to engage with him and give him what he wanted. It was harder to mobilize substances of scientific interest. From the beginning, Fore had limited the circulation of persons as goods, only now they threatened to withdraw all personalized valuables from exchange. The movement of a dying person from her home to Okapa had always generated resentment and anger, and the scientists soon stopped trying to do this. It was never possible to take people with advanced kuru out of the region for further study, despite numerous attempts. A few early cases were examined and tested at Port Moresby, but Fore usually resisted their mobilization. At first, Gajdusek had promised Australian investigators a live kuru "case" that they might study in their clinics. In March 1957 he proposed

> sending an ideal case to Brisbane, Sydney or Melbourne for study in a unit such as Dr. Wood's Clinical Research Unit. This would yield, in the long run, far more information and far more reliable results at a far smaller expense than all sorts of half-hearted efforts at getting experts and equipment into the highlands . . . Now I am not suggesting accepting a classical early case on the Clinical Unit ward for autopsy purposes but rather for clinical study and evaluation.[57]

Kuru victim on makeshift stretcher. Photograph courtesy Peabody-Essex Museum,
Gajdusek Collection

It was already clear to Gajdusek that the Fore would not permit a person to
die outside the region. The transfer never took place, partly because his relations with the Melbourne scientists rapidly broke down, but mainly because
the Fore would not let him circulate sick people.[58] Like the scientists, they too
were trying to mark the boundaries of exchange regimes and determine the
proper limits of consumption and distribution.

In such circumstances, it may at first seem curious that Gajdusek began to
"adopt" Fore and other Melanesian boys. Starting in 1962 with Mbaginta'o,
a Kukukuku lad living at Agakamatasa, Gajdusek took these boys to the United
States and ensured they went through the local high school, and sometimes
on to university. He preferred the boys from Agakamatasa and from Moraei,
across the Lamari River, taking seventeen or so across the Pacific over the following twenty or more years. Rowdy, disobedient, cheerful, they came to live
in the scientist's rambling house in the Maryland suburbs. The collecting of
adolescent Melanesian males amplified Gajdusek's earlier efforts to "save"
postwar German boys like Wolfgang. It reflected too his belief in individually
directed compensation for debts to whole communities, a belief he had first
formulated in New Britain. Most of all, it became a means for him to create
a family and to ensure some sort of social reproduction. None of this was

alien to the Fore. They also took in orphans and fleeing children from other communities; they respected personal, rather than collective, agency in exchange; and they frequently adopted or affiliated outsiders and made them kin. Healthy, robust Fore had often visited adjacent people. With the coming of colonialism, they traveled to Kainantu and Goroka, and many young men were conscripted into the Highlands Labour Scheme. Pragmatically, most Fore and Kukukuku chose to regard Gajdusek's adoption of their boys as similar to these acceptable or even desired movements and displacements of people, and to separate such activities from his initial attempts to take away their sick. Perhaps expecting the boys to come back eventually, these highlanders were prepared to review and shift the boundaries of approved consumption. As Masasa said in relation to the lads who accompanied Kaoten to America: "He paid their school fees and built a road for them."[59] Through education Gajdusek built a road for the boys, just as through laboratory work he built a road for the specimens he collected. Moreover, through mundane gift exchange, the Fore had built a road for Kaoten too. In taking these roads, all parties were becoming enmeshed in other exchange regimes, but their earlier relations and commitments continued to move them.

WE HAVE FOLLOWED THE SCIENTIST as he sought to "attain his type of value, to follow his line of social ambition." We have seen him—as Malinowski saw his Trobriand "savages"—"led on to perilous and difficult enterprises by a tradition of magical and heroical exploits."[60] Yet in reading Gajdusek's early published papers on kuru, we learn little about the social relations of the scientist in the contact zone. His interactions with the Fore suffer a total eclipse in the pages of the *New England Journal of Medicine*. However, through reading Gajdusek's journals and letters and talking with those who were involved, it is possible to glimpse the material cultures of kuru research, to gain an impression of the heterogeneous, puzzling, and often indeterminate exchanges taking place between locals and strangers around Okapa in the late 1950s and early 1960s. Thus may one come to appreciate some of these magical and heroical exploits.

I have tried here to recast the research field as a contact zone, a place where persons previously separated geographically and historically come together and elaborate a means of communicating. Like others, I use the term to "foreground the interactive, improvisational dimensions of colonial encounters so easily ignored or suppressed by diffusionist accounts of conquest and domination."[61] That is, I have tried to draw attention to confusion of relations in

Four boys and four girls with kuru. All died within six months. Photograph courtesy Peabody-Essex Museum, Gajdusek Collection

which Gajdusek became entangled—a confusion allowing no simple assertion of dominance and control—when he set out to investigate kuru. One might also imagine this site as a trading zone where scientists interact pragmatically and strategically: such a trading zone is "an arena in which radically different activities [can] . . . be locally but not globally coordinated." The evolution of new scientific languages—which happen also to be colonial languages— from pidgins to creoles, permits communication between different scientific cultures and some limited commensurability in practice, some common understanding of value.[62] From this perspective, we can view Gajdusek and the Fore operating in a contact zone, which is at the same time, and perhaps more revealingly, a trading zone.

It is important to realize that it was not only language that was developed in this trading zone. Evidently some linguistic mixing or creolization did take place, but so too did the detachment, transformation, and displacement of persons and things—categories that often were not readily distinguished. Scientists and Fore alike engaged in gift exchange, barter, and market transactions; they dreamed of easy appropriation of others' goods and persons, had reveries of consumption without reserve; and they imagined things that became persons and persons who might become things. Through these confused, perplexing and often unsatisfactory transactions, persons and bits of persons sometimes shifted form, gained informational density, and moved from one regime of value to another. Each party in these scientific exchanges was seeking to make visible the other, to find out what, or who, was valuable in the eyes of the other, and thus bring the other into being and into a relationship. To ask what the Fore thought of scientific research, or to raise the question of informed consent, is to miss the point: it was what they made of the investigator as a person that mattered. The contact zone around Okapa was a site for the redistribution of trust, affect, and desire, for the making of persons and new structures of feeling. Some Fore came to develop a research subjectivity and to experience a heavily medicalized colonialism; others simply became intimate with Kaoten. Gajdusek became kin at Agakamatasa, a sort of scientist and inadvertently a colonial emissary. Encounters in this particular contact zone were making persons visible in ways they had not anticipated.

It was Kaoten who met the boy Kageinaro at Agakamatasa in the early 1960s:

I greeted him but was a bit surprised by his lack of vocal response, by his unusually dirty appearance, his shredded loin cloth, his rather restrained stance, and lack of usual boyish response . . . And as though electrocuted, I suddenly realized that another of my boys had kuru . . . Here was Kageinaro, the wiry, noisy, impish, always naughty, and constantly playing youngster of my last trips, now a fattening adolescent, restraining his speech, holding himself stiffly, awkwardly, and diverting his eyes in the same tragedy which Azibara presented to me on my last sojourn here. I was spellbound with discouraging realization thrust so forcefully upon me, that I was back in kuru country, and that there is no life here without the ever-threatening death from kuru!

It was Kaoten again who wondered, "how many of the boys will pass to a similar fate?"[63] A few months later, Kageinaro was dying—"starving and thirsting to death in a hideous state." By then, Gajdusek was wondering about an autopsy. "It is a curse of such magnitude," the scientist wrote, "that I am ashamed of my feeble efforts and appalled by my own callousness."[64]

Historians and philosophers of science are fond of talking about translation, creolization, and the production of limited commensurability between the social worlds and theoretical paradigms of science. They tend to ignore the achievement of limited *incommensurability* with other, local social worlds, a differentiation—perhaps just the achievement of a callousness—that makes some things recognizably scientific and some people identifiably scientists.[65] Gajdusek knew that at some point he must cut the local network, attempt to mobilize inalienable possessions as his own, and limit his intimacy with the Fore—if he were to become a big man in science. In making persons into scientific objects he was producing an incommensurability that allowed them to be removed from the local regime of value and circulated in another. The Fore were right when they discounted the possibility that Gajdusek was like the sorcerers they knew: but meanwhile he was working his own magic, trying to make persons into things, into scientific goods no longer commensurate with Fore estimates of personhood. Thus it was through their involvement in the scientist's magic that many Fore first experienced a profane colonialism.

Chapter 5

HEARTS OF DARKNESS

IN FEBRUARY 1958, VIN ZIGAS DROVE UP from Kainantu to visit the Fore again. Barely a month earlier, Carleton Gaj-dusek had finished his first period of field work and left New Guinea, heading back to the United States. "I came to Okapa with the idea of obtaining a few more brains as per our arrangement," Zigas wrote to his American collaborator.[1] On this occasion, he experienced little trouble obtaining a brain—Aoga, the girl Gajdusek had first examined eleven months earlier, died soon after his arrival, and her relatives allowed the autopsy to take place. "Aoga died," Zigas reported, "and after four hours' hard work, brain, cord, eyes and viscera were removed."[2]

But Zigas encountered competition for possession of these valuable specimens. Already Australian and American scientists were bickering over the harvesting and disposal of the body parts of kuru victims like Aoga. Professor Norrie Robson, from Adelaide University, had slipped into the region as Gajdusek was leaving and began collecting genealogies and trying to establish the genetics of kuru. Robson demanded that Zigas send Aoga's brain and three other brains in the Okapa refrigerator directly to Adelaide. "I fought," Zigas reported to Gajdusek, "but was told straight to stop interfering." Robson advised the excitable Balt—as Zigas called himself—to "relax" and implied that "kuru research would be stopped for a while." But Zigas proceeded to ignore Robson, who seemed to him a stubborn, cunning, egotistical Scot, an entirely unsympathetic figure, aloof and irritable.[3] He sent Aoga's brain instead to Gajdusek's new laboratory at the National Institutes of Health (NIH) in Bethesda, Maryland. "Dear Carleton," Zigas confided, "I am unable to express my despair and feelings."[4]

Later, Zigas told Gajdusek that in the eyes of the Australian investigators he was no more than a "dumb, blasted foreigner."[5] They "do not want to include you or me on the papers, want all the additional information they can get, plan to use the materials we gave them and those that were 'stolen' . . . When we and they talk of collaboration and cooperation it appears that they expect it all from you and me but have no thought of reciprocating."[6] Zigas repeatedly warned his friend not to trust the Australian scientists and to treat any of their assurances as "bullshit."[7]

To his affable mate Jack Baker, Gajdusek lamented that the last three of Zigas's autopsies were "stolen from under his nose even against his complaints." He told Baker that "we have discovered a number of astounding new histopathological facts, and the material I had prepared is apparently NOT being used for anything but the prestige of others . . . Vin in his wrath sent on Aoga's brain, eyes, etc., here in spite of 'orders' to the contrary."[8] Not long after, Gajdusek asked Baker for "all the brains you have on hand, as well as eyes, cords, and other tissues that may be available."[9] The *kiap* managed to dispatch four more brains but pleaded with his friend not to tell any Australians: he was in enough strife already. "Your queries re. the whereabouts of the brains you left in pickle are answered by yourself," he told Gajdusek. "Robson liberated them." Baker missed their old camaraderie, the intense companionship of Gajdusek and Zigas, and disliked the more respectable Adelaide scientists. "If Moke [Okapa] used to resemble the rumpus room in an insane asylum, it is now like a refrigeration chamber in a morgue."[10]

John Gunther was encouraging scientists from the Adelaide Medical School to take an interest in kuru and supplant Gajdusek and Zigas, whom he regarded as erratic and unreliable. His own kuru ambitions thwarted, Macfarlane Burnet joined Gunther in urging Robson and his earnest colleagues to nudge aside the flamboyant and annoying American. For Robson, kuru presented a fascinating set of epidemiological and etiological problems, and an opportunity to undertake original research in an exotic place. On returning to Adelaide, he set Henry Bennett, the dapper young professor of genetics, to work analyzing Fore genealogical data, hoping to find clues to some obscure hereditary cause of the illness.

As an Adelaide medical graduate, Roy Scragg, the new director of public health, maintained close ties with the new Australian investigators—though he remained ambivalent about Gajdusek, hoping that all the scientists would eventually learn to play well together in the field. Growing up in a Seventh

Day Adventist household, Scragg had encountered many missionaries to New Guinea and decided to go there once he finished medical school in 1947. "I just thought it would be interesting to go to for a couple of years," he told me over a cup of tea in his house on the beach in Glenelg, South Australia. Not from a medical family, unlike most of his cohort of graduates, Scragg's career was not mapped out for him. His clinical experiences in Bougainville drew him toward social and preventive medicine. With Robson's encouragement he conducted demographic studies of fertility in the 1950s, thereby attracting Gunther's attention and securing a post in Port Moresby. Despite his affiliation with Gunther and the Adelaide team, Scragg developed a grudging admiration for Gajdusek, who seemed to share his enthusiasm for New Guinea. "Oh, keen young man, full of interest," was his gruff description of the American adventurer. "Hyperactive, extremely hyperactive."[11]

Like Gajdusek, the Adelaide investigators ruled out infectious and toxicological explanations of kuru. The disease seemed to favor women and children and run in families, so hereditary transmission appeared most likely. Gajdusek too had wondered about a genetic cause, though he was never wholly convinced. Later he insisted he had come up with the idea that kuru was hereditary first, even if it was an inadequate explanation: "This was our clear hypothesis as early as March 1957 and it is stated in all of our kuru publications."[12] Burnet also had assumed "the likely answer is a genetic anomaly, possibly with some environmental factor determining its expression."[13] But Gajdusek's attitude became increasingly ambivalent. "If genetic, the Fore are doomed," he wrote to Scragg late in 1957. "Thus, it behooves us to deny the genetic evidence and keep trying to locate toxic, deficiency, or post-infectious factors."[14] Since the Adelaide investigators seemed more emotionally prepared and better equipped to master the genetics of kuru it made some sense in the circumstances for them to take over the study of the disease.

By now utterly committed to the Fore and entangled in their lives, Gajdusek resisted any displacement. Despite the intermittent efforts of Australian authorities to delay or prevent his return to New Guinea, the resilient American continued to make his semiannual pilgrimage to the Fore, collecting more blood and brains, reconnecting with old friends, undertaking prodigious hikes, and passing the time at Agakamatasa. Even when he was not among the Fore, Gajdusek managed to get various affiliates, such as Zigas, to conduct autopsies and send him the desired organs and tissues. His persistence infuriated the Adelaide team and its sponsors, leading to heated outbursts and moral out-

Visiting Adelaide medical team. Donald Simpson is at right. Photograph Courtesy Peabody-Essex Museum, Gajdusek Collection

rage on all sides. When colonial authorities attempted to limit his activities, to channel his research into an apparent backwater, Gajdusek proved expert at circumventing their restrictions. As he said at the time, he was pleased to be a thorn in their side.

HAVING ENDORSED THE ADELAIDE INVESTIGATORS, Scragg initially appointed Zigas to assist them at Okapa, but it soon became obvious that the medical officer remained loyal to Gajdusek and could not abide Robson. Gunther told Bennett, "It is evident that Dr. Zigas could not be relied upon to give his whole hearted support to the Adelaide group."[15] Sensitive to his predecessor's opinion, Scragg instructed Zigas to return to Kainantu and provide aid to both research teams, while Andrew Gray, a medical officer whose fidelity to the Public Health Department was unquestioned, would instead be stationed in Okapa, helping visiting Australian scientists.[16] Later in 1959, Robson sent some young Adelaide doctors and medical students to the region to continue clinical studies and collect genealogical information. Clive O. Auricht stayed with Gray at Okapa; Bronte Gabb and later Donald J. Perriam decamped to Purosa. Gabb in particular became popular among the south Fore: they called him *liklik masta* in Pidgin, since he was short, and *pitasol*, which means "next?" in Fore, because he kept asking them who was born next in the family.[17] The Adelaide investigators fanned out across the region, compiling their own genealogical maps of the Fore.

Early in 1959, Robson traveled to New York City where he went to the Rockefeller Foundation to enlist its support. To his dismay, he found that Robert Morison, the Foundation's director of biological and medical research, had "obviously been thoroughly indoctrinated with the Gajdusek view of the New Guinea situation," so much so that he regarded the Adelaide studies as "superfluous." Robson, though, managed to convince Morison that his team had the only promising lead on the cause of kuru.[18] Soon afterwards, Morison agreed to give the Adelaide group a small grant to continue their work.

Then Robson caught the train to Bethesda to discuss a possible division of research labor with Joe Smadel at the NIH. He reported to Gunther that "spheres of activity have been defined as between our group and the National Institutes." Gajdusek would continue to study his various tissue and blood specimens in the laboratory while the Adelaide team, with Rockefeller support, would pursue its genealogical survey, seeking the most likely hereditary cause of the disease. If time permitted, Gray and Auricht might supply Gajdusek with further specimens.[19] At Bethesda, Robson noted that Gajdusek was "generally highly regarded as a brilliant though difficult young man. He is much admired for his energy, productiveness (i.e., papers) and the fact that he has, for his age, achieved really remarkable things in a number of fields." To Robson, Smadel appeared the "brusque, high-pressure business type, dogmatic—thinks highly of Gajdusek but really knows very little about the whole business in detail." So far Gajdusek had little new to show from his laboratory studies, just some abnormal blood protein patterns of dubious significance. His genealogical studies, Robson lamented, consisted only of "family histories, in varying completeness, of each case of kuru studied"— and no comparable studies of normal individuals. Smadel assured him that Gajdusek's data would be made available to the Adelaide investigators, who might add them to their central file.[20]

Scragg—an inveterate optimist—was delighted; he saw no obstacle to both research teams operating in the region at the same time. Gajdusek would "collect material and in particular, post-mortem material . . . these post-mortems could be undertaken with the assistance of the native staff at Okapa though not more than two per week should be undertaken." Before going to the south Fore, Gajdusek would have to gain Gray's approval. The American might build a laboratory at Kainantu, though not at Okapa. Scragg expected that Gray,

Auricht, and Gabb would work solely for the Adelaide team and not assist Gajdusek. He believed that these arrangements should keep everyone happy.[21]

Predictably, the apparent reconciliation was short-lived. When Gajdusek returned to Okapa in August 1959 he encountered Robson, Bennett, and Gray there. "I made the grave error," Gajdusek wrote, "of talking openly, hiding nothing of my hopes, ideas, plans and impressions about kuru."[22] As they listened, the Adelaide investigators became alarmed. "It at once became apparent that Dr. Gajdusek's intentions went far beyond those previously stated and agreed," Bennett and Robson informed Donald Cleland, the administrator of Papua and New Guinea. The American intruder wanted to take twenty or more kuru sufferers to the hospital at Kainantu and to conduct more therapeutic trials in the field. He demanded laboratory facilities at Okapa. Worse still, "he wished to have an open brief to range freely throughout the whole area to indulge in what he described as medical 'snooping.'" Bennett and Robson felt this would prove too disrupting to the Adelaide investigations. "Dr. Gajdusek's continued presence in the area is not compatible with the continuance of our work."[23]

The meeting with Gajdusek at Okapa had been disastrous. Also in attendance was R. A. Fisher, the aging British geneticist who was working in Adelaide with Bennett, once his student at Cambridge. He loathed the pushy American. Later he told Burnet: "I met the celebrated Gajdusek in New Guinea last winter, and after hearing him talk, which he does readily, I think I never met a man I should trust less." Gajdusek seemed able to bluff the locals, but "his ideas are always second-hand, and often only half understood . . . He is very much what in his country they call a 'con man.'"[24] Having long admired Fisher's scientific work, Gajdusek was taken aback. Fisher "was ratchety, crotchety, and had an ugly disposition, very anti-American, and anti–Vin Zigas and myself." The great English biostatistician was "neither open-minded, receptive, nor intellectually interesting."[25] Such obvious animosity between Fisher and Gajdusek deepened the Adelaide team's suspicion of the American.

Late in 1959, Gray intensified his efforts to control Gajdusek's access to the Fore. Baker had told Gajdusek that Gray was "a gentleman and eager to cooperate with you and all comers," but the young South African medical officer proved unsympathetic to the American scientist's constant demands for more kuru material.[26] He referred to the American as a "man whom none of us trusts, likes or respects," contrasting the "truly scientific nature" of Bennett's approach with "the 'medical snooping,' hit-or-miss methods and propaganda of a Gajdusek."[27] He tried to prevent unauthorized collecting of specimens and autopsies, and refused to allow Zigas to take kuru sufferers to the Kainantu Hospital for further study.[28] In December Gray wrote to Scragg to confirm that the aggressiveness of Gajdusek and Zigas had led to a "complete breakdown of confidence between the Fore people and ourselves in Okapa." He observed "no kuru victim and no relative is happy at the thought of investigations. Both victim and relative are, moreover, afraid of the ultimate post-mortem." Their apprehension had focused on the possibility of an autopsy distant from the victim's home—a legacy of the first few months of investigation—but it seemed now to blur into a more diffuse wariness and distrust. Gray recommended that the investigators abandon all post-mortem dissection.[29] H. N. White, the assistant director of public health, agreed. "Sufficient material should have been assembled by this time for full-enough investigation to be made and it is most important that the Fore people are not unduly disturbed."[30] Although aware of Gray's criticism, Gajdusek continued to admire the frustrated medical officer, whom he observed at Okapa morosely reading La Rochefoucauld and Baudelaire in French. The American later wrote in his journal: "I find I still like Andrew very much, tending to feel compassionately helpless to do anything or suggest anything, in his dejected presence."[31] Before long, Gray left New Guinea, decrying kuru research efforts, and disappeared.

"Much time and energy have been wasted in attempts to resolve futile interpersonal squabbles," Robson told Scragg. "The whole project has been marred by jealousy and by unscrupulous and unethical behaviour."[32] As Donald Simpson put it to Zigas, "Robson and Gajdusek are not well fitted for cooperation, and their splendid temperaments would exhaust any intermediary."[33] Constantly battling Gray and various Adelaide emissaries, Zigas began to despair. "The present situation is very unpleasant—shall I say, just shitty," he wrote to Gajdusek. "I've never met such bastards and such a piece of bastardry. Please excuse my jumpy thoughts at the moment but what else can a man do when his spirit is completely emasculated." He told Gajdusek that the Adelaide group was "spitting not in your occiput but in your face."[34] He wanted his friend to take them on. But Scragg intervened, reproving Zigas and telling him that "research workers must learn to work with other people even though the laboratory may be . . . the whole of the Fore area."[35] While Gajdusek may on occasion have imagined himself as Lord Jim, it seems the Australian colonial authorities and attendant scientists were coming to regard him as Mr. Kurtz.[36]

As usual, Gajdusek found supporters in the United States, sometimes beyond Bethesda. Kenneth "Mick" Read, one of the pioneering Australian anthropologists in the high valleys, wrote in November 1959 from his new post at the University of Washington–Seattle, to W. C. Wentworth, a campaigning member of the Australian parliament, protesting against efforts to limit Gajdusek's access to the Fore. He reported that it was common knowledge in America that parochialism and professional jealousy plagued the kuru investigations. Read knew Gajdusek as a "most intense young man of boundless energy and drive and devotion to his work. He seldom sleeps more than a couple of hours per night and never seems to relax. He is easily irritated by administrative conventions. He is domineering and impatient of social intercourse." Having established a certain rapport with the young scientist, Read wanted to defend his right to conduct further research in the highlands. Specifically, he urged Wentworth to intervene so Gajdusek might study the genetics of kuru as this was a more promising lead than any other.[37] Gunther responded to Read's letter, clearly vexed at the anthropologist's interference. Somewhat abashed, and perhaps apprehensive about his own access to the highlands, Read later wrote to Wentworth assuring him that he had not meant to stir up trouble. "Dr. Gunther goes a long way towards satisfying me that Gajdusek is to blame for the unfortunate climate of the kuru investigation." Read now claimed that he had simply hoped to "clear the air."[38]

DURING THIS PERIOD Bennett was developing an intriguing genetic explanation for kuru. In 1958 he and his colleagues advanced tentatively the theory that kuru is controlled by a single mutant gene, "K," which is recessive to its allele (or alternative form) "k" in males and dominant in females. That is, males with both K versions of the gene are potential victims of kuru while those Kk or kk are normal. In contrast, females with both K versions are victims in childhood, those with Kk are potential sufferers later in life, and only those kk are normal. Presumably the heterozygotes—those Kk—derive some tremendous survival advantage, otherwise the gene would not be so prevalent.[39] In order to confirm this hypothesis, the Adelaide emissaries continued through 1958 and 1959 to collect as many Fore pedigrees as possible, starting around Okapa and extending to the borders of the region. Acquisition of specimens became subordinate to genealogical work, yet such activities were never abandoned.[40]

Soon Bennett was able to predict with some accuracy who among the Fore would succumb to the disease. In January 1959 Bennett had told Cleland that "our genetical studies to date must be regarded as preliminary ones and our hypothesis has been proposed but tentatively at this stage. However, the striking agreement between observation and expectation on the basis of this genetical model is most encouraging."[41] By September he and Robson were more confident, claiming that kuru "now certainly does present a unique problem in human population genetics. No other case in the world is known in which such a genetically determined disease of this nature is present in a population with a comparable frequency."[42] Gajdusek and Zigas conceded that "genetic predisposition is strongly suggested . . . but the ethnic-environmental variables that are operating on the pathogenesis of kuru have not been determined."[43] Gunther recalled that Bennett's hypothesis was accepted with caution as the best available and never really received strong scientific affirmation.[44] From the beginning, some geneticists wondered how such a lethal gene could become so common, especially when so many sufferers died before reproductive age.

When I went to talk with him in Adelaide in December 2005, Bennett insisted that he would meet me at the airport. He was unmistakable: a rangy, courtly figure, dressed in whites for the summer heat, discreetly waving his walking stick. Over the following days he entertained me at the Adelaide University faculty club and drove me recklessly around the city in his old Volvo. He told me that Robson had pushed him into kuru research and then more or less abandoned him to it. Knowing now how the kuru story turned out, he wished he had never become involved. As he was a geneticist, he sought a plausible genetic explanation for the disease, but he may have been less committed to his theory if he had known more about family patterns of cannibalism. No scientist was thinking seriously about that in 1958 and 1959. To my surprise, Bennett told me how much he had enjoyed his two visits to the Fore, despite being carried out with dysentery on the second occasion. He learned some Pidgin and was happy to distribute trade goods in return for food, accommodation, a few specimens, and ever more pedigrees.[45]

ASSAILED WITH COMPLAINTS from the Adelaide group about Gajdusek's apparently insatiable need for more clinical and postmortem material, Scragg decided to establish a kuru research committee, chaired by Burnet, to coordinate and control the scientific investigations. When the committee met in Sydney in December 1959, it declared without qualification that "the genetic basis is essentially correct."[46] The theory convincingly explained the limitation of the disease to the Fore, its association with certain families, and its preference for women and children. Burnet and the other Australian members of the committee also argued for a "more disciplined and integrated pattern of research than in the past." In particular, they expressed concern that "serious friction had occurred in the past through Dr. Gajdusek's assumption that all aspects of native life were available for his study irrespective of the interests and feelings of others." They suggested that Frank D. Schofield, recently appointed as assistant director (research) in the Public Health Department, should take over responsibility for kuru investigations. Again, the Adelaide team was allocated genetic and demographic studies, while Gajdusek was allowed restricted access to specimens and limited opportunities to conduct clinical trials among the north Fore.[47] That is, Bennett would study the presumed cause of kuru while Gajdusek determined the nature of the pathological process.

After the meeting, Burnet wrote to Smadel deploring yet again his protégé's "naïve and overbearing attitude and . . . complete insensitivity to the usual decencies of scientific behaviour." But he continued in a more conciliatory vein:

> As you know well, [Gajdusek's] unique personality can irritate academic types like [Gray] Anderson and Bennett. Neither of the latter has been wholly reasonable and the Administration on its side has shown little interest and no consistent line of policy in the matter . . . I found Gajdusek very reasonable and willing to accept our recommendations in toto and once again I rather fell under his spell when he got on to his real hobby-horse—child development in primitive communities.[48]

Smadel welcomed the compromise. To Burnet he wrote: "I think we should create a New Guinea Peace Prize and make you the first recipient. The division of the country into 2/3 for the Australians and 1/3 for Gajdusek and his cohort was worthy of Solomon."[49] Scragg let Gunther know that Gajdusek had "agreed to toe the line," though he must have realized how ephemeral that commitment would be.[50]

Astounded by the ferocity of the disputes, Schofield was dismayed when he heard about his new role supervising kuru research: as an English outsider he wanted to let Australians and Americans fight their own battles. Bored with clinical studies in London, Schofield was lured to Papua and New Guinea late in 1959 by the prospect of conducting research into major public health problems, especially malaria and neonatal tetanus. At his job interview, Burnet had mistakenly assured him that he would be insulated from the bickering over kuru.[51] Gajdusek wrote to Schofield warning him: "I would not want you to get 'buried' in kuru politics and I fear it is now inevitable for anyone involved not to escape the conflagration and a scorching."[52] During lunch in Brisbane, Schofield told me that he eventually came to like the American scientist and respect his dedication to research. He met him and Zigas at Okapa and later, with Schofield's wife, they partied together at Clarissa de Derka's apartment in Moresby. "I found Carleton very communicative and interesting," he recalled. "And not the slightest bit worried I was from the administration." As the flies buzzed around us in the Brisbane heat, the elderly Englishman, still very proper and precise, with a trace of his Harrovian education lingering, also made clear his annoyance with Gunther, whom he regarded as rather too obstinate and unsparing. An independent spirit, Schofield proved loathe to enforce the wishes of Gunther and Burnet. In any case, he insisted, as "assistant director (research)" he actually "wasn't a director *of* anything." He could not command anyone in the field: "I wasn't directing, it's a *silly* word!" He heatedly disowned responsibility for kuru research. Like so many others, Schofield later chose to follow Gajdusek to the NIH for a brief period—but to work separately on tetanus, never on kuru.[53]

If kuru was hereditary, as Bennett claimed, then it would be necessary to take measures to prevent the "kuru gene" from spreading. As he proposed early in 1959, "control of reproduction may be the only way in which the incidence of such a disease might be limited."[54] At the kuru research committee meeting in December, the Adelaide scientist recommended that "an effort be made to prevent the spread of the genetic taint by creating a kuru reserve from which no emigration was allowed." But members of the committee thought this constituted denial of the rights of the Fore, and in any case, probably was not feasible.[55] Robson made it clear he did not share his colleague's opinions on the need to isolate the Fore. Scragg noted their discord, telling Gunther after the meeting that "Robson is happy but Bennett who disagrees with Robson still wants 1st the people treated like cattle and 2nd Gajdusek kept out."[56]

Sharing some of the eugenic enthusiasms of his mentor Fisher, Bennett continued to campaign for a *cordon sanitaire* around the Fore. After spending five weeks among the Fore, he wrote to Cleland in February 1960 expressing concern that so many men were moving out of the region to Goroka and beyond. It was "alarming" that more than one thousand men—all "capable of transmitting the kuru gene"—had already left. He was "astonished" to find that a hundred or more Fore were recently recruited into the Highlands Labour Scheme and sent away. Bennett urged the colonial authorities to keep a "close watch on any migration of these people," and to consider creating a Fore reserve "taking account of the features of Aboriginal reserves which exist in Australia."[57]

On April 21, 1960, Paul Hasluck, the Australian Minister for Territories, approved restrictions on Fore movement out of the region and plans to return those who had already departed. Gunther believed that "the disease is such a foul one, it seems to me that we have an obligation to prevent its spread." It was only though administrative action that Fore had been able to leave their high valleys without fear of death; now administrative action might render them as stationary as before.[58] But the etching more deeply of such stigma onto the Fore appalled Gajdusek and Zigas. "The government of Australia has declared a 'genetic quarantine' of the kuru region," they wrote in the *American Journal of Tropical Medicine*. "This eugenic policy, pragmatically unenforceable, is without precedent anywhere."[59] Theodosius Dobzhansky, a leading population geneticist, was passing through Port Moresby when he heard about it. Indignantly, he jotted a letter to *Science*. Without challenging the genetic explanation of kuru, Dobzhansky described the "unprecedented" eugenic policy as a "severe restriction to be imposed on a whole tribe."[60] In practice, though, colonial authorities did little more than exempt Fore from the Highland Labour Scheme; they could never effectively prevent their movement in and out of the region. As doubt about the genetic hypothesis gathered momentum, administrative efforts to enforce the genetic quarantine soon lapsed. In 1962, an ad hoc committee on medical research in New Guinea recommended that "'quarantine,' which is ineffective, resented by the people, and is making registration difficult, be quietly discontinued."[61]

THROUGHOUT 1958 AND 1959, Gajdusek struggled to collect specimens from the Fore, overcoming increasingly fraught relations with the families of kuru victims and constant recrimination from the health department and other investigators. His single-minded pursuit of samples of Fore tissue was remarkable. In March 1959, he urged Zigas to keep on "collecting pathological specimens, blood specimens and all other environmental and clinical specimens." Gajdusek interpreted the Adelaide preoccupation with genealogical studies as "a complete green light for the continuation of our international collaborative venture in environmental, etiological studies." He was prepared to leave demographic and genetic analysis to Robson and his colleagues, so long as he could conduct metabolic and electrolyte studies and therapeutic trials at the Kainantu hospital and collect specimens in the field. "The division of effort thus leaves the pathological, clinical, laboratory, and ethno-environmental studies to ourselves and our collaborators— and this is our major interest in any event." Moreover, if anyone in Adelaide wanted "portions of the sera and pathological specimens we can collect, we can easily distribute them without breaking up all the other studies under way. There should certainly be enough material available for every laboratory that is interested in doing constructive work on the problem."[62] For Gajdusek and his associates, the acquisition of more specimens, their distribution to various laboratories, and their intensive scientific study seemed the only available routes to the real solution of the problem.

Once back in the highlands, Gajdusek quickly set about trying to realize his bold plans. In June 1959 he reported that "we are enlarging a KURU hospital here [at Kainantu], bringing new patients in and already a series of KURU and control autopsies have been done here and specimens are awaiting shipment to the US and laboratories around the world."[63] Tireless, Gajdusek was digging in, anticipating research for another five years at least. He demanded a "full laboratory approach to the problem and complete freedom to collect specimens, patients, autopsies, and data as we wish." As he told Richard L. Masland, the director of the National Institute of Neurological Diseases and Blindness (NINDB), which was Gajdusek's new location at the NIH: "Vin and I are desperately trying to collect cases and materials. Three brains go off today."[64]

During the 1960s, Gajdusek repeatedly tried to play the NIH card in New Guinea, hoping it might trump whatever else was held against him. Evidently, his U.S. institutional patronage, with control of an intramural laboratory, was his best argument for continued involvement in kuru research, his strongest shield against the attacks of Burnet and Bennett. He therefore sought to magnify the NIH role in kuru investigations. In 1960, for example, he made a plea to Masland and Smadel:

> Since we have been criticized for having "so small" a field operation—in spite of many an Australian attempt to curtail our work in the past—it will be most politic and helpful and a very much needed gesture to have other NIH experts arrive to see and study kuru a while at Kainantu. I beg that Len Kurland [an NIH epidemiologist] plan to stay a week with us at Kainantu.[65]

Woman with kuru helped to her feet. Photograph courtesy Peabody-Essex Museum, Gajdusek Collection

On occasion, the colonial authorities cautiously welcomed the modest NIH investment in kuru research, though more often they felt embarrassed and resentful. Unlike his senior colleagues, Schofield was readily drawn into the orbit of the NIH, optimistically assuring the Americans in 1962 that he could "guarantee our cooperation for at least 2 years from now."[66] He argued for a contract which would commit the NINDB to paying the Papua and New Guinea government for "logistical support" of kuru research. But Burnet protested against any formal arrangement. "The contract puts Gajdusek in a position to make himself an insufferable nuisance," he warned T. K. "Terry" Abbott, the acting director of health. "I doubt if there is any way to terminate Gajdusek's activities but I should certainly not be a party to regularizing and increasing them."[67] Abbott agreed: "Until we find ourselves lacking in competence to manage our problems we should not make this place a free-for-all among our more wealthy uncles."[68]

Many Australian scientists continued to express skepticism if not outright hostility toward NIH involvement. Bennett was always "a little uneasy at the prospect of Gajdusek and the NIH buying their way in."[69] By 1962, however, the Adelaide scientists had mostly lost interest in further kuru investigation. They had devised the best genetic model they could, and intellectually their research program was nearly exhausted. Burnet, however, was gradually renewing his enthusiasm for treating New Guinea as a vast laboratory for studies of disease and human ecology. He proposed establishing an institute of human biology in the territory and gave notice to Gunther and Scragg that NIH competition would threaten the project. Throughout the 1960s Burnet attempted to repulse any intruders into his arena. Gajdusek complained that Scragg, through the Papua New Guinea medical research committee, the successor of the kuru research committee, was making "extensive efforts to curtail, or even stop, all research work by myself and my many collaborators."[70] Frank Fenner, the chairman of the committee, responded, pointing out that "neither the committee nor the administration can be expected to endorse a blank cheque when the welfare of natives is involved, as it must be for every New Guinea project."[71] Burnet and Fenner eventually warned Paul A. Siple, the Antarctic explorer and scientific attaché at the U.S. embassy in Canberra, about his menacing compatriot. Soon Siple was able to assure the two scientists that "we have taken steps to control the activities of Dr. Carleton Gajdusek . . . I believe the situation is now under control."[72] The U.S. State Department would forbid his traveling to New Guinea without approval from Moresby. To Scragg, Siple wrote more ominously, telling the director of health that "we have one last heavy hand approach we could use if necessary."[73] But Gajdusek, of course, was never deterred. As Burnet remarked to Robson, the intruder was "quite impervious to reprimand and even military orders—but he does add something unique and nonconformist to the American scientific world."[74]

ALTHOUGH THE PATHOLOGICAL FEATURES of kuru were now well recognized, laboratory investigation seemed still to offer few clues to its cause. Fore tissues and blood turned up in laboratories around the world, but to no avail. No one could suggest any alternative to the Adelaide group's genetic hypothesis, though few expressed complete confidence in it. Laboratory scientists could find no environmental trigger or toxin, no metabolic or hormonal anomaly, no obscure infection. More than two years after his entry into the Fore region, Gajdusek remained frustrated in his effort to make his mark as the discoverer of the true cause of kuru. Despite his extraordinary energy and perseverance, his research was going nowhere, his specimen collecting appeared ever more futile.

Then in July 1959, William J. Hadlow wrote to Gajdusek pointing out the similarity in appearance of human brains with kuru and sheep brains with scrapie. Their common features struck Hadlow, an American veterinary pathologist assigned to an English agricultural field station, when he happened to view a traveling exhibit of enlarged photographs of kuru brain tissue at the Wellcome Medical Museum in London.[75] For hundreds of years, scrapie had been recognized in sheep. They became excited, rubbing their fleece against trees and fences, trembling and staggering; sometimes they showed a high-stepping gait or hopped as rabbits do. Like kuru it was a degenerative disease of the nervous system, leading inevitably to death, and its cause was unknown. Postmortem examination of the brain revealed characteristic vacuoles, or tiny spaces, within the cells, sometimes giving the tissue a spongy appearance. In the middle of the twentieth century, veterinary scientists had conducted experiments successfully transferring the disease to healthy sheep and to goats, though the incubation period was lengthy, with signs of the disease developing only after many months or even years. Scrapie thus became known as a transmissible spongiform encephalopathy caused by a mysterious "slow virus."[76]

This odd affliction of sheep and goats would give Gajdusek a model for explaining the transmission of kuru and, finally, an experimental program at Bethesda. He hoped to conduct long-term inoculation experiments in chimpanzees—"even in the face of overwhelming evidence for completely genetically mediated pathology"—to determine if they eventually succumbed to kuru. It was at least a new use for brain specimens.[77] "I suggest we make cautious long-term plans in this direction since no one else in the world of medicine seems to be doing so," he wrote to Masland and Smadel. "I admit my skepticism—but—who can say?"[78]

post-Sputnik.[9] In the late 1940s there were only a few foreign research awards; by the early 1960s there were almost a thousand. Much local, intramural research also drew on foreign resources, addressed overseas problems, or involved distant collaborators. At the NINDB, Pearce Bailey, its director until 1959, supported field studies of retrolental fibroplasia (an eye disease of the newborn) and employed Leonard Kurland, an epidemiologist, to investigate the geographical distribution of amyotrophic lateral sclerosis (ALS), a motor neuron disorder.[10] Richard Masland, Bailey's successor, promoted training programs for foreign researchers, especially those from Eastern Europe, and encouraged Gajdusek's studies in the Pacific and elsewhere.[11]

Yet Gajdusek could not conform to an institute that still focused mostly on laboratory work in Bethesda. In order to allow him to travel, he received an extramural salary even though he ran an intramural laboratory. Since the NIH were no longer concentrating on "applied" investigations, particularly if they concerned the health problems of another country, it also was necessary to configure his research as "basic" inquiry. While in New Guinea he might emphasize the plight of the Fore, in Bethesda he always represented his work as a fundamental study of the nervous system.[12] The NIH during this period showed remarkable bureaucratic flexibility and a commitment to shielding, if not nurturing, talented scientists, most of them young men from elite eastern universities. In some ways it was a challenge to fit Gajdusek into the NIH; in others, he seemed preadapted, ready to flourish.

When he returned from New Guinea in early 1958, Gajdusek found some space in Building 8, a neocolonial red-brick structure adjacent to the director's office at Bethesda, to sort through his case records, tissue samples, and correspondence. Soon he acquired an exceptionally patient and efficient secretary, Marion Poms, who ordered his life and work for the following three decades. She proved adroit at finding him and communicating with him at the most isolated locations, whether a Pacific island, the Siberian tundra, or highland New Guinea. Crucially, she also displayed a knack for resolving personal disputes between Gajdusek and his collaborators and friends, for providing a warm social gloss to his blunt interactions with others.

At first, his "laboratory" at the NIH functioned more as a data clearing house than a site for hematological and other pathological investigation. Gajdusek relied on Igor Klatzo for analysis of the brains that he sent from New Guinea and turned to Kurland for epidemiological advice—both of them at the NIH but outside his small group. Roy Simmons at the Commonwealth Serum

Laboratories and Cyril Curtain at the Baker Institute, back in Melbourne, continued to do most of the blood work.

Soon Gajdusek began to attract to his laboratory some of the young men he had met through scouting or encountered in New Guinea. Barry Adels arrived first, in the summer of 1959. "I would be a hell of a taskmaster," Gajdusek told Adels, who was embarking on his own medical career. "But if you really want to get a taste of how dull the compiling of laboratory data and its analysis is, and of what the analysis of field epidemiological records is like, I can give it to you."[13] Later he explained in more detail what went on at Bethesda:

> The work on hand is largely data processing and, as with any scientific data processing, complex matters of assorting, assigning, codifying and tabulating, and of establishing the rules of classification and the units and symbols, form the major part of the work. Once these are done, the mathematical manipulation of figures is a rapid, easy and pleasant "play."[14]

In practice, though, Adels quickly became preoccupied with the logistics of specimen preservation, identification, and transportation. By then Gajdusek was back among the Fore, demanding instructions for the preparation of tissue samples and blood tests and attempting to send fresh material to Bethesda and other laboratories. Adels supplied him with equipment, reagents, and technical advice, receiving in return more brains, blood, epidemiological information, quotations from T. S. Eliot and W. B. Yeats, and a discourse on Dada. Delivery failures and problems with refrigeration of specimens worried them both: early on, Adels suggested using dry or crushed ice in the newly developed Styrofoam containers, which seemed to work well enough, though the packages could still go astray.[15]

In the fall of 1959, Jack and Lois Larkin Baker arrived in Bethesda on a sort of extended working honeymoon. Jack helped to sort out maps and census reports while Lois set up the "laboratory and specimen receipt center." She was, as Gajdusek reported to Masland, an "astounding technician," whom Joshua Lederberg was trying to lure to Rockefeller University. Even if she lacked a college diploma, "she can manage any and every technical problem we may decide to enter into," he wrote.[16] A few years ago, when I asked Lois Baker what it was like to set up a laboratory in a foreign country and an unfamiliar institution, she quickly dismissed the question. "Well, there's nothing easier when you don't have to worry about the money," she said. We were drinking beer in the stifling heat outside their house, next to the caravan (RV) park at Bribie Island, Queensland, which they managed after Jack retired from New Guinea in the late 1970s. "You just said, 'Well, we'd like this and this and this,' and this this this turned up! So it wasn't hard at all." The opportunities at the NIH presented a stark contrast to her experiences at the Hall Institute. "The United States had the money to do the research which Australia didn't have. And I don't know where [the Australians] were going to get it from." After eight months at Bethesda, though, she and Jack returned to New Guinea to his new post in Chimbu, and sadly, she never again ventured into a laboratory.[17]

FASCINATED BY THE ANALOGY TO SCRAPIE, Gajdusek realized he needed to begin a series of inoculation experiments.[18] First, he went in 1961 to see Bill Hadlow in England to obtain some goat scrapie material, which he then secretly brought into the United States.[19] With J. Anthony Morris at the NIH, he began inoculating mice and observed that within months they were manifesting signs of the disease.[20] Next, he would try in the same way, using Fore brains, to give chimpanzees kuru.

Early in 1962, Smadel assigned a young virologist, Clarence J. "Joe" Gibbs Jr., to assist Gajdusek in his experiments. A reserved and mild man, Gibbs had worked previously on arthropod viruses with Smadel at Walter Reed. When he told Smadel that he planned to accept a Rockefeller Foundation fellowship to study these viruses in São Paulo, the reaction of the "boss" was "immediate and violent and in his inimitable fashion he pointed his finger in my face and said: 'Goddamn it Gibbs, you're not going to Brazil!'"[21] Instead, Gibbs would join Gajdusek, studying the transmission of scrapie and kuru. Having met Gajdusek at Walter Reed, Gibbs was at first a little apprehensive. Writing from the bush, from a place he described as the "southwestern section of Kuk-type people," Gajdusek tried to reassure him, explaining that his laboratory of "child growth and development and disease patterns in primitive cultures" was

very nebulous and "unstabilized" on paper and I want it that way . . . I want it to stay small, intimate, and "non-committed" to any one line of study or attack. Since it is a very personal section and remarkably "free-wheeling" I try to do everything to preserve that freedom . . . I want our approach and laboratory to stay thus fluid and I approve of a parasitic approach to the vast sprawling NIH rather than any attempt to build our own empire and facilities.

Joe Gibbs at the National Institutes of Health, Bethesda, Maryland. Photograph courtesy Peabody-Essex Museum, Gajdusek Collection

It was a place where one might study anything "from sex to serology, from sea urchins to starlings, from steroids to scrapie." Above all, Gajdusek continued, "I think we should not do what others are doing, want to do, and will do, but rather stick to things others either do not want to do or bother to think of doing."[22] Somewhat to his own surprise, Gibbs liked the idea of this sort of unstructured, independent research.

In practice, Gajdusek could readily sketch out the broad parameters of his investigations. "We have immediately on our hands a vast wealth of possible studies among the best 'sentinels' in the world, the primitive communities still existent . . . We get into the most remote and wildest regions," he told Gibbs, a quiet, retiring man. Now warmed up, Gajdusek expatiated on his research:

We work on neurological disorders clinically, and have a number of new "infectious" (probably) neurological disorders under investigation or needing such study, and we have the full responsibility for the most fascinating branch of virology in medicine and perhaps the only human model with a good chance to be a case like temperate phages and the lysogenicity story— kuru and its analogies to scrapie. Thus, slow and chronic and latent virus infections and the possibility of gene-borne viruses in man will all be our province.[23]

Although not inclined to traipse through the bush, Gibbs found the notion of a "slow virus," and the speculation on an infectious cause for a genetic disease, intriguing. While Gajdusek cavorted with the Kukukuku, Gibbs would stolidly conduct inoculation experiments back in the United States. He recalled Smadel, not long before his premature death, telling him "in very fatherly terms: 'Joe, I want you to undertake this work and I guarantee that within five years you will either have golden positive or golden negative results— and I am sure they will be golden positive.'"[24]

Gibbs took charge of the new animal house at the Patuxent Wildlife Research Center in Laurel, Maryland. In a cinder-block, uninsulated building, he led efforts to transmit kuru and other subacute progressive degenerative diseases of the nervous system to chimpanzees, monkeys, and rodents. He inoculated the nonhuman primates with pieces of fresh brain from victims of kuru, CJD, ALS, the parkinsonism-dementia syndrome of Guam, multiple sclerosis, Alzheimer's disease, and many more neurological conditions whose cause was still unknown. In 1962, Gajdusek wrote to Frank Schofield, warning him that "we are immensely interested in obtaining appropriate inoculum material from autopsies."[25] Previously, he had sought to preserve and fix brain tissue so its structure would be legible to pathologists in distant laboratories; now he wanted the freshest possible material. "The kuru brain is going into *every-thing*," he told Schofield a month later. "We need further such specimens."[26] In January 1963, after three years reflecting on analogies to scrapie, Gajdusek finally inoculated intracerebrally a newborn rhesus monkey with brain homogenate from Eiro, a kuru victim.[27] On August 17, along with Morris and Gibbs, he proceeded to inoculate intracerebrally the first chimpanzee, which they called Daisey, with 0.2 mLs of a 10 percent homogenate of brain from Kigea, a young Fore girl who died of kuru. On August 21, a second chimpanzee, initially named George, later Georgette, had brain homogenate, again from Eiro, injected into its frontal cortex. By the end of 1963, Gibbs could boast of inoculating seven chimpanzees, seventy-five smaller nonhuman primates, and more than ten thousand mice, with homogenate from kuru, scrapie, and

Primate house in later years. Photograph courtesy Peabody-Essex Museum, Gajdusek Collection

other diseased brains. Now it was just a matter of giving it time, even though the prospect of waiting years left Gajdusek alternately appalled and irritated.[28] When not inoculating animals, Gibbs passed the ensuing months patiently and methodically protecting them from other infections, such as tuberculosis, and fending off inquiries from the U.S. Department of Agriculture, which feared inadequate containment of scrapie at Patuxent.[29]

In attempting to control the new field of human slow virus investigation, Gajdusek was staking a claim on a territory far more abstract than the Fore region. Here his research group—now renamed the Laboratory for Study of Slow, Latent, and Temperate Virus Infections—had few competitors. At the Rocky Mountain Laboratories, Hadlow and Carl M. Eklund conducted scrapie and mink encephalopathy transmission experiments mostly with rodents, but lacking Gajdusek's field connections and clinical engagement, their access to human pathological material was limited.[30] A symposium at the NIH in 1964 consolidated Gajdusek's leadership in the search for human slow viruses. As Gajdusek and Gibbs observed at the time, if any one of the "wide range of subacute and chronic affections of the central nervous systems of unknown etiology" were to prove the result of a slow virus infection, "we shall win a significant advance in our understanding of diseases of the human brain."[31] The meeting brought together most scientists—predominantly veterinary pathologists—studying viruses with properties of latency, masking, slowness, or temperateness. A few human viral diseases, such as herpes and rabies, shared qualities of latency and masking, but scientific attention previously had focused on the pathogenesis of acute infections. Meanwhile, some veterinary pathologists had taken an interest in the slow viruses afflicting the brains of other animals: at the 1964 meeting scientists could present findings on scrapie in sheep, goats, and mice; Aleutian disease and transmissible encephalopathy in mink; and visna in Icelandic sheep. A fascinating pattern was emerging: these degenerative neurological conditions evidently were transmissible, even if no agent had yet been isolated. The question was whether Gajdusek could fit kuru and other diseases of the human nervous system into this etiological framework. "Kuru remains today," he told delegates, "an unsolved challenging problem, the solution of which cannot fail but to give us many new leads in neurophysiology and neuropathology, in the study of the whole range of heredofamilial degenerative diseases of the brain, in human genetics of isolated groups, and even in the pathogenesis of senile changes."[32]

ALWAYS IMPATIENT AND LONGING FOR WILD PLACES, Gajdusek spent little time in the Bethesda laboratory and the Patuxent animal house. In the Fore region, his collaborators—first Vin Zigas and later Michael Alpers, a young Australian doctor—were doing more and more specimen collection for him. In Bethesda and Patuxent, others had done most of the laboratory work from the beginning. Gajdusek would suddenly appear, coming down from the sky in a helicopter or plane in Okapa or Bethesda, and then take control of activities for a time, talking incessantly, throwing out new ideas and provocations. Other times he spent on the road or off the beaten track, often in Micronesia or Siberia, directing the comings and goings of his associates from a distance, with Marion Poms helping to move people and things along.

Gajdusek wanted his friends and assistants from New Guinea around him when he was in the United States, so he invited Zigas, the Bakers, Schofield, and Alpers to the NINDB—he even found a temporary post for Clarissa de Derka at the National Library of Medicine. It was important for him to give the American laboratory scientists an experience of life in the field, among the Fore. Usually they returned with a stronger personal commitment to the research and with a keener appreciation of the value of even the most battered or contaminated specimen. Having struggled along the trails south of Okapa, they generally became less sensitive to a hard-won tissue sample's imperfections. Richard "Dick" Sorenson, another of Gajdusek's young friends, started at Bethesda late in 1961 and spent most of the next decade between laboratory and field, where he studied the changing ecological relationships of the Fore and made countless documentary films, many focusing on child care and personality development.[33] Not long after Paul Brown began working with Gajdusek at the NINDB in the early 1960s, he too was hauled into the Fore region, where he conducted autopsies and collected more specimens. Even Gibbs was compelled to endure a few weeks among the south Fore, though he was, as expected, thoroughly miserable the whole time, yearning for the animal house at Patuxent.

Visiting Gajdusek's NIH laboratory in the 1960s, a journalist from *Look* magazine found herself entangled in a skein of papers, equipment, books, journals, and boxes. The scientist, amazingly impulsive and loquacious, used a cluttered passageway as his office. The main lab was "a confusion of psychedelic posters, maps of New Guinea, photos of primitive people, dog-eared paperback editions of Camus and Hermann Hesse, conventional laboratory equipment and scientific journals." In one corner, "fuming like a witch's cauldron," stood a tank of liquid nitrogen. Melanesian children and other visitors crowded the rooms. Somewhere a researcher was meditating, while another practiced yoga. Bob Dylan's *Highway 61 Revisited* was playing, loud. The journalist overheard conversations on "American imperialism, the superiority of the new morality, the American achievement hang-up, the comparative merits of different cameras, the cornering powers of various sports cars, and bluegrass music."[34] Now and then Gajdusek would blow up. This imbroglio did not match the journalist's preconceptions of a normal scientific laboratory.

"We used to answer the phone, 'Dr. Gajdusek's office,'" Judith Farquhar recalled. She was one of many anthropologists Gajdusek collected at Bethesda, though she frequently thought "handmaiden" better described her job. She remembered:

There were the precious Revco freezers in which a phenomenal variety of specimens waited to be useful, the freezers themselves demanding meticulous record-keeping from us. There were the cinema and still photographic records from Papua New Guinea, Micronesia, and other parts of the Pacific, all requiring assembly, copying, documenting, and special storage. And there were Carleton's journals, growing all the time with descriptions of other fields, labs, scientists, museums, villages, mountain trails, and palaces; bursting with analyses of research reports; speculations and hypotheses; facts historical, natural, and metaphysical; gossip and scandal; and embarrassing psychoanalysis of all of us.

In the late 1960s, the crowded laboratory was accumulating an "ever-thicker global network," a "global 'family and friends' community." The ambition to collect and archive materials far exceeded the capacity of Gajdusek or anyone else to make use of the stockpile. "The outside world that flowed through these few rooms is a testament to a certain refusal to reduce the import of kuru and its kindred diseases," Farquhar believed. "The variables of the phenomenon were not under control." Kuru came alive to her in the specimens of frozen tissue, the field reports, the constant stream of visitors, and the animals in Patuxent. "Despite its rigorously local character as a human epidemic, kuru after 1957 quickly became a global phenomenon and the Bethesda lab, handmaidens and all, was one of its key nodes."[35]

The laboratory remained a troublesome place for Gajdusek, a site where old friendships could be broken and were rarely strengthened. Early on he wrote to his mother about his interactions with laboratory colleagues:

> I like them all and work well with them all, but my method of working, implying equal accord between the various personalities, only seems to stimulate difficulties. Everyone from you to my newest friends and co-workers takes a proprietary interest in me and my work that the others who work with me find difficult. I admit that I encourage this, for I appreciate such enthusiastic dedication. However, it can create problems and, just as overly devoted love, I find it disquieting and, at times, restricting.[36]

When Adels returned to Bethesda in 1966, Gajdusek expressed his "firm conviction that NO ONE should stay around me for more than a few years, and forseeing this, I warn you from the start!" He complained that "we are too big and too rushed, to be sure! I would welcome the day when I was again ALONE."[37]

Most of the time, Gibbs and other laboratory workers showed amazing forbearance and tenderness. But not everyone could shrug off Gajdusek's episodes of shouting, his insistent demands, his occasional tantrums. The disputes with his young acolyte Sorenson became increasingly intense and demoralizing. Gajdusek's "desire to have the illusion of personally manipulating everything," his "tendency to play personalities," came to grate on the budding anthropologist.[38] The laboratory had revealed "the other Carleton," who issued the "occasional erratic demands coupled with petty threats alternating with effusive affection and the unfulfilled direct and implied promises of forthcoming bounty." Sorenson complained of "tirades concerning the work and fancied fears concerning the personal dedication of the staff," and he deplored "rapid changes, often in midstream, of program and objectives, motivated by transient but powerful emotional states."[39] Later, Gajdusek wrote to Adels to tell him, "I am amazed that to date it has been only Dick who has reacted with such destructiveness to my changeable whims, loves and lusts."[40] When the laboratory became too stressful or tense, Gajdusek might still resort to the New Guinea highlands. After a particularly difficult period in 1967, he retreated to Moraei, in the Kukukuku region. "Personally, I'm again having fun," he wrote, "productive fun—and that reduces the need for Wagnerian display."[41]

AS THE DEMAND FOR FRESH MATERIAL INCREASED in the 1960s, the logistics of specimen acquisition, registration, and transportation became even more pressing. Gajdusek often told Frank Schofield what he coveted:

> Well documented specimens from living patients who are WELL-studied and followed from a clinical investigative point of view. Urine specimens, bloods, CSF, throat washings, blood specimens and tissue culture and frozen biopsy specimens are all needed on patients, "pre-patients," etc., for BOTH microbiological and biochemical work. Likewise, autopsy material is needed for BOTH approaches in the form of live cells for culture, quickly frozen material as well, and, finally, specially collected material for microbiological histopathology and for biochemical histopathology in the form of whole CNS [central nervous system] specimens, pituitary and adrenal specimens, etc., etc.

Gajdusek hoped for an improved "flow of properly collected and handled and transported specimens from the field as a long-term matter over the years." His goal was to muster more than one thousand blood samples and some twenty or so autopsies each year. This would require a small field laboratory south of Okapa, where he and his collaborators could prepare and register specimens under standard conditions. Additionally, he urged Schofield to maintain the Kainantu laboratory specially for "transport in and out of materials and specimens."[42] To Scragg, Gajdusek wrote: "We are, obviously, immensely interested in obtaining appropriate inoculum material from autopsies . . . We are ready to handle as many specimens as we might receive."[43]

The challenge of preserving and transporting serviceable specimens was always formidable. Frequently Gajdusek despaired:

> I am torn between the desire to get autopsies, and the full knowledge that there is no way possible of getting the adequate specimens we need this year—24–48 hours of no cooling, doubtful cooling at Okapa, no cooling without immense imposition to everyone en route to Kainantu, and probably complete failure of refrigeration from there on to Melbourne . . . Thus, without the logistic support needed, the enthusiasm for field heroics fails me.[44]

And then again, an opportunity would come his way, an expedient found, a pragmatic solution determined—and his enthusiasm returned, stronger than ever.

Since the colonial authorities were making it so hard for Gajdusek to base himself in the highlands, he frequently relied on collaborators to obtain and dispatch the specimens he wanted, whether for distribution to other scientists or for use at Bethesda and Patuxent. With Zigas transferred out of the field, and Gray and Auricht, both Adelaide acolytes, resistant to his demands, Gajdusek soon came to depend on Michael Alpers, a renegade Adelaide medical graduate, to conduct autopsies and gather specimens. In 1960, Norrie Robson, the Adelaide dean, had suggested sending this "most intelligent young man" to Okapa to help with medical research there.[45] Twenty-six years old and recently married, Alpers readily disclosed his enthusiasm for anthropological studies and offered hints of a literary mindset. "I am interested in the social background of the people," he wrote to Scragg, "and the folklore associated with the disease."[46] After a few months training at the Summer Institute of Linguistics in Ukarumpa (in the eastern highlands), Alpers traveled up to Okapa and hiked through the Fore region. He decided to base himself, with wife and baby, in Waisa, a hamlet off the main track to Purosa, in a relatively isolated area with high kuru incidence. The community was eager to keep him there, so they built a house for his family and killed some pigs for him. Soon after his arrival, he reported to Scragg: "Dr. Gajdusek is in the area, but I have only seen him for a short period . . . He is certainly a most stimulating and energetic man."[47] Alpers's reaction to the American intruder must have chagrined the director of public health, his employer.

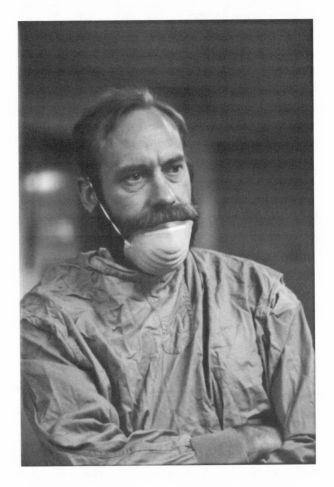

Michael Alpers,
1960s. Photograph
courtesy Peabody-
Essex Museum,
Gajdusek Collection

Before he left Adelaide, Alpers had been warned about Gajdusek. "The main thing was to be very careful about Gajdusek because he was a *dangerous* person," Alpers told me, as we sat talking on a chilly day in 2002, surrounded by his New Guinea library in Fremantle, Western Australia. "He spent a lot of time among 'those Kukukukus' and they were 'terrible people,' and you just had to be 'very cautious.'" But on meeting the notorious scientist, "we immediately hit it off, just from that first evening . . . In terms of his interests and my interests, it just *clicked*."[48] The same day, Gajdusek noted, "I have opened to him most of my pet ideas, suspicions, 'leads,' etc."[49] Later, he described the lean, bearded young man as a "quiet and conscientious doctor who may quietly go about digging up really valuable data."[50] The two scientists, both unflagging walkers, bonded more closely early in 1962 on a hard trek deep into Kukukuku territory—it was, Alpers recalled, a "marvelous jour-ney."[51] The Australian was reading Gibbon's *Decline and Fall of the Roman Empire,* while the American dipped into Proust's *À la recherche du temps perdu.* They thoroughly impressed each other.

Although Alpers initially decided to concentrate on collecting clinical and epidemiological information, Gajdusek had little trouble persuading him to conduct autopsies to obtain fresh inoculate. Alpers, however, soon realized the sensitivity of the Fore to relentless demands for new specimens and came to appreciate their fears of losing control over the bodies of their loved ones. From the beginning, he therefore made sure that as he followed the decline of the kuru sufferer, he came to know the family well, soliciting their friendship and trust. At Waisa he was "essentially already part of the family . . . the people were very open to that."[52] He was learning to speak Fore, running a popular clinic, participating in initiations, and contributing to bride payments and fu-neral exchanges. Eager to retain their white man, the Waisa people plied his family with food from their gardens. More than any other scientist, Alpers be-came incorporated into the community, adopted and cherished. To this day the people of Waisa maintain his house and wonder when he and his family will return permanently to their home.[53]

When Michael came to live in the village, a strapping young man named Pako came and introduced himself, offering to help. "Michael didn't rush into examining people with kuru," Pako told me. "He established a 'foundation' so that people could be close with him. He gave them food and sat with them, and after some time, everyone was pleased." Pako found the white man in-teresting. "He tested the sick and we followed him around." It was not long before they regarded him as a "Waisa man," before Waisa became his "asples." Pako called him his "age-mate" and friend.[54] As we sat looking out through the mist toward the trees, the old man remembered how Michael "worked all the time. He gave food, tinfish and rice, to his workmen and provided them with bridewealth." However, Pako continued, "some people are wondering what he is doing now, why hasn't he come and made a medicine to help us?" In any case, he should return soon to Waisa to live out his last years.[55]

"The first time was, I guess, the worst—because I wasn't sure whether it was going to work." But to Alpers's surprise the autopsy went smoothly. As he recalled:

> I was able to get the skull cap off, exposing the brain, while someone held the head. I had pre-labeled sterile vials in an insulated container—I would

take the selected tissue, pop it in, put that back, and so on. Then I took the brain out, got the cerebellum out, got down to the brain stem and then removed the brain. We had a bucket of formalin ready and just put it in there. Then I put the skull cap back, sewed it up, and basically handed over the body to the relatives, who were there giving advice.

Quickly, Alpers made his exit in order to get the specimens into the Okapa refrigerator as soon as possible. He called a plane to pick up the materials on the following day at the airstrip nearby at Tarabo. Packed in dry ice, the brain samples were flown first to Lae, then through Melbourne to the United States. After dispatching them, Alpers returned to the village "for the feasting and so forth . . . and then [to] give gifts to the family and be part of that process." He paused, smiled wryly, and said: "The number of blankets I've distributed amongst the Fore over the decades is astronomical."[56]

In June 1961, Alpers was called to the hamlet of Miarasa to examine Eiro, a boy of twelve years, well built and "rather sulky." The lad could not stand and shook uncontrollably: the diagnosis of kuru was inescapable. For more than a year, Alpers continued to visit him, observing his deterioration. Soon the child was weak, unable to speak; eventually, he stopped moving and just grunted softly, emitting a "general putrefying smell." Gently but insistently, Alpers talked with the relatives about an autopsy, and toward the end they gave permission. He heard of the boy's death on September 10, 1962, and arrived in the village an hour or so later, relieved to find the corpse still warm. "Further persuasion was required before they would let me perform an autopsy," he noted in the case record. "But [they] eventually agreed—with the prospect of a little pay in the offing—to allowing his brain to be removed." Eiro's brother sat beside him, weeping, while the operation took place. "Villagers getting impatient at the end," Alpers observed, "as it was drawing on in the afternoon. Pay promised to them later." Within four hours of death, the boy's brain was in the Okapa freezer. It arrived in Bethesda, still frozen, ten days later. The following year, Gibbs inoculated the chimp Georgette with some of it.[57]

At Waisa, Alpers became fond of a cheerful, playful girl called Kigea. Then in February 1962, she complained of headache and painful limbs. Soon she became unsteady and began to shake. In August Alpers described her as a "thin, rather frightened-looking child holding hands firmly together . . . Speaks very little, and then softly and shyly." By November she needed support for walking, and her speech was slurred. A few months later, Alpers found her lying down, "composed. Gave a smile of recognition. Speech rather grunt-

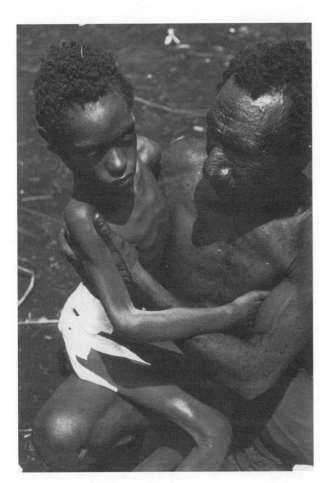

Kuru patient in
terminal phase,
held by father.
Photograph courtesy
Peabody-Essex
Museum, Gajdusek
Collection

like." In March Kigea was terminal, showing "just the faintest response of recognition and a slight attempt to grasp a proffered lolly [sweet]." Her father had kept vigil but was now "cross with her for taking so long about dying; and it is true there is very little recognizable as his daughter left."[58] This "wonderful young girl," Alpers told me, suffered a terrible, slow, painful death. Forty years later, in Fremantle, he became distressed as he remembered it.[59] The family had granted him permission to perform an autopsy. He took only her brain. Five hours later it was frozen in Okapa, and before long, a piece of it was put into the chimp named Daisey on the other side of the world.[60]

"Everything has been dependent on the fine specimens Michael Alpers got to us," Gajdusek reported to Schofield in 1963. "The frozen inocula are essentially our only hope."[61] Together with Roy Simmons, the hematologist in Melbourne, Alpers organized an efficient means of transferring frozen Fore brain samples from the highlands to Bethesda. Fresh from the corpse, the brain was frozen in Okapa and placed in dry ice later in Lae, for transport out of New Guinea. When he received the material in good condition, Gajdusek was impressed and relieved. "Michael got a good set of brain specimens off to us," he told Schofield. "They got to Roy [Simmons] refrigerated and from there to us frozen!!!!! It looks good."[62] As brains and other specimens were extracted and transferred out of New Guinea, more equipment, reagents, and personnel moved in. As these valuables passed from scientist to scientist in one direction, other goods, such as refrigerators, brain knives, and dry ice traveled in the reverse direction. Thus the portable canister of liquid nitrogen at Okapa was constantly replenished from canisters coming from Australia *via* Port Moresby, Lae, and Goroka. Scientists referred to the "brain shuttle" and the opposite "nitrogen shuttle." It is not altogether fanciful to compare this series of scientific gift exchanges with the *kula* of the Trobriand Islands. There, necklaces circulated clockwise across the sea, exchanged for armshells flowing counterclockwise. As they followed their *kula* path, shells accumulated value, and their transactors, a chain of trustworthy partners, acquired renown. After passing through several hands, the valuable eventually would return to the *kula* partner, just as Kigea's brain sample, moving counterclockwise across the Pacific, belatedly came back into Alpers's possession in Bethesda.[63]

IN 1963, AFTER ALPERS LEFT NEW GUINEA to work at the NIH, Gajdusek sent out Paul Brown to conduct more autopsies. Schofield urged Brown to come as soon as possible, fearing that "there won't be all that number of kuru people dying in his short stay to get brains from."[64] Brown, a twenty-seven-year old virologist mostly interested in measles research, was somewhat apprehensive about venturing into the last unknown to obtain human tissues. But when he arrived in Port Moresby, tired after days of flying, Schofield and de Derka made sure to entertain him. He found the Hungarian *femme fatale* "terribly attractive and thoroughly continental. She is quite a woman," he wrote in his journal.[65] On landing in Goroka a few days later, Brown was met by Dick Sorenson, who immediately whisked him up the road in his creaky Volkswagen to Purosa to perform an urgent autopsy, only to find the woman still alive. She soon died, and the neophyte Brown did the postmortem while the corpse was warm. They "laid her out on a table for the cutting and sawing, with her husband in tears and other relatives holding her head and otherwise assisting." At the time, Brown observed:

> The husband was most particular that every scrap of meat—his word—be put back in the body. Dick returned her to her hut, where the mother was waiting and groaning, and also quickly noted the loose skull cap and pushed it back and forth with more groans and crying. Talk about a macabre scene.[66]

Sorenson noted in his journal that "the natives" had been "loathe to have a stranger, Paul Brown, do the cutting."[67] Lacking fluent Pidgin and feeling disoriented among the Fore, Brown wondered how he was going to develop the rapport necessary to get permission for future autopsies.

Brown visited as many advanced cases of kuru as he could find, hoping to persuade the relatives to allow postmortem dissection. In desperation, he decided to try to "purchase" autopsies with a "couple of pounds and a couple of blankets. I don't much like the idea, but with only 3 autopsies possible in the next month . . . we might do worse than cater to their quite obvious motive."[68] Despite references to payment and purchase, the itinerant scientist realized he was still engaged in a complex series of gift exchanges. "It wasn't really 'pay,'" Brown told me early in 2006. "The Fore people have a very strong sense of economic balance. If I were to give you four pigs, you would be obligated and feel guilty not providing me with something the next time we saw each other. There was this sort of balance sheet."[69] When Igi'erakaba died at Wanitabe, her husband, after much argument and negotiation, consented to a limited autopsy. Sorenson observed at the time that the man was "quite unprepared to have a complete stranger come in at the last minute and remove half of his wife's brain."[70] Brown gave the husband four pounds and some salt. Thereafter, the people at Wanitabe became "loathe to show us their kurus."[71]

Just before he left the region, Brown managed to perform another autopsy, this time on Kabu'inampa who died at Purosa. Within an hour he had her brain in liquid nitrogen, and a few hours later it was on the road to Okapa. "We were told that we had taken a lot of meat from the body," Sorenson wrote, "and that we had to give meat in replacement."[72] According to Brown, her husband seemed determined to bargain hard: "Not content with a 3 lb. canned ham, he complained we had taken a lot of meat; so we gave him other of our

Dick Sorenson filming treatment of a Fore child, 1964. Photograph courtesy
Peabody-Essex Museum, Gajdusek Collection

own tinned meats. After this, rice, blankets, laplaps, and heaven knows what else, he demanded tobacco. Then he went back to his hut and cried with the other relatives. Funny people."[73] Within a month of his arrival, Brown flew out to Lae with a heavy cold and no regrets. "I am very anxious to get home," he wrote, "for I have seen the Highlands, and the Fore, and they do not hold me."[74]

Alpers was dismayed when he heard that Brown was merely flying in and out, just dropping in to get a few autopsies of people he did not know. He protested to Gajdusek, claiming that these activities—"commando work," he called it—would imperil the trust he had built up with the Fore. "Paul came in on a parachute and had a list of patients I'd seen and sort of raced around and did a few autopsies . . . Carleton was anxious to get more brains, basically. But we had enough in reality," Alpers told me.[75] Brown felt uncomfortable about the whole enterprise, but he had only recently joined the Bethesda lab and wanted to make a good impression, to appear dutiful and obliging. Although he returned to his measles studies and later investigated CJD, not kuru, his New Guinea field work at least reinforced his sense of the worth and scarcity of the specimens arriving in Bethesda.[76]

Mert Brightwell, the pugnacious new *kiap* at Okapa, surmised that Gajdusek was utterly irresponsible and his collecting out of control. Believing that these apparently cavalier autopsies constituted interference with a corpse, he demanded that none be performed without instructions from the coroner—that is, without Brightwell's own endorsement. Since brain specimens must be fresh, his directive seemed to present a serious impediment to the scientists' activities. Gajdusek's first meeting with the *kiap* was strained. Arriving in drizzling rain and low clouds at Okapa, the American inadvertently interrupted an assembly of police. Brightwell was furious and told the intruder that from now on, he was in charge. Gajdusek responded bluntly: "To glibly relinquish a right that medical men had won over the centuries to a bunch of half-baked 'kiaps' was certainly not my intention."[77] Later, Brightwell asked Schofield what the health department was doing to "keep out any crackpot with a whim to try something that may be of no scientific value and in other ways detrimental."[78] In the early 1960s, the shadow of Brightwell would loom over kuru research. A corpulent, Falstaffian figure, a good cook and heavy drinker fond of classical music and Victorian novels, Brightwell fascinated and sometimes intimidated the scientists. "Your name is mud around here," he wrote, rebuking Gajdusek. "Although the medical officers of the Territory may be suckholing after NIH gold, of which they stupidly regard you as the sole dispenser, and others cannot see through the fog of verbiage that you delight to wear for confusing purposes, there are others who have never been fooled by any of this and are not prepared to quietly tolerate the impositions and folly."[79] But as Gajdusek learned, Brightwell growled more than he bit.

During his brief visit, Brown watched the *kiap* carefully, noting "his habit of oscillating his large, bloodshot, and watery eyes . . . when thinking of things to say. He waddles around the house in stocking feet muttering half phrases of 'bloody well' and 'are you eatin'?,' and generally plays the role of district officer to the hilt . . . He may still be trying to size me up."[80] Yet Sorenson and Alpers came to like and respect Brightwell. Sorenson felt that "his tendency to dwell on what is wrong with everyone he knows is a way of defending his own acute sensitivity to slights and fear of possible slights."[81] According to Alpers, he was "tough and he used to throw his weight around and had this sort of blunt and blustery attitude to 'the natives.' But he was very interested in what was happening . . . and when it came to the crunch, he was helpful. He always did what he said he would do."[82] For Alpers, and later Brown, he agreed to provide approval for a coronial investigation before death, so long as he was convinced the relatives had no objection. "Things seem to be on a pretty good footing at the moment," Schofield reported to Scragg. Alpers and Brightwell "have reached a mutually satisfactory arrangement on post-mortems at present, so Alpers would like things kept that way."[83] But it was feared Gajdusek's incursions would trigger explosions.

HAVING OVERCOME THE MULTIPLE CHALLENGES to obtaining and transporting kuru brains, Gajdusek pondered how to distribute them strategically. From the start, gifts of Fore tissues to Australian investigators might repair, albeit temporarily, his relations down south. In return for specimens he often received equipment, reagents, advice, medications, and access to the field. Equally important, tissue offerings to pathologists and others in the Northern Hemisphere enabled him to forge useful relationships, gather information, and expand the kuru research network. Such debts generally were recognized in scientific authorship. Eventually, of course, more and more kuru material was reserved for inoculation at Patuxent, and thus removed from circulation, but some remained available for gift exchange with other scientists whenever Gajdusek deemed this advantageous. Blood was more readily partible and divided up than brains. Gajdusek boasted of his special skill in distributing blood:

> Since early in 1957 we have worked out a complex system of couriering and special handling of blood specimens, such that a single large bleeding from a patient or control subject may eventually be used to supply information from the laboratories of the large group of collaborators who are working with us . . . In many cases well over a dozen laboratories are benefiting from a single bleeding of a patient. I am rather proud of the remarkable economy in the research and effort which this complex personal organization between myself and my co-workers has made possible.[84]

Gajdusek's entrepreneurial genius is undeniable: he anticipated the extractive perspective of fellow scientists and set about manipulating the emerging economy of *his* kuru blood and brains.

Fore brains and blood, and sometimes other parts of their bodies, had become Gajdusek's valuables. He framed these things astutely, emphasizing their scarcity, the effort required to mobilize them, their exotic character, and relevance to contemporary medical science. The flouting of conventional indigenous boundaries of exchange, the disrespect for standard transactional limits, surely added value to these things, even as such nonconformist circulation might put the transactor in moral peril. Confident in the power of these farfetched material objects to produce desired outcomes, in their potency to transform others, Gajdusek was practicing a civilized form of fetishism. He obscured the original social relations of these specimens, so he could reify and mobilize them—thus fixing on a brain his attempted disavowal of the complicated cross-cultural encounter through which it was acquired.[85]

In managing the subsequent movements of the new kuru things, Gajdusek gave social identities to others, forged new relationships, became trustworthy, and reproduced value in science. As these novel things circulated, they acquired the power to create social relations, to move other persons, to make persons visible in surprising ways. For the recipients, these objects were personified around the figure of Gajdusek. That is, these things, once Fore, were now Gajdusek's kuru valuables.[86] In this sense, the generating and reproducing of value in science consisted in a motion, in the excitation and bustle of a network, rather than in recognition of the preordained quality of things.[87] These transactions became a means of eliciting and realizing the personal capacities of others, of making them manifestly scientific collaborators. "A large group of world experts in their specific fields in Australia, America and Europe stands ready to work with us on these specimens," the scientist assured Scragg.[88] Gajdusek was making himself into a big man in biomedical research, carefully managing his scientific gift exchanges, regulating the brain wealth of selected laboratories, and thereby accumulating social and intellectual debts. In the drive for credit, whether among Melanesians or scientists, it is necessary to keep extending and confirming networks through strategic exchange. As an anthropologist observed, "for big men it is important both to have large networks and to manage them well."[89]

Gajdusek felt on surer ground in these exchanges with scientists than with Fore, but sometimes mistakes were made and negotiations could be delicate. Hearing that some brains were allocated to Bethesda, a Melbourne neuropathologist complained that it "does seem reprehensible to send specimens to two places without informing the other. I am baffled by it all and obviously do not understand all the facts."[90] Such errors in scientific decorum were more readily corrected than those made among the Fore. Gajdusek quickly apologized, inviting his erstwhile collaborator to "visit our Kuru Research Center and help us with this fascinating problem in which your pathological investigations are already of great assistance."[91] "Although I have attempted to deal directly with all our collaborators," Gajdusek told Smadel after evidence of Australian irritation filtered through to Bethesda, "prestige and publicity considerations have brought in numerous 'intermediaries' at many stages . . . Yes,

Young boy with kuru helped by friend.
Photograph courtesy Peabody-Essex Museum, Gajdusek Collection

Joe, Australian feelings have been hurt by not having everything on kuru studied in their hands . . . However, all collaborators we have had are fully credited in reports as they come out, and joint studies are all based on extensive correspondence."[92] Those Australians and Americans who supported him often received Fore brains in appreciation. "We have given one [brain] to Prof. Sunderland of Melbourne, who visited us; and another of our posts, including the complete spinal cord, went off to Adelaide with Simpson who was here with us."[93] Distribution of specimens thus strategically distinguished collaborators from competitors, even within the same medical school. "We have on frozen file," he wrote to Frank Fenner, "aliquots of all the specimens collected—even those from 1957—and . . . I should certainly be interested in collaborating. They . . . have already been distributed far and wide for many studies."[94]

While other scientists recognized Gajdusek as personally bestowing the valuable kuru brains they received, he and other field workers never managed to deny completely the Fore provenance of such material. In making a brain into a thing and rendering it scientifically serviceable, they found it still attached to a person they once knew, not an object but a quasi-subject trailing an individualized Fore aura. For Gajdusek and Alpers, this brain tissue was part of Tasiko or Kigea; that inoculum was still part of Eiro. "How strange to fondle the brain of the infant I so well remember," Gajdusek mused, "of the mother I shall never be unable to recall."[95] Meanwhile Alpers scrutinized the chimps, waiting to see if they started to behave like Kigea and Eiro, the children fixed in his memory. For these implicated scientists it was not easy to mark a boundary between the affective bonds of the field and the supposedly more desiccated emotional texture of the laboratory. Efforts to make Fore persons into alienated kuru things in the laboratory repeatedly failed to truncate the social and emotional ties Fore had forged, to jettison the claims that Fore made on the scientists as persons. Such relational claims were turning the scientists into new persons, composite persons, fashioning identities more complex than the name "scientist" might usually imply. While scientists tried to make kuru things, Fore were generating persons, collecting white men, drawing them time and again back into their field.[96]

"AFTER READING JOE'S FASCINATING ACCOUNT of the chimps," Gajdusek wrote to Alpers from an atoll near Guam on August 28, 1965, "I am so excited that it is hard to exaggerate my enthusiasm."[97] At Patuxent, Daisey and Georgette, the chimpanzees inoculated more than two years earlier with brain tissue from Kigea and Eiro, had begun to stagger and tremble. Gibbs was unfamiliar with evidence of kuru. At first, he wondered if they were developing an acute infection, but the signs of their neurological disorder became more severe and intractable by the day. As their condition worsened, Gibbs told Alpers and called in other physicians with neurological training. After examining the animals, Alpers, by then the best chimpanzee neurologist in town, wrote his report, and "dashing it off in the usual fashion I made my clinical summary at the end, almost mechanically: 'clinical impression— kuru.' The word leapt from the page." Gibbs and Alpers bravely summoned Gajdusek back from his Pacific wanderings. "If the diagnosis of kuru had first to be discerned through the dry sequence of clinical notes," Alpers recalled, "it was magnificently displayed in the performance of the first chimpanzee when Carleton first came to see her. For on this particular morning the resemblance to the human disease was uncanny."[98] They began testing the chimpanzees for every imaginable nutritional deficiency and poisoning but found nothing. It looked as though kuru was directly transmissible.[99]

Chimpanzee with kuru. Photograph courtesy Peabody-Essex Museum, Gajdusek Collection

Having come to know them well over the years, Gibbs felt sad when the time came to "sacrifice" the chimpanzees.[100] Carefully he and Alpers conducted the autopsy. Elisabeth Beck, the chosen neuropathologist, flew in from the Maudsley Hospital, London, to advise on the preparation and fixing of the brain tissue. As she left, with brain in hand, she predicted that someone would receive a Nobel Prize for this breakthrough. It took several weeks before the pathologist believed the brain tissue was ready for examination. "When the microscopic sections came through," Beck recalled, "they presented the most amazing picture."[101] She immediately reported that the appearance of the chimpanzee brain was indistinguishable from the kuru brains she had seen. The following day, Gajdusek, Gibbs, and Alpers drafted an announcement of their findings and submitted it to *Nature*.[102]

Late in 1965, before the neuropathology was determined, Gajdusek began writing to Burnet and other Australians, telling them that the inoculated chimpanzees had demonstrated a neurological disease "astoundingly akin to kuru."[103] The news stunned even the more skeptical of Gajdusek's rivals and the more cynical bureaucratic worthies. "I have heard gossip of Gajdusek's infection of chimpanzees," John Gunther wrote to Schofield. "Does this mean that Gajdusek and Mac Burnet now agree that we have a virus? If the Australian and American groups came up with this we have an amazing achievement."[104]

A year earlier, Burnet had proposed an infectious origin for kuru and scrapie, within definite genetic parameters. Narrowly hereditary explanations of disease always made the disease ecologist uncomfortable. In 1964, then, he speculated that a transmissible agent like hepatitis virus might cause kuru in those who were genetically susceptible. The postulated virus of kuru—and of scrapie—was evidently a very unusual one, and quite elusive. Like a hepatitis virus it must be capable of maintaining low-grade persisting infection. Moreover, it must at some stage find and infect a highly vulnerable mutant cell in the body, which would then release huge quantities of virus, setting up a chain of infection, a lethal irreversible process. Making a "foolhardy attempt to follow a tenuous chain of logic into completely unknown territory," Burnet argued that Fore uniquely possessed nerve cells genetically susceptible to such infection with circulating virus.[105] But the transmission of kuru to chimpanzees suddenly deflected attention from human genetic predisposition. For a while it focused the minds of scientists on the character of the virus, not on the peculiarities of the host.

The trend toward an infectious explanation of the disease countered the prevailing tendency to seek genetic causes or correlates of illness. Since the discovery of the double-helical structure of nucleic acid and the mechanism of its replication, medical scientists often seemed consumed with enthusiasm for genes and their products. Even Burnet in the 1960s admitted the future for infectious disease research looked very bleak indeed.[106] Kuru stood out against the growing indifference to microbes, yet it was not alone. On his "tumor safari" in Africa, Denis Burkitt discovered and mapped the prevalence of a new form of cancer from which the causative Epstein-Barr virus was isolated.[107] Later in the sixties, Baruch Blumberg and colleagues happened to find a protein in the blood of patients with serum hepatitis that reacted with the blood constituents of an Australian Aborigine. This "Australia antigen" enabled detection and charting of the persisting infection that became known as hepatitis B.[108] All the same, viruses remained a minority interest during this period.

Convinced they had finally discovered the cause of kuru, Gajdusek and his colleagues set out to confirm its transmission and to find more human slow virus diseases. By 1967, the disease had been passed three times serially from chimp to chimp, thus proving its infectious character. Eventually, inoculated monkeys also succumbed to a syndrome resembling kuru.[109] Then Gibbs and colleagues reported the successful transmission of human CJD to inoculated chimpanzees as well as to monkeys, cats, mice, and guinea pigs—thereby recalling Igor Klatzo's original observation of the pathological similarity of kuru and CJD brains.[110] Later, Gibbs and Colin Masters found that Gerstmann-Sträussler-Scheinker syndrome, an exceptionally rare subtype of CJD transmitted genetically, could also spread through infection.[111] During the 1960s and '70s, a vast array of brain specimens from patients who died from neurological diseases of unknown cause went into dozens of species of laboratory animals. It soon became clear that medical procedures might also be spreading these agents: researchers confirmed that patients could contract slow viruses from intracerebral electrodes, dura mater (brain lining) implants, corneal grafts, human cadaveric growth hormone, and gonadatrophin.[112] Patuxent became a factory for the transmission of putative slow viruses—making it, as Gibbs boasted, an exciting new frontier for infectious disease research.[113]

The mechanism by which the agent spread in nature was still unknown. How was the slow virus transmitted among the Fore? What, if anything, prevented it from infecting their neighbors and visitors? No laboratory would provide an answer to these questions. Instead, anthropologists and epidemiologists now clamored to explain patterns of the disease in the eastern highlands of New Guinea. Thus the current of research was again pulling everyone back toward the field. To understand the behavior of the slow virus in the wild, we must follow this backwash, retracing the activities of anthropologists and epidemiologists in the kuru region.

Chapter 7

WE WERE THEIR PEOPLE

EVER SINCE RONALD AND CATHERINE BERNDT
had ventured into the northern reaches of Fore territory in the early 1950s, anthropologists shaped the meaning of kuru. The Berndts sought to present kuru in its proper social and cultural context, within a matrix of beliefs about sorcery and patterns of exchange relations, conflict, and competition. Carleton Gajdusek, too, often regarded himself as an amateur anthropologist, an expert on Fore and Kukukuku mores, especially on the topic of sexuality. Once kuru turned into a medical problem, anthropological studies generally became more applied; that is, they were more commonly directed to elucidating the possible cause of the disease in the wild. Therefore a shift occurred around 1960 from a relatively autonomous social anthropology to an explicitly *medical* anthropology, toward social inquiries serving to elaborate and refine medical explanations of the phenomenon. The Fore region thus became during the 1960s a testing ground for the nascent specialty of medical anthropology. If anthropology were to have any medical significance, surely it would be here.[1]

As early as 1958, the medical investigators were amused to find doddery Reo Fortune rummaging through Fore fireplaces, selecting various stones and putting them in his sack. Vin Zigas watched the anthropologist "madly gathering rocks and cooking stones." "He appears to me a peculiar type," Zigas told Gajdusek, "and I fear, in time, will prove a nuisance more than a help."[2] Hearing about the announcement of a new disease, Fortune had decided to visit the Fore after further compensating the Kamano man he shot in the 1930s, on his way to catching up again with the sorcerers of Dobu. In Okapa, he and Zigas "talked non-stop for two weeks on many subjects of mutual interest."[3] Fortune later claimed to have assisted with autopsies, but mostly

he checked Fore pedigrees and looked for toxic rocks. He studied siblings in 150 families where at least one sister or brother was suffering from kuru. J. H. Bennett and his Adelaide colleagues found the deaths of young persons with kuru concentrated in siblings born of mothers also affected, but Fortune observed that most of the siblings he investigated were born of kuru-free parents. He too therefore challenged Bennett's genetic hypothesis.[4] But he failed to identify any toxic causes of the disease and left the region disheartened. His health was deteriorating, and the twitches and tics that had tormented him since the 1930s grew worse. Concerned, he wrote to Gajdusek: "There is a chance it was a touch of kuru that developed with me about eight months after leaving the field—ataxia, now tremor."[5] Gajdusek did his best to rally the aging Cambridge don.

Lacking personal anthropological ambitions, the Adelaide team readily understood that they needed to enlist ethnographic help to sort out Fore kinship patterns and health beliefs. In 1960, Bennett asked W. R. Geddes, the professor of anthropology at Sydney University, whether any of his graduate students were interested in conducting their fieldwork among the Fore.[6] To John Gunther, Bennett wrote: "Further anthropological study of the kuru population is an important and urgent requirement."[7] The Adelaide geneticist advocated in particular a detailed, comparative study of regional folklore relating to kuru, but he also hoped for assistance collecting reliable genealogical data to counter Fortune's claims. Geddes quickly recommended a young couple, Robert and Shirley Glasse. Robert was a New Yorker, disenchanted with cold-war politics, who worked among the demanding Huli in the southern highlands and completed his anthropology Ph.D. at the Australian National University (ANU). Shirley had just finished her diploma of anthropology at Sydney. John Barnes, the professor of anthropology at the ANU, warned Bennett that Robert Glasse possessed the "reputation of being a rather difficult person to cooperate with in the field." But then, his subjects were "as surly and uncooperative a bunch as you could find anywhere in the highlands." Barnes observed that although Robert was "unlikely to make any sensational advances in his subject . . . his findings would be quite sound and reliable."[8] The couple intrigued Bennett when he interviewed them in Adelaide. "[Robert] Glasse was rather quiet but he asked a number of thoughtful questions," the fastidious scientist reported to Gunther. "He introduced the subject of Gajdusek by saying he had met him in Canberra several years ago and had been very impressed by him. I hope I have succeeded in disillusioning him on this." Mrs. Glasse seemed to Bennett "a very alert young lady, an enthusiast with much good sense." She was the sister of his old friend Ken Inglis, a historian whom he had known at college in Melbourne.[9] This kinship proved enough for Bennett: he immediately offered them research fellowships, initially with Rockefeller Foundation funding, and told them to pack for a trip to the Fore region.

IN JUNE 1961 AT OKAPA, the Rabelaisian assistant district officer, Mert Brightwell, gruffly told Robert and Shirley Glasse they could borrow his Land Rover to search for a place to live. As the vehicle rumbled over the mountain divide toward Wanitabe, in the south Fore, the noise reverberated through the valley, drawing a crowd. "We saw a large group of people by the road," Shirley recalled, so "we stopped to have a conversation . . . We said we were sort of looking for a place to stay. And they said, 'Well, stop here!'" The local women clustered around Shirley, rubbing her breasts and gesturing as though to take the flesh into their mouths, as a supplication for friendship and intimacy. Everyone wanted to participate in the new custom of shaking hands too. The Wanitabe people promised these strangers they would provide them with a house. "It was in an area that we knew was a hot spot for kuru, and people were very welcoming and said, 'Come in.'"[10] I was talking with Shirley Lindenbaum (as she became later in the 1960s) on a bright spring day in 2005, looking out over Central Park from her apartment in New York City. Although still active in the Graduate School of the City University of New York, she now mostly worked from home, surrounded by Indian and New Guinea artifacts. Despite having spent most of the past fifty years abroad, her lilting Australian accent was still strong.

Initially the local people "were very interested in our cargo," Shirley told me. "They were fascinated by all the stuff they were seeing coming in from outside." The village had been the site of a failed cargo movement a few years earlier, and its leaders felt vindicated with the arrival of the Glasses. Soon the locals would be calling the Glasses' field house the Wanitabe "store." In particular, they wanted salt, soap, safety pins for jewelry, blankets, clothing, tomahawks, and paper for cigarettes.[11] "We understood that we had entered into important exchange relationships entailing detailed reciprocity, the key to our acceptance as social beings."[12] The anthropologists also brought with them a large suitcase full of shillings, and they occasionally dispensed these coins in return for services. The Wanitabe people learned they could use the money to buy goods at the Okapa trade store.

It did not take the locals long to catch on to why the visitors were there:

Shirley Glasse at Wanitabe, 1961. Photograph courtesy Shirley Lindenbaum

the Fore began to address them as *stori masta* and *stori missus*. Together the Glasses interviewed the men, with Shirley taking notes; while Shirley, sweet and quizzical, alone talked with the women. Bob spoke to the men in Pidgin, and Shirley quickly acquired some familiarity with Fore words and phrases. All parties assumed that telling stories was just another of the exchanges that were taking place. "We didn't start asking about kuru to start off. We wanted to establish ourselves as being interested in the whole culture, because we didn't know how the kuru story was going to work out."[13] Unlike the medical investigators, these new arrivals never demanded blood or autopsies. In any case, everyone with kuru was over with the Gimi people, neighbors of the Fore, receiving special treatments. The Glasses did not see kuru at all for the first few weeks.

"So we were theirs," Shirley remembered. "We were *their* people." The village affiliated them, incorporating them into its social life. Shirley, known as "Shorlay," found she had sisters and cousins, and eventually a cheeky boy, Kivengi, called her "mother." When the rain began late in the afternoon, people came to the Glasses' house to roast their sweet potatoes in the pot-bellied stove and to tell stories. One day Shirley made coffee for them, since it was now growing nearby, but no one liked its taste. The Fore gave the anthropologists fresh vegetables, which accumulated in the house until the donors returned to eat their gifts. "Our outer room became a place that people plunged into—it filled up with bodies." Shirley would also sometimes sit with the women who were menstruating in their special, isolated hut. From the beginning, she and Bob—called "Bapu"—contributed to bride payments and funeral transactions. (At such ceremonial exchanges, Fore were perplexed when the couple preferred pig meat to its fat or "grease" and neither made any speeches.) Whenever a child was born, they killed one of their chickens for the mother. Shirley even tended a garden. "I didn't do heavy work in the gardens, but I poked around. And I sat under trees a lot while they poked around!"[14] More than any of the medical investigators—more than Alpers even—the Glasses became part of their Fore community.

Two adolescent age-mates, Inamba and Patali, soon offered assistance to the Glasses. The year before, they had walked to Kainantu to work with the police and acquire some Pidgin. "We went around together and talked to old men and women and gathered their stories," Inamba told me in 2003, not far from Wanitabe. "Like you are doing now, like that." The sprightly old man explained to me that Yona became Shirley's father at Wanitabe, and since he was also Inamba's classificatory father, they all belonged to the same family line. Shirley "adopted" a daughter, Inamba remembered, and when the time came for her child to marry, she decorated her with a *pulpul* (a genital covering) and "ate" the bride wealth. Shirley "wasn't afraid and didn't come and go," Inamba remarked. "She spent time with us and 'storied.'" Although on occasion she treated malaria and scabies, the new affine seemed particularly interested in determining the relations of her kin and collecting memories of kuru. Sometimes Shirley and Bob did travel—perhaps south to Purosa, or over to the Gimi—to find out more about patterns of kuru and the significance of its sorcery. A retinue of Wanitabe adolescents accompanied them as the anthropologists had become Wanitabe people too. This entourage helped establish contact with related groups at Purosa and Agakamatasa, but caused trouble among traditional enemies like the Waisa people, among whom Alpers was living.[15]

Kivengi remembered Shirley giving him a *laplap* and asking him to look after her garden to keep the chickens out. "She liked me, she said I was a good boy and I listened." Kivengi did not regard her as a missionary or a scientist or a patrol officer—like the other white men he had seen—but simply as a "parent." In the afternoon, she "would cook food and divide it and give it to us."

The boy listened as Fore men and women told her their stories. In 1963, when she and Bob went back to Australia, he "was sad, saying: 'Why did she come to stay with us?' And then she left." Kivengi grew up and worked with missionaries, and then on a coffee plantation, but he always tried to keep in touch with Shorlay.[16]

Inevitably, Shirley and Bob soon became intimate with kuru. A young girl call Nogiya used to help Shirley in the gardens. "I loved her," Shirley recalled. "She was animated and funny and had a wonderful laugh." Then suddenly Nogiya showed the first signs of kuru, just before Shirley left to spend some time talking with the scientists in Adelaide. "She was very sick when I came back . . . I visited her and I wept, and she wept. Then the people commented: 'She wept!'" The Fore expressed surprise and concern. "I realized that it was something they hadn't expected from white folks." Shirley also told me about a woman friend who started to succumb to kuru close to the time the anthropologists were leaving Wanitabe permanently after living there two years. "She struggled out of her house and she came up the field and she was hanging on to the fence post outside my little kitchen garden because she wanted to say goodbye. She knows it's the end and I know it's the end . . . I felt emotionally impaled. So I wept when I left the field 'cause I figured that I wasn't going back again."[17] Shirley paused for a while, looked across Central Park, then said she had never anticipated such emotional involvement.

As anthropologists, the Glasses displayed a more cautious approach to publication than the medical scientists. They needed first to immerse themselves in Fore culture, to observe and participate in mundane and ceremonial events in Wanitabe. Above all, they tried to develop rapport with their new relatives so that they could gradually learn from them. In his introductory report to Bennett in 1962, Robert Glasse began with a geographic and demographic overview of the Fore before proceeding to a conventional account of local economic relations, agricultural practices, and histories of conflict. There was little on sorcery and cannibalism. Glasse focused on Fore kinship patterns, which impermanent alliances and shifting affiliations rendered so confusing. The shallowness of Fore genealogical reckoning presented a vivid contrast to the Huli obsession with lineage. It made collection of Fore pedigrees exceptionally trying and unreliable.[18] By this stage, with simple hereditary explanations looking increasingly implausible, the anthropologists must have wondered if such research was worth the effort.

Robert Glasse separately reported in 1962 on the spread of kuru among the Fore. All highlanders seemed convinced the disease was a manifestation of sorcery, but the Glasses relied on medical evaluations of the condition, pointing out that it was a neurological degeneration, the cause of which was still unknown though genetic hypotheses were current. Robert Glasse was especially keen to find out how long kuru had prevailed among the Fore. People told him this thing came from the north a few generations earlier, but the inhabitants of the alleged origin sites proved understandably mistrustful and suspicious of any inquiries. Further discussion did, however, confirm that the first afflictions near what was now called Okapa occurred as late as the 1920s. From there, kuru rapidly spread southward, with fears of the new form of sorcery diffusing to the outer limits of settlement. A few deaths in one village would be followed by a few deaths in the next. "People were alarmed," Glasse reported. "When they heard of a case they travelled long distances to watch the victim's behaviour." Between 1925 and 1930, some Purosa women began to succumb to kuru sorcery. But more isolated hamlets suffered no kuru deaths until the 1950s. Everyone agreed that the arrival of Europeans seemed to have magnified the problem.[19]

The consistency of Fore dating of kuru deaths was remarkable. The Glasses were collecting hundreds of mutually reinforcing stories. Siandibi, a man from Purosa, probably in his fifties, was typical. "There was no kuru in the time of our fathers," he told the anthropologists. "Before, we fought with bows and arrows; we killed with yanda and karena sorcery and with tokabu. Kuru came only recently." Siandibi first saw kuru when his daughter, aged about thirteen, was ready to marry: it killed her before she wed. "Later," he said, "when white men came to Moke more women began to get kuru at the same time, and men began to die from it. The first time [an adult] woman got kuru here we stared at her and watched her shaking. She took a stick and hit us and we thought she was playing . . . Then she died and we knew it was kuru and a man was responsible." Similarly, Ogana, an elder at Wanitabe, told the Glasses: "There was no kuru in my youth. Before I was initiated [at ten years] there was no kuru." Around the time Ogana married, in his early twenties, his mother died of kuru. "Men said at first: 'Is it sickness or what?' Then they said that men worked it . . . Before we fought with bow and arrow; then kuru came and killed off the women one by one. Now kuru kills everyone. Where will it end?" Time and time again, old men assured the Glasses that no such scourge had existed in their youth.[20]

But if kuru was so recent, and so rapidly dispersed, a simple hereditary explanation seemed harder to justify than ever. At best, one might postulate genetic susceptibility to some environmental factor spreading through the region in the 1920s. How, the anthropologists wondered, might Fore have exposed themselves to the inciting agent in this period?

Robert Glasse could not stop thinking about cannibalism. Shirley Lindenbaum recalled him reading aloud over breakfast an article in the May 18, 1962, issue of *Time* magazine on cannibal flatworms that seemed to acquire the memories of those they consumed. "After reading it," Shirley said, "Bob began our discussions about cannibalism."[21] It had, of course, been such a striking feature of Fore ritual. But the anthropologists heard of few instances in the 1960s and witnessed none. Their friends and neighbors, though, spoke readily about their earlier participation in cannibal feasts. Only the corpses of those loved ones who died of dysentery were refused. Cannibalism was primarily the practice of women and children—most men believed it made them weak and vulnerable to enemies. The maternal line of the dead person claimed the corpse, cooked it, and then distributed it. The widow received a dead man's best parts, his buttocks and genitals, while his sister ate the brain, and female maternal kin took the remainder, feeding scraps to their children. A dead woman's parts often were allocated more idiosyncratically, though her brain tended to go to her son's wife. Yona, Shirley's honorary father at Wanitabe, recalled as a young boy consuming flesh from a kuru victim that his mother gave him. The corpse of a kuru victim, with its layer of fatty tissue from long recumbency, was especially tender and appetizing. But after initiation, like most other men, he refrained from eating any human flesh. Bob soon deduced that patterns of kuru resembled patterns of ritual cannibalism. Perhaps predilection for eating the flesh of kuru victims accounted for the high incidence of kuru among adult women and children of both sexes? Moreover, Fore usually told the Glasses that their mothers and grandmothers had begun the practice, often within living memory. It was a new fashion that spread from the north, as did kuru. "No doubt it is too late for a direct test of association between cannibalism and kuru," Bob lamented in 1962. "The present evidence is suggestive, but very far from adequate."[22] Still, it was an intriguing notion.

The Glasses took every opportunity to expatiate on the possible association of cannibalism and kuru to *kiaps* and medical investigators. When Burnet heard about it, the scientist retorted that the suggestion was "incredible."[23] But Bennett, losing faith in his genetic theory, was secretly more receptive. "Perhaps the causative agent of kuru can survive gastric and intestinal juices and produce symptoms after absorption," he wrote in 1962 to a British expert on scrapie. "Do you think it possible that scrapie could be artificially induced in this way? Could you possibly test this by feeding scrapie brain, both raw and semi-cooked, to some sheep from a scrapie-free group?" Bennett regretted it was impossible to do such experiments with "human material."[24] His correspondent replied, telling him it was already established that feeding a soup of scrapie brain to goats and sheep gave them the disease. There was no reason why cannibalism might not transmit an unknown kuru agent.[25] Bennett kept this exchange, and doubts about his genetic theory, to himself.

Notions of cannibalism came to animate many scholarly arguments in the early 1960s. The lists of imagined, documented, and denied cannibals grew ever longer during this period. Cannibal metaphors also proliferated. Indeed, scientists and engineers routinely "cannibalized" artifacts and materials. In part, this fixation on cannibal appetite reflected more general postwar concerns about excessive or transgressive consumption. The figure of the cannibal often seemed to epitomize the unsatisfied, acquisitive subject in a consumer society, as the end point of improper or uncontrolled desire.[26] No wonder that anxieties about cannibal appetite came to inflect disease explanations too.

It proved impossible to keep cannibalism out of the etiological reckoning. On encountering the Fore in March 1957, Gajdusek had briefly wondered whether eating human brains might lead to kuru sufferers setting up an autoimmune reaction against their own brain. After all, the scientist had recently developed the autoimmune complement fixation test at the Hall Institute in Melbourne, so the process was fresh in his mind:

> Early in kuru investigation we entertained such an outlandish hypothesis as the possibility that, in the course of ritual cannibalism . . . the infantile gut may have permitted the passage of un-denatured homologous brain antigens, which might have initiated a state of auto-sensitization. Kuru symptomatology might then have been initiated by further ingestion of human brain at a later date by the hypersensitized individuals.[27]

But kuru brains showed no evidence of any immune response, and no autoimmune antibodies to brain tissue were detected. Moreover, many kuru sufferers denied ever participating directly in cannibal feasts. Gajdusek came to dismiss the association of cannibalism and kuru as distastefully vulgar and demeaning to the already over-exoticized Fore. It echoed the snide remarks about "those terrible cannibals" he overheard in the bars of Kainantu and Goroka.[28]

Some other medicos remained fascinated. When Andrew Gray, the medical officer at Okapa, came to doubt Bennett's genetic hypothesis, he wondered about possible environmental or behavioral causes of the disease and for a moment seized on cannibalism. Late in 1960, he read the speculations of Ann and J. L. Fischer, two biological anthropologists from Tulane University, who reviewed the scientific findings on kuru and discounted simple hereditary explanations, favoring instead either a direct "nutritional etiology" or the transmission of some agent through diet. The Fischers postulated the Fore habit of eating corpses might pass on a toxin or infection. If women partook more frequently in cannibalism, then the greater prevalence of kuru among them was explained.[29] "To investigate the matter," Gray wrote to the director of public health, "a team of married observers, preferably with medical degrees as well as training in anthropology, is required."[30] Similarly, around 1960 Vin Zigas began telling Frank Schofield and others that he was sure kuru was somehow related to cannibalism. "I agreed that it would be a good idea," Schofield recalled, "to get some social anthropologists, including a female one, because the men told us that they didn't practice cannibalism, but the women did when their relatives died."[31] The Glasses' studies therefore confirmed what many others had long suspected and feared; as anthropologists they were licensed to amplify the persistent whispers of cannibalism's medical significance.

The Glasses observed the "kuru region" and its inhabitants changing rapidly during the early 1960s. During these years the Fore gave up covering their bodies with pig fat and abandoned nose piercing and hair braiding; instead they donned *laplaps* and secondhand European clothes. Body paint and twist tobacco declined in value as trade goods. Houses came to assume a rectangular, not oval, shape; hamlets consolidated into villages. Aid posts proliferated, dispensing basic medications, and quickly yaws and leprosy came under control. In 1962, a new government hospital opened at Okapa, and Lutherans established a kuru hospital and orphanage nearby at Awande. Many Fore now attended Seventh Day Adventist, World Mission, and Lutheran church services, and some of their children could go to school to learn English. Coffee plants spread across the land, and as trade stores opened at Okapa and Purosa, money began to circulate. The road from Goroka improved, and cars could penetrate as far south as Purosa.

Identification of possible behavioral correlates and risk factors for kuru was only a small part of the anthropological enterprise. Shirley Glasse decided to look more closely at the social consequences of kuru and to elucidate Fore beliefs about sorcery; that is, their thoughts on transgression, blame, and threat. While Robert Glasse reported on local memories of kuru and the possible connection of the disease to cannibalism, Shirley focused on how the loss of so many Fore women and children was affecting the community. The *luluai* (or

Gimi curer visits Waisa, 1961. Photograph courtesy Shirley Lindenbaum

community headman) of Wanitabe, for example, had lost his mother, his half-sister, three wives, one son, and one stepson. Poraka, a man from Kamira, witnessed the deaths of three wives, two sisters, two sons, one son's wife, and three daughters. Shirley could list dozens more examples of loss. Many of these men began caring for their young children and laboring in the gardens. Ten years before, most had prided themselves as warriors, now they performed women's tasks. They spent more time with the women who remained. Accusations of sorcery and oratory on the causes of the decline of the Fore people were among the few remaining means for these men to express hostility and to strive for dominance over others. As kuru gained a hold on the people, sorcery divination, declamatory speeches, resort to curers, and informal courts, all came to permeate Fore life. The inhabitants of the region were thus seeking to control or deflect an uncontrollable phenomenon.[32]

From the beginning, the magical thinking of the Fore captivated Shirley

Glasse. The medical officers who passed through the region tended to regard sorcery accusations as mere superstition, though sufficiently dangerous to the peace to warrant suppression. Shirley, however, wanted to understand the social and psychological functions of sorcery beliefs. In 1962 she observed:

> Fore gain relief from anxiety by their belief in sorcery: it gives them a course of action when faced with catastrophe. Yet their inability to avert or counteract it gives rise to further stress. They meet this stress by re-affirming their belief in sorcery.[33]

In a series of articles and later in her book, *Kuru Sorcery,* Shirley continued to examine Fore attribution of disease to malign human agents and disturbed social relations. Nothing shook the conviction of her friends and neighbors at Wanitabe that other men made kuru—if it were a sickness someone would have cured it and it would be gone. Belief in sorcery allowed Fore plausibly to identify a culprit, usually of inferior status, and to propose a solution. The durability of kuru despite the constant campaign against the men who made it failed to cast doubt on the existence of sorcery: rather it implied that these efforts to root out sorcerers must be redoubled. Among Fore, sorcery accusations served, often inefficiently, "to regulate relations between individuals who must cooperate and also compete." In practice, the obsession with sorcery "simultaneously registered and aggravated the social inequalities and demographic imbalances" of the area. The study of kuru sorcery thus permitted the anthropologist to compile "an epidemiology of social relations."[34] Much later, this arresting phrase would cause Shirley to remark reproachfully that contact with medical investigators had for a time affected her judgment on social matters.[35] But just as anthropologists might frequently resort to medical analogies, scientists in the region were forced to think more anthropologically.

IN NOVEMBER 1962, Michael Alpers wrote from his house in Waisa to Gajdusek, urging him to take seriously the Glasses' argument that kuru was a recent, emerging disease. The Australian medico's family and the anthropologists came together frequently for meals, drinks, and conversations on kuru, choosing to ignore the animosity of their Waisa and Wanitabe groups. Alpers, Shirley Glasse, and Brightwell—the Australians—all liked to party, and often they discovered unsuspected intellectual affinities after a few drinks. "I have been trying to persuade the Glasses to publish their work soon," Alpers wrote, "presenting it in as factual a way as possible, demonstrating the correlations and citing evidence of many diseases of the period before kuru which the informants remember well . . . However, they are very reticent about publication." Their anthropological observations prompted Alpers to wonder if the disease pattern was still changing, and how he might best track its incidence over time.[36]

The engagement of the anthropologists with some of the medical investigators was becoming ever more intense and productive. Clinical findings and laboratory experiments did not much excite the Glasses, but they soon realized ethnography might animate, and render more satisfyingly complex, the epidemiology of disease. That is, their oral histories and observations potentially added nuance and depth to the analysis of patterns of disease prevalence and incidence. Shirley told me that she and Robert came to understand that anthropological and epidemiological studies manifested a shared rationality, derived from "the same image of social relations," which aided communication. Epidemiology traces the health consequences of human behavior, while anthropology explains the context for behavior.[37] From the beginning, Alpers and, to some extent, Gajdusek demonstrated unusual openness to anthropological approaches and insights. "In our thinking there were no disciplines," Alpers told me. "We used what knowledge we had and where there were gaps we tried to fill them." He believed anthropological studies complemented his epidemiological work—which was, in any case, ethnically delimited. Translation was not a problem since the various investigators were speaking dialects of the same language.[38]

In the late 1950s, medical investigators had ascertained the boundaries of kuru, its distribution among Fore men, women, and children, and its annual incidence. As soon as he could confidently recognize the disease, Gajdusek tramped across the high valleys conducting a kuru census. In collecting Fore pedigrees, the Adelaide investigators also, in effect, surveyed the prevalence of kuru. Patrol officers and medical assistants made records of kuru incidence, kept in "village books," which scientists later carefully examined.[39] Everyone knew the disease was worst among the south Fore and that it mostly afflicted women and children. According to the Glasses, Fore were convinced that kuru was becoming more common, that its incidence increased after the arrival of Europeans. Gajdusek and Zigas, however, claimed the incidence of the disease had not varied over the years and cast doubt on local assumptions.[40]

Burnet and Gunther wanted an independent demographic analysis of the pattern of kuru incidence. They asked Norma McArthur, a brusque and efficient statistician at the ANU, to visit the Fore region and try to sort out the figures. Although McArthur had previously worked with Burnet, she evinced little confidence in the mathematical ability of medical scientists. Moreover, she regretted the time kuru was taking from her studies of Pacific depopulation. McArthur nonetheless diligently examined the epidemiological reports from the region, adjusting the estimated age distribution of the population in order to get the correct age-specific incidence of kuru. It appeared that kuru mortality for females increased steadily with age until forty-five years, after which it declined. There may have been some fall in kuru mortality among older males, but the levels were not statistically significant. The genetic explanation of the Adelaide group especially dismayed the demographer. In view of the numbers of children dying from kuru before reproducing, it was inconceivable that the lethal kuru gene could be maintained in the population at a frequency high enough to account for actual kuru incidence. Therefore, the Adelaide genetic hypothesis was untenable.[41]

Although he shared her skepticism toward the genetic hypothesis, Alpers felt McArthur was misinterpreting trends in kuru incidence—or perhaps simply failing to observe the more significant patterns. Proud of his own commitment to the Fore, he distrusted interlopers, outsiders like McArthur and Paul Brown who just passed through the region. As a neophyte epidemiologist, Alpers, with Gajdusek's encouragement, set about analyzing kuru mortality from 1957 to 1963 according to sex, age, village, and parish, using the records of the 1,450 cases accumulated in Bethesda. He was not surprised to find that overall mortality from kuru was declining a little. It was startling, though, to see that the disease was now far less common in children under the age of ten. Between 1957 and 1959, 116 children succumbed to kuru; yet in the 1961-63 period only twenty-eight died from the disease. Since the Fore population continued to increase despite the ravages of the epidemic, this decline would look proportionately more impressive if the total population was used as a denominator. Alpers, however, resisted figuring this trend as a rate of decline because he could not determine the population truly at risk. Nonetheless, expressed in terms of numbers of dead, it still represented a dramatic change. It struck the scientists that the fall in childhood kuru deaths was most extreme in those areas that had endured the longest contact with white men. "The changing pattern of kuru may be related, not merely in time, but through some causal link, with the increasing influence of civilization upon the people of the kuru region."[42] Alpers predicted that the next cohort of young adolescents would soon show the same decline, and then progressively older age groups, leading eventually to the disappearance of the disease.[43]

Field observations and laboratory experiments were converging. In Bethesda, Alpers alternated between statistical analysis of kuru incidence and care of the inoculated chimpanzees. By the beginning of 1966, it was clear that the pattern of kuru was changing and the disease was transmissible. How did these deductions fit together? One morning, as he wrote a paper in the make-shift office in the porch of his house, Alpers realized it was cannibalism that might spread the slow virus:

> Cannibalism as the single mode of transmission of the transmissible virus of kuru did make sense: it was suddenly all too painfully obvious. It was obvious because nothing further needed to be explained, and painfully so because of the agonies of uncertainty that existed when the explanations seemed so close at hand yet not quite there: until that moment when it all clicked into place. But what a moment![44]

Cannibalism explained why the disease was limited to the Fore, and the distinctive sex and age distribution. It solved the puzzle of how the epidemic began some forty or fifty years earlier, and why it was now declining among Fore children, who were growing up in communities free of the practice. It solidified the rather tenuous linkage of disease disappearance with exposure to "civilization."[45] It was almost *too* neat. And sadly, it seemed to confirm the speculations already bruited about in hundreds of bar conversations in the highlands.

OTHERS WERE ALSO HELPING to decipher the enigma of kuru. When Alpers went from Fore region to Bethesda late in 1963, his replacement in the field was Richard W. "Dick" Hornabrook, a New Zealand neurologist eager to elucidate the clinical course of the disease. Gajdusek had initially recommended Hornabrook, but it was Burnet who recruited him and became a persistent, if sometimes captious, supporter.[46] Alpers remembered showing the newcomer around the area. "I was hoping that Dick would follow on at Waisa, but he had other ways of doing things. He wanted to work more through the hospital and less out in the field." Hornabrook, with wife and children, felt more comfortable at the Okapa station, where he formally established the Kuru Research Centre, than in the villages. "He may have had good relations with a few people," Alpers told me, "but generally he wasn't a man of the people."[47] Soon after his arrival, Hornabrook dismissed the orderlies Alpers had used, seeking instead English-speaking casuals. He complained, too, that visits to nearby northern Fore hamlets were destroying his good city shoes. The punctilious neurologist preferred to augment his butterfly collection.

When Gajdusek passed through he observed that the "Okapa station is more a colonial enclave than ever—few natives hang about on it; it is ever more neat and clean and park-like and quiet . . . I would soon be very bored here."[48] His vexing interactions with Hornabrook depressed him. He repeatedly deplored the "atmosphere of tension and mistrust, of cliques and clans, of jealousy and dislike [that] flavors so much the European relationships of Okapa." It was yet another "case of 'kuruitis' and the personality troubles that afflict all kuru investigators . . . I am now not far from concluding that I must be the primary source of all infection—a fair conjecture admittedly!"[49] He quickly decamped to frolic among the Kukukuku.

Hornabrook did confirm that kuru primarily affected the cerebellum, and he assembled an extensive description of the disease's clinical course. Cerebellar involvement led to the disorganization of muscular contraction so characteristic of kuru. Typically, disorders of gait and sustained tremor, resembling shivering, marked the onset of affliction. In established cases, involuntary movements, twitching, flaccid muscles, and abnormal postures became conspicuous. With time, the body was no longer able to maintain the tremor. Hornabrook determined that involvement of the rest of the brain occurred only in advanced and terminal stages. He challenged the claim of Gajdusek and Zigas that intellectual activity was unaffected, describing a number of advanced cases in which sufferers became forgetful, disoriented, and apathetic. But above all, this was a disease of cerebellar dissolution, lasting as long as three years and leading inevitably to death.[50]

Unhappy at Okapa, Hornabrook became more reclusive and irritable over the next few years. He sensed Fore resentment of the repeated medical demands on them. It was hard not to sympathize with their feeling that the bleedings and autopsies over the past decade had been excessive and futile: certainly no one was offering any cure for the disease. Hornabrook resisted Gajdusek's requests for more blood and brains. He did not much like the American anyhow. He felt isolated and ignored, complaining: "I am particularly concerned with the fact that virtually all the important background information concerning kuru is in Washington and most of this has been obtained by Gajdusek."[51] After leaving early in 1966, he lamented that kuru research at Okapa was becoming a "lagging field," merely feeding more exciting developments at Bethesda.[52]

John D. Mathews, an ambitious medical graduate from Melbourne, took over from Hornabrook and, like his predecessor, based himself at Okapa with his young wife and children. As a student, Mathews had come to admire Macfarlane Burnet—who glowed with the luster of his 1960 Nobel Prize in Medicine or Physiology—and he began spending time at seminars in the Hall Institute. Hearing that the eager researcher was a mathematical prodigy, Burnet suggested he try to sort out kuru epidemiology. Soon Burnet was urging him to conduct further investigations among the Fore for a year or more. But Mathews, cautious at first, hoped to learn more about the problem before committing himself. In 1964, Mathews met Shirley Glasse when she was visiting her mother in Preston, a Melbourne suburb not far from where he had grown up. She told him about the possible link between kuru and cannibalism and shared with him some of the genealogies she and Robert had collected, even as she expressed doubts about the genetic explanation for the disease. Her stories appealed to Mathews's sense of adventure. He decided to visit Hornabrook for a few weeks over Christmas 1964. "Okapa is a delightful spot," he reported to Scragg. "The Fore people are very friendly and interesting, and I found the challenge of kuru very stimulating, so altogether I really enjoyed my work."[53] Mathews now wanted to return for a few years after finishing his internship. "He was looked on as brilliant," Ian R. Mackay recalled, "although Mac probably put him there because he was a useful chess piece on the board. Burnet was always putting his chessmen where he thought they'd be well placed."[54]

Mathews's arrival at Okapa, along with his research technician Ray Spark, coincided with the announcement of the transmission of kuru to chimpanzees. The young doctor desperately wanted to assist Gajdusek and his NIH colleagues to isolate and cultivate their slow virus. "This project is obviously the most important one possible for the Fore people," he wrote. "Unless the kurugenic agent is a fairly conventional virus from which a vaccine can be made, there is little possibility that there will be any major therapeutic advances in kuru."[55] In winter 2001, sitting in the director's office of the Australian Centre for Disease Control in Canberra, Mathews told me he felt it would be "churlish" to deny Gajdusek's repeated demands for more blood and brain tissue. "There was no one on the Australian end who could compete with the NIH resources that Carleton was able to harness."[56] The only hope was in Bethesda.

Rather tentatively at first, Mathews began cutting the kuru dead. In preparation, he spent hours with the dying victims and their families, trying to persuade them to allow the autopsy. Often he waited days, sitting in the village, sometimes playing a Beatles tape. *Sgt. Pepper's* jaunty melodies rang out across the eastern highlands. People asked Mathews why he needed more brains.

"Rapport with the local people is now fairly good," he wrote at the time. "Few patients refuse to be seen. About one in five or six requests for autopsy are granted; permission is more often given by groups with a stronger mission background." Frequently his efforts failed. "Understandably, some people have become tired of kuru research, and in several respects, attitudes have been of toleration only."[57] Mathews found that gifts of cash, about ten dollars, were now the most appreciated, since families of the dead could spend the money at the local trade store. But many bodies were not available at any price.

Concerns about ethics often intruded into Mathews's thoughts. It was, after all, the period in which many physicians began self-consciously to reflect on research and clinical ethics and to draw up explicit guidelines for proper behavior.[58] The fragmentation of the old moral economy of medical practice and the perceptible impact of market forces made the need for impersonal standards, supposedly untouched by contingency, seem ever more pressing. In particular, Mathews feared the requirement of voluntary informed consent for medical procedures was meaningless in a "stone-age society" without concepts of individual rights. Yet he hoped the problem might solve itself with time. "The dissolution of the concept of 'European' from that of 'compulsion' in the minds of local natives has proceeded rapidly, but is not yet complete." There were, however, signs that some prestige might now even accrue to Fore who rebuffed authority. Certainly, they did not hesitate any longer to refuse most requests for hospitalization, venepuncture, and autopsy. "It can be most unpleasant to attempt to persuade a relative to change his mind about permission for an autopsy," Mathews wrote. But he continued ambivalently:

In our society such an act would be in poor taste and would be considered to be unethical by many people. In Fore society, such an attempt is more acceptable. I have spent many memorable and not unpleasant hours haranguing the relatives of dying kuru patients.

Generally the young doctor believed the benefits of research in this setting outweighed any "departure from accepted ethical procedures in more sophisticated communities."[59]

"We are continuing close collaboration with Dr. Mathews in our attempts to obtain viruses directly from explants of kuru tissue," Gajdusek wrote in 1967. "These are urgently needed in the study, and Dr. Mathews is doing the necessary post-mortem examinations."[60] In contrast to Hornabrook, Mathews and his wife Coralie found Gajdusek enthralling. "The first time we ever met him we were at Okapa," Mathews told me. "We knew he was coming because the cargo was arriving. Things were falling out of the sky with his name on them." Unkempt and out of shape, Gajdusek turned up at midnight and did not stop talking until 4 A.M. While admiring the young Australian's "extensive cultural tastes," the infamous visitor issued instructions for specimen collection and disparaged local efforts at epidemiological analysis.[61] Then, and later, Mathews felt the danger of being "intellectually kidnapped" by Gajdusek. "He was very exciting. I mean, the people he knew, the things he knew—he was very intellectually engaging." Mathews stopped and reflected for a moment. "He's really a latter-day Richard Burton. He's out of his century, basically."[62] But it was never clear that Mathews's admiration was reciprocated. Decades later, the Australian scientist still wondered why Gajdusek rarely recognized his contribution to kuru research. Perhaps the Burnet association had tainted him? Perhaps it was because he published independently from the NIH group? Maybe—and Mathews was smiling as he said this—it was that he beat the great man at chess? In any case, Mathews believed Gajdusek was a poor exchange partner, failing to "recompense" him adequately, to bestow appropriate recognition. Mackay had warned him about the American's occasional, and often strategic, lack of finesse in scientific transactions. "You were either totally owned by him or you didn't exist," Mathews said.[63]

Feeling frustrated and neglected, Mathews decided to conduct his own experiments at Okapa. In June 1966, a contact in England sent him some freeze-dried scrapie material so that he could infect some mice and treat them with antiviral drugs. But the shipment disappeared between Lae and Goroka. In August, Spark asked quarantine officials about importing special mice and learned for the first time that introducing scrapie was prohibited in Australia and New Guinea. When the authorities discovered the attempt to bring in the scrapie material, they demanded an explanation.[64] Since Mathews could not give them the tissue, they accused him of hiding it and sent the police to search his home and office. "I was ignorant of the fact that it was a prohibited import," Mathews responded. "The implication of dishonesty is a little steep, I think. We have nothing to gain by hiding the material at all. The affair was due to my foolishness and no intent to deceive."[65] Scragg condemned the scientist's "irregular approach" of importing contraband.[66] Mathews remained deeply embarrassed and felt thereafter as though he were on probation.

It was about this time that patrol funds became scarce. Late in 1966, Mathews complained that he had not received any administrative backing for more

Land Rover and broken bridge, 1967. Photograph courtesy John and Coralie Mathews

than a month. He was now employing ten or so importunate *dokta bois,* including Inamba. "I have made do by finding money from my own pocket, and making our field assistants wait for their pay."[67] The lack of support from Moresby and Melbourne was galling. The following year there was little improvement. Mathews wrote indignantly to Scragg, alleging that he had "not had a fair deal from the department." Deploring the "apparent attitude that kuru research work is of no importance," he threatened to resign.[68] R. J. Walsh, the dean of medicine at the University of New South Wales, wrote to Scragg to try to smooth over the dispute. "I believe Mathews would not be the easiest person to get on with," Walsh claimed. "However, he is a rather brilliant lad in spite of his apparent personality problems."[69] Reluctantly, Mathews stayed until early 1968. "I was young and a bit bumptious," he recalled.[70]

On returning to Melbourne, Mathews again encountered Shirley Lindenbaum and they decided to write, with Robert Glasse, an article definitely linking the slow virus transmission with cannibalism. Earlier, Mathews had provided further epidemiological support for the Glasses' claim of kuru emerging within living memory. In 1965, he argued the "progress of kuru has followed the passage of some environmental agent, possibly infectious, through the Fore region."[71] Now certain of the disease's transmissibility and more than ever impressed with its correlation to cannibalism, the three sought to synthesize available knowledge. "I tried," Mathews told me, "to connect the anthropology and the epidemiology and the biology together, you know."[72] In 1968, the *Lancet* published their article on kuru and cannibalism. The authors believed the disease developed between four and twenty years after a person consumed poorly cooked human tissues containing the transmissible agent. Exposure to this agent, not genetic susceptibility, could fully explain the ethnic, familial, and gender patterns of the epidemic. They supposed the "kurugenic agent arose de novo in the Fore area from human, animal, or viral genetic material, and that its continued existence was ensured by cannibalism." Once someone carrying the agent was eaten, there was little chance of its extinction. But since cannibalism had ceased, the decline in kuru incidence should be rapid.[73]

Regrettably, not all the loose ends were tied up. Gajdusek and his colleagues had not yet demonstrated oral transmission to nonhuman primates of the mysterious agent; other possible routes of spread remained unexplored; and no one knew for sure the range of incubation period, or influence of genetic predisposition, on the expression of kuru.[74] Most disturbing of all, the germ just could not be found.

It was the ebb tide of kuru research in New Guinea. During 1968, after Mathews left, a Sepik research officer, Alphonse Kutne, tried to keep records of kuru and take blood. But he lacked research training and struggled to convince the Fore to cooperate. He pleaded with officials in Port Moresby for another job. "When I go around to the villages to record their names and do few simple things like collecting blood and all this. They just ignore me and the relatives would not give permission to do so. On one occasion I wanted to collect blood from a woman and the husband refused. He said you are just waisting your time. This kuru is done by sourcery and you will not collect blood from her [*sic*]." Sometimes the Fore just laughed at him.[75] In 1968 seventy-five kuru victims died; all were older than fifteen years and lived in the Okapa subdistrict. As Inamba recollected: "The doctors tried to beat the sickness, but then they just forgot it. The disease slowed down, the kuru slowed down, when they left." Like most Fore, Inamba assumed there was less sorcery occurring.[76]

ONCE CONCENTRATED ON KURU, medical research activities in the highlands were shifting toward other biological challenges and health problems. In 1965, Burnet renewed his call for a research institute that might study human adaptability and disease ecology in New Guinea. He suggested that investigations of human gene frequencies, growth and physique, nutritional status, and population dynamics might constitute Australia's contribution to the new International Biological Programme.[77] In 1967, the Papua New Guinea colonial government responded, establishing the Institute of Human Biology, based initially at Madang and then, from 1969, at Goroka. Hornabrook returned as the institute's first director. While continuing to report kuru deaths, he became committed to developing an extensive research program in human adaptability and to addressing more common disease threats such as malaria. The institute intermittently assisted visitors like L. Luca Cavalli-Sforza and James V. Neel, who wanted to "fill in the vast blanks on the genetic maps of Papua New Guinea"—though as Hornabrook scolded, these gaps were not nearly as large as his guests assumed.[78] All the same, the scientists were usually permitted to collect yet more blood. But until his retirement in 1975, Hornabrook resolutely managed to resist any potential competition from Gajdusek, severely limiting the American's access to the beloved "primitives." Generally, Hornabrook had little time for any of the interlopers. "There is a tendency for all different itinerant practitioners of one sort or another to arrive in Papua New Guinea, descend on a village without prior discussion, bleed, and then write papers," he wrote. "We have a long and bizarre history of this sort of thing."[79]

Hornabrook was pleased to be leaving just as Papua New Guinea gained independence from Australia. "It has been disappointing, indeed almost demoralizing," he wrote late in 1975, "to attempt to try and stem the tide of decay and degeneration which has set in here. All of my attempts at influencing and educating the new national bureaucrats and their political masters have met with little success." He observed the Institute of Medical Research (IMR), as it now was called, "visibly shrinking and evaporating in front of one's eyes."[80] National politicians distrusted general biological studies and demanded more practical investigations of disease outbreaks. All around him, Hornabrook saw inefficiency, corruption, intellectual indifference, and

Alpers in the field, 1972. Photograph courtesy Peabody-Essex Museum, Gajdusek Collection

"hostility to the white minority."[81] On his departure, he gloomily predicted the rapid end of scientific research in New Guinea.

But when Michael Alpers became director in 1977, he was able to maintain and expand the Goroka institute. Over the following twenty years, Alpers developed research projects in respiratory diseases, malaria, malnutrition, digestive ailments, sexually transmitted diseases (including HIV/AIDS), and women's health. The emphasis was on discovering and testing practical interventions against major disease challenges. All field staff were expected to "adopt an anthropological approach to their work; this means establishing a strong rapport with participating communities that is based on an understanding of their motivation and social organization; it means adopting an open attitude to their beliefs and customs; and it means asking open-ended questions about health-related behavior."[82] These were lessons he learned from the Fore.

Under Alpers's direction, the institute continued to support investigations proposed by the proliferating itinerant geneticists, though local discomfort with these projects sometimes was palpable. Typically, Cavalli-Sforza was the most insistent blood collector and the least willing to spend much time with local communities. In the 1980s he demanded to study the genetics of "populations other than Caucasians" living around Goroka: his target was a "tribe that has not been undergoing heavy recent out-breeding."[83] Thus the preliminary genetic sampling of the 1960s was scaling up to the human genome diversity project, which in the 1990s would attempt systematically to document the genetic heritage of isolated indigenous groups.[84] Not surprisingly, Gajdusek claimed priority in such sero-epidemiology, and he sought to become more closely involved in the growing study of primitive genes. "We want you to work over for us and with us," he wrote to Cavalli, "the vast genetic data we have accumulated."[85] Gajdusek expected the population geneticists to acknowledge his pioneering collection of indigenous blood and to negotiate with him for the use of his material. But these possessions did not interest them as much as he hoped, and so he always remained on the margins of the population studies. Alpers frequently wondered if the potential health benefits justified the demanding genetic investigations. Usually, the plan was to fly in on a helicopter, line up the community, take blood, and leave. He complained to one importunate geneticist that brief visits to study populations were "inconsistent with my own personal philosophy about how research should be conducted in developing countries." Therefore it put Alpers and his colleagues "in a false position if we are seen to be responsible for the work."[86] Still the geneticists came, even where they were not especially welcome.

While institute director, Alpers took the opportunity to return regularly to his house at Waisa and chat with his old friends, the kuru reporters—many of them former medical assistants. They were all getting frailer as the century advanced. "Kuru does remain an interesting topic," Alpers mused, "but there is nobody left who is working on it but me."[87] By 2000, when he retired, there was plenty of time to sit on the porch of his Waisa home and talk, as only a few kuru deaths occurred that year.[88]

Chapter 8

STUMBLING ALONG THE
TORTUOUS ROAD

"A BEAUTIFUL STORY IS UNFOLDING," CARLE-
ton Gajdusek wrote in 1972, observing from Agakamatasa the continuing,
predicted decline in kuru deaths. He also regretted that kuru field research
was now "so 'tied up' and tidy that it is boring and unchallenging."[1] For a
while, Gajdusek had continued to reject the association of the disease with
cannibalism. Michael Alpers recalled him saying the explanation was so
bizarre and exotic no one would ever believe it.[2] Yet he may really have been
more concerned that some people were only too ready to endorse the theory.
By the 1970s, however, Gajdusek felt sure that cannibalism was responsible
for the epidemic, whether through oral transmission or, more likely, contam-
ination while preparing the bodies of the dead. It seemed little remained to be
done in the highlands. In the 1970s, his grouchy colleague Paul Brown was
"sorry to hear that Carleton has developed an immune tolerance to New
Guinea—but it was probably inevitable as more and more 'other people' penetrated into the hinterlands, 'spoiling' the people and in fact causing divisions
in loyalties, etc., etc. I really hope he can find some sort of serenity from time
to time in something."[3] Now it was necessary to identify the agent in the laboratory, which most likely would happen in North America. But as Gajdusek
attempted to meet this new challenge, he found no serenity.

Late in 1976 his life became even more complicated. "I just learned that I
have won the Nobel Prize," the fifty-three-year-old scientist wrote in his journal on October 14. He was spending the week in his childhood home in
Yonkers, surrounded by eight adolescents from the Pacific. "Major reaction:
worry! Can I manage to complete the work I am trying to do, serve the boys
well, and retain humility and creativity?" he wondered. "I must figure out some

Gajdusek with Marion Poms after learning he won the Nobel Prize, 1976. Photograph courtesy Peabody-Essex Museum, Gajdusek Collection

way to escape this notoriety and regain equanimity in solitude." But already reporters and television crew were blocking Palmer Street. That afternoon, having just heard of the award, Gajdusek was whisked away to Bethesda. The car of the director of the National Institutes of Health picked him up at National Airport and took him to an assembly of institute staff. As the new Nobel laureate walked to the podium, which was littered with microphones, lights, and wires, everyone in the hall stood and applauded. Marion Poms, his loyal secretary, could not stop crying. "The frenzy the Prize has caused in our laboratory is huge," Gajdusek observed that day.[4]

Together with Baruch Blumberg, whom he hardly knew, Gajdusek received the prize in physiology or medicine for discovering new mechanisms for the origin and dissemination of infectious disease. They joined Saul Bellow and Milton Friedman as part of a clean sweep in Nobel prizes for the United States that year. Not surprisingly, the outgoing U.S. president, Gerald R. Ford, declared these achievements a cause for national pride, a refreshing relief from the aftermath of Watergate and defeat in Indochina. In the 1960s, Blumberg isolated the "Australia antigen," which made possible diagnosis of hepatitis B, even though the virus still was not cultivated.[5] Gajdusek, of course, had discovered the first human disease caused by the putative unconventional "slow virus," a completely new type of infectious agent, also still hidden. But he certainly did not do this alone, and he was especially embarrassed that Joe Gibbs and Vin Zigas were not also honored. It could have been worse. "I have feared that Hadlow and some of our . . . scrapie colleagues might be my 'running mates' if I ever received the award and that, to the exclusion of Vin and Joe, would have been less tolerable." Nonetheless, all those left out made sure not to express openly any displeasure or bitterness toward Gajdusek's success. If they felt resentment, they kept it to themselves. In any case, the Nobel laureate was already hoping that unraveling the structure of the kuru and scrapie agent might soon warrant another prize in which more of his laboratory colleagues could share.[6]

During December 1976, Gajdusek and his entourage, including the eight irrepressible Micronesian and Melanesian boys (the "Mikes" and "Moros" as he called them), enjoyed the round of banquets, receptions, and lectures in Stockholm. Quickly Gajdusek came to appreciate his new status. On December 13, the hall of the Clinical Center of the Karolinska Institute was packed with admirers as he delivered his Nobel lecture. "Kuru," he began, "has led us . . . to a more exciting frontier in microbiology than only the demonstration of a new mechanism of pathogenesis of infectious disease, namely the recognition of a new group of viruses possessing unconventional physical and chemical properties and biological behavior far different from those of any other group of microorganisms." These unconventional viruses evidently also caused Creutzfeldt-Jakob disease (CJD) and familial Alzheimer's disease and might even be responsible for multiple sclerosis, Parkinson's disease, Huntington's chorea, and many other chronic degenerative conditions. Thus, in elucidating the cause and pattern of spread of an exotic disease in an isolated population, Gajdusek had expanded the scope of infectious disease research. Still, many important aspects of etiology and epidemiology remained unresolved. These unconventional viruses therefore continued to tax even the Nobel laureate's imagination.[7]

THE BASIC PROBLEM was that no one could find the cause of kuru. Through the 1960s and 1970s the infectious agent continued to defy efforts to cultivate and characterize it. For a "virus," it was proving remarkably enigmatic. "It is now essential to try to grow the virus directly from the tissues of kuru victims," Gajdusek had written in 1967. Yet researchers in his Bethesda laboratory employed conventional viral culture techniques to no avail. New electron microscopes revealed no organism in affected brain samples. At

Gajdusek with the "Mikes" and "Moros" on the way to Stockholm to collect Nobel Prize, 1976. Photograph courtesy Peabody-Essex Museum, Gajdusek Collection

Awande, near Okapa, Gajdusek set up an "elaborate tissue culture and virus laboratory wherein live tissue cultures of small bits of tissues obtained both from kuru victims on autopsy, and from autopsies of Fore natives dying of other causes, are being grown."[8] But the virus remained mysteriously immaterial.

Since scrapie proved relatively easy to induce in cheap laboratory animals such as mice and hamsters, it became a model for the investigation of human slow virus conditions like kuru and CJD. British veterinary scientists already knew the scrapie agent resisted heat and formaldehyde. It was extremely small and survived more than two years in dried brain tissue.[9] During the 1960s much more was learned about this particular slow virus, even though no one was yet able to grow it. At the Hammersmith Hospital in London, Tikvah Alper began to doubt it was even a virus, at least in the sense, which became conventional in the 1950s, that it was an infectious agent composed of nucleic acid and protein.[10] In association with colleagues from the Institute for Research on Animal Diseases at Compton, England, she found it was smaller than any known virus, perhaps too small to include any strands of DNA or RNA.[11] Moreover, the émigré South African radiobiologist determined that tissues containing the agent remained infective even after exposure to ionizing radiation, which disrupts the structure of nucleic acid. Frequencies of ultraviolet light that destroy nucleic acids also did not render brain homogenates any less infective, while other frequencies that deform proteins seemed to reduce their virulence.[12] Perhaps, then, the scrapie agent did not require nucleic acid for replication—perhaps it was an autocatalytic protein? Some other scientists ridiculed Alper as the Wendell Stanley of her generation, recalling the Berkeley biochemist who claimed wrongly in the 1930s that tobacco mosaic virus was solely a self-replicating polypeptide.

When Francis Crick asserted in 1970 the "central dogma" of molecular biology, he argued that "information" flows only from DNA to RNA to protein and not in the reverse direction. Each gene consists of two strings of DNA nucleotide building blocks, with each nucleotide base combining with its adjacent complementary one to form a base pair. The sequence of these DNA base pairs determines the nucleotide sequence of single-stranded RNA transcribed from it. The RNA strand then acts as a messenger, carrying genetic information to the cell's protein synthesis complexes, known as ribosomes. At these sites the RNA nucleotide sequence serves as a template, ordering amino acids into proteins and thereby translating genes into functional products. The process seemed remarkably regular and tightly ordered, though Crick did wonder about the "chemical nature of the agent of the disease scrapie."[13] Could mere protein carry information for its own replication?

"Protein, so far as we know, does not replicate itself all by itself, not on this planet anyway," declared Lewis Thomas, the president of New York's Memorial Sloan-Kettering Cancer Center and a popular writer on science. "Looked at this way, the scrapie agent seems the strangest thing in all biology." A great enthusiast for modern medical investigation, Thomas used the possible existence of a pathogenic protein to argue for more basic research:

> To be able to catch hold of, and inspect from all sides, a living, self-replicating form of life that no one has so far been able to see or detect by chemical methods, and one that may turn out to have its own private mechanisms for producing progeny, novel to earth's life, should be the chance of a lifetime for any investigator. To have such a biological riddle, sitting there unsolved and neglected, is an embarrassment for biological science.

It remained, he lamented, a complete mystery.[14]

Gajdusek was baffled. His experiences at Caltech had made him skeptical of explanations that posited self-replicating proteins, but nonetheless the notion fascinated him. "The kuru and scrapie viruses may well prove to be the first infectious 'micro-organisms' without nucleic acids for their genetic information," he mused in 1976. He still favored the theory that they would contain some very small nucleic acid. But "if they are instead replicating proteins or membranes without DNA or RNA, we will have opened up a very new chapter of microbiology."[15] The prospect excited him. The imminent Nobelist now wanted everyone in his laboratory "working creatively on the molecular biology and biological properties and physical properties of scrapie."[16] He felt torn between natural iconoclasm and habitual practice. "It is my romantic beast that keeps me secretly 'hoping' that no nucleic acid will ever be found, yet I keep trying to find nucleic acid."[17]

The "slow virus" refused to give itself up. During the 1970s and 1980s, a number of research teams in the United States and Britain continued to try to purify and characterize the scrapie agent. Gajdusek competed with veterinary scientists at Compton and the Moredun Research Institute near Edinburgh, Scotland; he closely watched Carl Eklund and William Hadlow at the Rocky Mountain Laboratories in Montana as they studied the pathogenesis of scrapie; and later he contended with a pushy upstart, Stanley B. Prusiner, perched precariously at the University of California at San Francisco (UCSF), who began to argue that the "slow virus" was actually an infectious protein, which he creatively chose to call a "prion." By the 1990s, the talk was all about these prions, and attention to the slow virus diminished. The shocking epidemic of bovine spongiform encephalopathy (BSE), or "mad cow" disease, in Britain during this period, along with the consequent rise in CJD incidence, greatly amplified interest in the strange, anathematized infectious proteins. At the end of the century, as controversy raged, some British scientists decided to venture once more into the Fore region to reinterpret kuru as a prion disease, abandoning altogether the increasingly discredited slow virus. Yet again, scientists would go to the Fore people for answers.

IN THE EARLY 1960S, as an undergraduate student at the University of Pennsylvania, Stanley Prusiner demonstrated a knack for chemistry. An ambitious young man from the Midwest, he began assisting scientists at the medical school with their experiments. Then, as a medical student at Penn, Prusiner conducted his own investigations on the metabolism of the fat cells of hamsters as they arose from hibernation. Fascinated, he started to consider a career in scientific research. After completing his medical degree, Prusiner moved west to become an intern at UCSF. From Parnassus Heights, the young doctor could look down on the remnants of the previous year's Summer of Love as they unraveled in Haight-Ashbury and Golden Gate Park. At the end of his internship, Prusiner accepted a post at the National Institutes of Health, where he worked with the biochemist Earl Stadtman, studying the enzyme glutaminase in *E. coli*. There he learned to develop bioassays, purify molecules, record experiments, and write scientific papers.

In 1972, Prusiner started his residency in neurology at UCSF. The specialty seemed to him more intellectually invigorating than any other, and more likely to offer opportunities for research. Within a couple of months, he admitted a woman, already diagnosed with CJD, complaining of progressive memory loss and failing coordination. Supposedly a slow virus, with some striking biochemical attributes, was responsible for the disease:

> The amazing properties of the presumed causative "slow virus" captivated my imagination and I began to think that defining the molecular structure of this elusive agent might be a wonderful research project. The more that I read about CJD and the seemingly related diseases—kuru of the Fore people of New Guinea and scrapie of sheep—the more captivated I became.[18]

The mysterious character of the infectious agent tantalized Prusiner—just as, fifteen years earlier, the Fore themselves had enthralled Gajdusek.

Prusiner believed his biochemical training gave him an advantage in discerning the structure of the agent. Initially he spent much spare time reading everything he could find on the puzzling slow virus. In 1974, Prusiner became an assistant professor in the neurology department at UCSF, supported with "soft money"—short-term grants—like so many new faculty members in the medical school. But his efforts to achieve recognition as a slow-virus investigator frequently were rebuffed. When the NIH rejected his first grant proposal, disparaging his qualifications for research in infectious diseases, he decided to take a course in virology. Sensitive to his status as an outsider, Prusiner made contact with Eklund and Hadlow in Montana and argued that his expertise in biochemistry would help advance their quest for the scrapie agent. Thus began his uneasy relationship with the fractious community of slow-virus researchers. Over the next few years, Prusiner repeatedly traveled to Montana to acquire scrapie-infected brain tissue, which he worked up chemically back in San Francisco, before returning with partially purified homogenate for inoculation into laboratory mice to determine its infectivity.

Often genial, sometimes prickly and aggressive, Prusiner never hesitated to "sell" himself. The young scientist therefore thrived in the enterprising culture developing at UCSF during the 1970s. A decade earlier, Clark Kerr, the president of the University of California, had promoted his vision of a university of individual faculty entrepreneurs.[19] Now UCSF was transforming its staid medical school into a center for the study of basic molecular mechanisms of "higher organisms" such as humans. Like its rival medical school at Stanford, UCSF soon became the matrix for the new biotechnology, emphasizing biochemical approaches to understanding physiological processes and practical solutions to problems of pathology. Scientists at UCSF readily linked this ambitious research agenda with the commercial interests of the specialized legal firms and venture capitalists of nearby Silicon Valley.[20] By the end of the decade, Herbert Boyer, a professor in the ramifying Department of Biochemistry and Biophysics at UCSF, had created Genentech, the first therapeutic biotechnology company.[21] Prusiner found the times and circumstances to his liking.

During the 1970s, Prusiner dedicated himself to purifying scrapie-infected tissue in order to isolate the cause of the disease. The work was labor intensive and time consuming. First he separated the different fractions of brain and spleen homogenate of infected mice, using a centrifuge. Then each fraction was inoculated into mice back in Montana to see if it was infective. The most virulent fractions were further subdivided through centrifugation, then each finer fraction was inoculated into more mice. Every time, he waited up to a year for scrapie to show. Eventually, he isolated a small fraction that manifested 50–80 percent of the infectious activity of the original diseased tissue. After performing a battery of biochemical tests, he too concluded that the partially purified scrapie agent contained no nucleic acid.[22] But the cost of the experiments was not sustainable and the results came too slowly. "We rapidly went through our ten thousand mice," Prusiner told Gary Taubes in 1986. "Even if we were handed money on a silver platter we couldn't go on like that."[23] In 1978, the NIH shut down scrapie research at the Rocky Mountain Laboratories, deliberately forcing Prusiner to find other means of testing for infectivity. Fortunately, it propelled him away from mice and toward hamsters, which conveniently had a much shorter incubation period for scrapie. He also stopped using "end-point titration" which meant waiting until fifty percent of the inoculated animals developed the disease. Instead, he tried an "incubation-time assay," calculating infectivity by gauging how quickly the animals became sick.[24] What the new test lacked in accuracy it made up in speed.

Often Prusiner tried to approach Gajdusek and his group at Bethesda, but they spurned him. One of them, Paul Brown, found the ambitious biochemist particularly irksome and ridiculous. Undaunted, Prusiner visited the Bethesda scientists and offered to work with Gajdusek, provided he dismiss the others.

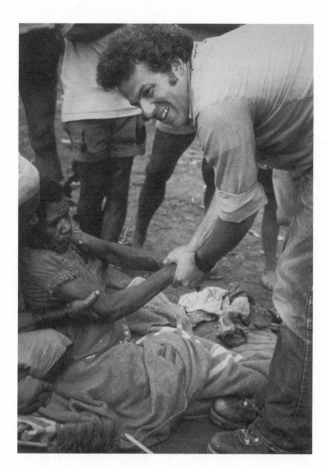

*Stan Prusiner
taking the pulse of
kuru victim, 1980.
Photograph courtesy
Peabody-Essex
Museum, Gajdusek
Collection*

To Brown, the usurping Californian appeared surprisingly ignorant of the biological and social intricacies of slow-virus research.[25] Prusiner only became more insistent. He obtained an invitation from Michael Alpers, then director of the Goroka Institute of Medical Research, to the Fore region in 1978, where he set about examining nine kuru sufferers.[26] "I was a pretty good neurologist at that point," he recalled. On this visit, Alpers and his family looked after him in Waisa, where he became known as Prusli or Bruce Lee. In 1980, Prusiner intruded again into what was unquestionably Gajdusek's territory and studied another seven cases. Before fleeing to Agakamatasa, the older scientist arranged for a few boys to assist the interloper. But Prusiner followed, struggling along the slippery mountain tracks, the boys hauling him up the mountain. "I'm not a great walker," he conceded years later.[27] Once at Agakamatasa he succumbed to a gastrointestinal disorder. Inamba watched the two scientists arguing a lot in Gajdusek's house. "We were in continuous bull sessions for all that time, discussing the future of kuru and CJD and scrapie work," Gajdusek later told Richard Rhodes. After a few days Inamba carried the ailing white man down the mountain to Purosa.[28] Prusiner's "hectic life" prevented him from returning.

In 1982, Prusiner published a paper in *Science* describing his latest laboratory findings. All tests indicated that the partially purified scrapie agent lost its power to cause disease when exposed to most reagents and enzymes that disrupted protein structure, while it remained impervious to nucleic acid destruction. Prusiner radically declared he had discovered a novel infectious entity. Although he did not at this stage completely rule out nucleic acid involvement, it seemed the agent displayed some unique qualities. He decided to give it a name, even though its structure was still conjectural. He called it a *prion*, signifying a proteinaceous infective particle, with the vowels reversed for euphony.[29] He insisted on pronouncing it "pree-on." "Prion is a terrific word," he later claimed. "It's snappy."[30]

As Prusiner flaunted his talent for self-promotion and public relations, his competitors seethed. Two months before publication in *Science*, UCSF had managed to place an article on the new infectious entity prominently in the *San Francisco Chronicle*. "They put my picture and prions on the upper left-hand corner of the front page on Friday," Prusiner boasted. "Reagan was on the right."[31] This made one of his colleagues, Frank Masiarz, so uncomfortable that he left the laboratory soon after: "I said there's no point in creating a name for something we don't even know exists yet. But Stan's ego is unusually important in the way he interprets data. He tends to jump to conclusions, to give no credence to the facts that would discount his preferred interpretation of the results."[32]

"Of course he upset everybody," Michael Alpers recalled. "Stan used to upset everybody for all kinds of reasons." Unlike most other slow-virus hunters, Alpers retained some sympathy and respect for Prusiner. In particular, he admired the biochemist's persistence and resilience. "Everyone whom I spoke to, except the people working directly in Stan's lab, were abusive of what he was trying to do. I mean it's ridiculous, but that happens in science. Stan had to put up with it."[33]

In Edinburgh, Alan Dickinson and his colleagues at the Neuropathogenesis Unit (formerly the Moredun Research Institute) were outraged. For years they had meticulously studied the interactions of the scrapie agent with its various hosts, describing twenty or so different "strains" of the disease, each with a distinctive incubation period and neuropathological lesion.[34] Dickinson suggested that the small infectious entity might consist of a protective coating of host protein around a tiny piece of nucleic acid, which coded for the different strains. He called this agent a "virino," a facetious reference to the neutrino, an elusive subatomic particle.[35] When one of the Edinburgh scientists, Richard Kimberlin, read Prusiner's *Science* article, he responded that the reagents and enzymes that destroy nucleic acid did not inactivate the agent simply because the protective protein coat prevented access to the nucleic acid core. Moreover, how could mere protein contain the biological information needed to produce distinct strains of the disease?[36] The veterinarians pitted their genetic interests against Prusiner's biochemical enthusiasms.

Prusiner and his colleagues persisted in their efforts to define chemically their prion. Later in 1982, David Bolton, one of the lab's postdoctoral fellows, applied the enzyme protease-K, which destroys some protein molecules, to partially purified scrapie material from infected hamsters. Then he separated out the remaining intact proteins using gel electrophoresis, finding a distinct band in the gel that was absent from uninfected hamsters. This protease-resistant protein, named PrP, seemed a likely candidate for the prion, the infectious protein.[37] Once he purified PrP, Prusiner wanted to know if any host genes or viral genes coded for it—if so, it was not a transmissible pathogenic protein working alone, he reasoned. He sought help from Leroy Hood at Caltech, who worked backward from a small segment of the amino-acid building blocks of PrP to make a nucleotide probe that identified a clone from a cDNA library. With Charles Weissmann and Bruno Oesch at the University of Zurich, they used the radiolabelled cDNA as a probe to search for a corresponding segment of DNA in hamster cells. To Prusiner's great surprise, they found the gene coding for PrP in both scrapie-infected and healthy brains. This implied that the host genome *normally* produced PrP.[38]

For six months Prusiner was silent, brooding over his lost pathogenic protein. Then he learned that only the PrP in scrapie-infected tissue was resistant to protease-K: in uninfected tissue the same molecule was susceptible to the enzyme. Moreover, the characteristic minute fibrils found in infected brains, called scrapie-associated fibrils (SAF), seemed to consist of this protease-resistant protein. Prusiner and his colleagues concluded that the PrP gene coded for a protein that could exist in two forms, one normal and the other pathological, depending on its conformation, on how it was folded.[39] In the mid-1980s, Prusiner renamed PrP the prion protein, and the SAF became for him the prion rod. He was now investigating the transformation of normal cellular PrP (or PrP^c) into scrapie PrP (or PrP^{sc}). Not only had he salvaged the prion, he made it more interesting biologically. It was now a protein whose shape connoted infectivity, a rogue molecule able to propagate its pathological configuration on contact with molecules of the same composition in the host. The malformed protein grabbed onto and recruited the normally shaped one, setting off a chain reaction in the body. As the pathological form spread, it produced vacuoles and progressive degeneration in brain tissue, with accumulation of long filaments of amyloid, leading eventually to the death of the host. Its presence in an innocuous form in cells could explain the lack of any immune response to its harmful isoform. But for most scrapie researchers, the notion of an infectious protein still seemed farfetched, if not heretical.

In his studies of prion function, Prusiner was becoming ever more reliant on molecular biologists who could identify, clone, and splice genes. With the help of Karen Hsiao, a new postdoctoral fellow, Prusiner and his colleagues managed to sequence the PrP gene from cases of Gerstmann-Sträussler-Scheinker (GSS) syndrome, a rare hereditary form of CJD, and another presumed slow-virus disease.[40] Hsiao found a mutation of the human prion protein gene in victims of GSS, causing the protein to fold abnormally. She cloned the mutated gene and injected it into mice embryos, creating a transgenic mouse with the defect, which soon succumbed to a neurological disorder. Prusiner and Hsiao declared that GSS was an inherited prion disease, not the result of a slow virus affecting susceptible families.[41] Evidently, the pathological protein might be hereditary as well as infective. Over the next decade, multiple mutations of the normal PrP gene were discovered, each producing clinical signs and brain pathology. Some scientists, especially those beyond the small group studying the scrapie agent and "slow viruses," began to feel that Prusiner was on to something significant. But when he and Hsiao repeatedly failed to transmit the disease from transgenic GSS PrP mice to normal mice, others raised doubts.[42]

The evidence for a pathogenic protein was quickly accumulating. In the early 1990s, Weissmann worked out how to inactivate the PrP gene: mice with the inactivated gene, the PrP-knockout mice, appeared healthy and normal. When inoculated with scrapie, they stayed well for more than a year, while infected mice with the active gene manifested disease within four months.[43] Prusiner therefore claimed that without the presence of the normal protein, scrapie prion was unable to replicate. The prion found no material suitable for transformation in the knockout mice.

When participants in the 1993 meeting on prion diseases, sponsored by the Royal Society of London, heard about Weissmann's results, many were impressed. The scrapie agent really did seem to be an infectious, deforming protein. Alpers recalled the tremendous impact of the announcement on his beliefs about the cause of transmissible spongiform encephalopathies.[44] Even Gajdusek became far more doubtful about the role of nucleic acid, though he refused to stop calling the agent a "slow virus." "I use the term facetiously," he told Georgina Ferry. "Pasteur would have applauded."[45] As he argued, "The potent abstract concept of a virus as a self-specifying transmissible entity requiring the machinery of the host for its replication did not specify any specific structure." Indeed, he noted that "computer viruses" contained no nucleic acid, yet no one disputed the terminology. He also observed that nucleic acid *was* involved—only it was presumably host nucleic acid producing normal PrP, ripe for malformation, not foreign viral nucleic acid.[46]

The rest of the Bethesda group would also divulge, when pressed, their conversion. "I would say that the prion hypothesis looks very good in the absence of any other candidate," Brown conceded in 1996. "But it ain't formally proved." Joe Gibbs, too, murmured his assent:

When you try everything possible using the most modern technology that you have to your hand and you still can't show anything other than the protein associated with infectivity, I think you have to admit that the prion hypothesis is very strong. I think, frankly, that you have to keep an open mind and continue to look. But it's now—what?—thirty years almost. And nobody's come up with a substitute for the protein yet. That's pretty strong evidence.[47]

Others, however, remained fiercely skeptical.

Prusiner's critics could readily point to anomalies and gaps in his research. Some still claimed that PrP might act merely as a receptor molecule for a small, and as yet unidentified, virus.[48] For years, others wondered why no one could convert PrP^c to PrP^{sc} in a test tube, that is, without the involvement of cells. Evidently the switch from one conformation to another occurred in laboratory animals, but it proved difficult to make PrP^{sc} through chemical means. Moreover, the synthetic protein, free of nucleic acid, embarrassingly failed to generate infection.[49] Strain variation was even more troubling. The Edinburgh scientists demanded to know how a single protein could cause the many distinct strains of scrapie. If deformation of host PrP protein was responsible for disease transmission, how might this strain resilience be explained? Surely the infectious agent must include an informational molecule like nucleic acid?[50] Despite his experimental successes, many of Prusiner's adversaries maintained their indignation, causing "prion" meetings sometimes to dissolve into acrimony. One disgruntled scientist called the prion hypothesis the "cold fusion" of infectious diseases research. Prusiner responded that the putative virus was "like the nineteenth-century ether."[51]

Gajdusek and his collaborators had inserted their clinical observations of kuru and other transmissible spongiform encephalopathies into larger biological, epidemiological, and anthropological frames in order to make sense of the diseases. In contrast, Prusiner was resorting to molecular analysis of the presumed agent, hoping to explicate the biochemical logic of these conditions. The "slow virus"—or the "virino"—was inherently a biological entity, implying a certain interaction between host and parasite, while the "prion" was a chemical object separable from the interactions of life forms. The slow virus roamed the field, while the prion paraded around its laboratory niche. Since the 1950s, the "molecularization" of life and its disorders had grown in appeal. The discovery of the double-helical structure of nucleic acid in 1953, with the immediate linkage of chemical pattern and mechanism of heredity, gave particular impetus to this research trend in the biomedical sciences. For many investigators, attempts to reconfigure clinical problems biochemically displaced efforts to understand human illness in broader biological perspective.[52] Thus Prusiner's studies of the composition of the disease agent, whatever the quibbles of his critics, seemed especially compelling. Molecular biology did not move Gajdusek, but he was obviously in the dwindling minority. Increasingly, his anthropological enthusiasms, clinical commitments, and biological sensitivity appeared eccentric and recondite, while Prusiner's single-minded quest for the pathogenic molecule now promised to reshape knowledge of the cause and treatment of these conditions.

The fashion for molecular analysis was embedded in a new set of economic relations, an emerging transactional order that sometimes alienated and discomforted Gajdusek and his generation—just as it sometimes attracted them too. Prusiner and his colleagues were participating eagerly in the biotechnology industry developing in the San Francisco Bay area. Depending initially on short-term grants, Prusiner needed an efficient laboratory, carefully defined projects, and quick results. His management style favored contractual relationships with potential competitors such as Hadlow and Weissmann, a clear division of labor, and limitation of access to potentially valuable information. Prusiner resisted the sort of exploratory exchanges of research materials and ideas that Gajdusek once used to perform so adroitly. Instead, he privatized and commercialized the products of his laboratory, turning molecules into standardized commodities and filing more than one hundred patent applications. He hired a public-relations consultant to market him and his work. Rather than using exchange to make himself visible as a person in the network of slow-virus researchers, Prusiner set himself up as an owner and purveyor of molecular objects. His company, InPro Biotechnology Inc., based in South San Francisco off Highway 101, was founded to "optimize and commericalize prion diagnostic and disinfectant products. Its motto is "saving lives through proteomics."[53] Weissmann also took to patenting results, including the PrP knockout mice, expecting a financial return from their circulation. In the late 1970s, he contributed to the founding of Biogen, a major biotechnology company, and in 1997 he organized Prionics Inc. to develop more diagnostic tools.[54] Human tissue thus was acquiring a different sort of social currency: its exchange once created a community of scientists, now its transaction extended the market. To old-timers like Gajdusek, the refusal to engage in reciprocal exchange, to participate in the old gift economy of science, brought to mind Fore sorcery. Like alleged sorcerers, many of the younger generation of commercial scientists started off as marginal men attempting to overturn social hierarchies through magical performance, thus circumventing more conventional methods of obtaining advantage through manipulating exchange networks. Of course, to those more accustomed to the new way of doing science, these researchers were just ordinary molecular entrepreneurs.

IN 1985, FARMERS IN SOUTHERN ENGLAND began to report unusual behavior in their cattle. A few cows became excitable; they lost weight and staggered; eventually they fell down and would not rise again. Called to Pitsham Farm in Sussex, the local veterinarian saw some of these "mad cows" and noted their tremor and lack of coordination. He examined those that died and submitted their tissue samples to the main English veterinary pathology laboratory. To the pathologist, the brain looked surprisingly spongy, like the brains of sheep infected with scrapie—only this was a cow's brain. Meanwhile in Kent, a veterinarian was examining some nervous, aggressive, and shaky cattle at Plurenden Manor farm. They were slaughtered and their brains sent for pathological assessment. Again, the lesions resembled those of scrapie. By 1986, veterinary pathologists were talking about "bovine scrapie," or bovine spongiform encephalopathy (BSE), a new ailment of cattle.[55]

Epidemiologists tried to track the emerging disease across Britain. Its similarity to scrapie led them to assume it was infectious, even though they could not initially find the causative agent. Soon BSE was cropping up in dairy herds throughout the country. By the end of 1987, reports of more than four hundred cases had accumulated. Strikingly, farmers had fed all the victims meat-and-bone meal (made from the remnants of a variety of stock) after weaning in order to enhance growth and milk production. Some veterinary scientists suggested that this foodstuff might have amplified any scrapie agent present in "rendered" sheep sources and enabled it to cross the species barrier. Others, blaming the epidemic on forced bovine cannibalism, speculated that the meal contained infectious prions generated through the spontaneous mutation of a cow's own PrP gene.[56] In 1988, the British government banned feeding ruminant-derived feed to other ruminants. It ordered the slaughter of all infected cattle and their incineration in vast bovine pyres. Still the cases proliferated, indicating either a long incubation period or other, as yet unrecognized, routes of infection. In 1989, ten thousand cows developed the disease; in 1990, nearly twenty-five thousand succumbed.[57]

Would the bovine epidemic affect human health? Though not uncommon in British sheep, scrapie was never transmitted to humans. In 1988, Richard Southwood, an Oxford zoologist, convened a working party to assess the risk to the people's health. Although the group suspected the consumption of flesh and offal from cattle presented little danger, it recommended a ban on their use in baby food. But the public outcry continued. Toward the end of 1989, the Ministry of Agriculture, Fisheries, and Food stipulated that brains and spinal cords from cattle must be excluded from all food. Within a month, the newspapers were reporting that a cat had died from "feline spongiform encephalopathy," heightening concern that the BSE agent was hazardous to other species.[58]

It was hard to know what BSE might look like in humans, though something like CJD seemed a fair guess. Despite arguing the risk of transmission to humans was remote, the Southwood Committee urged the establishment of a CJD surveillance unit. Based in Edinburgh, the unit determined that between 1970 and 1989 no one in Britain under the age of thirty contracted CJD; yet between 1994 and 1996 ten cases of a novel form of the disease were reported in young people. Generally, the new variant started with psychiatric symptoms, progressing slowly to tremor, failure of coordination, and death. The brains of its victims showed the vacuolation characteristic of the spongiform encephalopathies, though mostly in the basal ganglia and thalamus, not the cerebellum as in kuru.[59] Mice inoculated with brain homogenate from variant CJD (vCJD) victims developed the same lesions.[60] In 1996, Kenneth Calman, the chief medical officer in Britain, announced that BSE was responsible for the epidemic of vCJD. However, he could not predict the severity of the problem. Virtually everyone living in Britain between 1980 and 1996 was exposed to the BSE agent, but it was not yet clear how susceptible the population was or how long the incubation period might be. Some speculated that the disease toll would be a few hundred, others warned of a million dead. By the end of 2003, the authorities knew of almost 150 definite or probable cases of vCJD.

The British and international papers were full of the lessons of kuru. Journalists regularly conjugated vCJD with kuru, mad cows with New Guinea cannibals, and pathologies of progress with diseases of primitives. Both vCJD and kuru shared an insidious, indestructible, invisible pathogen; both diseases emerged from intraspecies recycling of brains; both, so it seemed, demonstrated the consequences of tampering with nature. They both lurked in emblematic, and supposedly strengthening, foods: the roast beef of old England and the bodies of Fore loved ones.[61] Indeed, many commentators asserted that vCJD really *was* kuru: the behavioral changes, progressive disorders of gait, and brain lesions of vCJD resembled kuru more than ordinary CJD.

The analogy with kuru guided social critics down a path to a place where alarmism and agitation mixed freely with monitory advice. According to the London *Times,* kuru declined only when primitive tribesmen were "dissuaded from feeding the tastier parts of their enemies to their women folk." The lesson from kuru, then, was to stop "enforcing cannibalism on cattle."[62] When the British government forbad the feeding of ruminant protein to other ruminants, the *Mail on Sunday* declared: "The lessons of kuru had been learnt."[63] These new diseases were "enough to turn a New Guinea cannibal vegetarian," Sue Arnold wrote in the *Observer.* She went on to describe kuru. "The natives caught it from eating their ancestors. It was predominantly the women who died from Mad Granny Disease, because the men ate first, leaving the women the brains and spinal cord."[64] The lesson of kuru, according to the *Guardian,* was that cannibalism "leaves the brain riddled with holes."[65] "The disease known as 'laughing death,'" the *Economist* warned its readers, "seemed to have been spread by ritual cannibalism: the Fore honored dead relatives by eating their brain or smearing it on their bodies."[66] In the United States, *Newsweek* reported that mad cow was the "creepiest in a family of disorders that can make Ebola look like chickenpox."[67] Thus a disease of civilization, a disorder of industrial agriculture, was refigured as the return of the primitive.

Talk of prions was in the air as New Labour rose to power in Britain. In the mid-1990s, the prion seemed to insinuate itself everywhere. Politicians and journalists freely scattered Prusiner's linguistic coinage across the realm. Scientists and farmers alike spoke in the idiom of infectious proteins. There remained some resistance to the novel pathogens in places like the Neuropathogenesis Unit in Edinburgh, but these holdouts were increasingly beleaguered and starved of funds.

Then in 1997 Prusiner's molecular conjuring was rewarded with the Nobel Prize in medicine or physiology, thus confirming prion supremacy. The award recognized his discovery of an "entirely new genre of disease-causing agents," an addition to the list of infectious agents including bacteria, viruses, parasites, and fungi. Normally innocuous cellular proteins (Dr. Jekyll, as the Nobel press release put it), prions converted their structures into pernicious conformations (Mr. Hyde) that damage nerve cells, leading over a long period of time to CJD, kuru, GSS, and fatal familial insomnia in humans, and scrapie, BSE, chronic wasting disease of deer, and transmissible mink encephalopathy in other animals. It appeared prion diseases might be inherited, transmitted through infection, or occur spontaneously. This "sensational hypothesis" also promised to enrich understanding of other types of dementia, such as Alzheimer's disease.[68] In his banquet speech, Prusiner recalled the difficulty in convincing other scientists that prions even existed. It was the "mad cow" epidemic, and the subsequent emergence of the new variant of CJD in humans, that finally persuaded scientists and the public to take infectious proteins seriously. "Yet the principles of prion biology are still so new," he said, "that some scientists and most laymen, including the press, still have considerable difficulty grasping the most fundamental concepts." Nonetheless, "the story of prions is truly an odyssey that has taken us from heresy to orthodoxy."[69] In the *New York Times*, though, Lawrence K. Altman described Prusiner as a "maverick scientist in San Francisco whose discoveries about infectious particles called prions have been criticized by other researchers as unproved." He reported that many scientists still doubted that proteins alone could cause disease.[70]

Still glorying in his Nobel Prize, Prusiner testified the following year to a committee of inquiry the British government set up to investigate the response to the BSE outbreak. While prepared to expatiate on the chemistry of the prion, he was clearly uneasy when pressed on local epidemiological concerns. The committee wanted him to explain the origins of the British outbreak, but this was far outside his expertise. Regrettably, he could not help them and told

them he must leave soon for another meeting. Prusiner did, however, find time to discuss his efforts to develop diagnostic tests and therapies for prion diseases. At the end, he was asked whether he would ever contemplate eating British beef again. Prusiner vigorously shook his helmet of curly gray hair, but admitted this was merely an emotional reaction. "When there is a disease like BSE," he said, "things do not sound very appetizing. But at a scientific level, I cannot give you a scientific basis for choosing or not choosing beef, because we do not know the answers."[71]

In his lectures, Prusiner sometimes called on Kurt Vonnegut to explain the prion. As a youth, the scientist had read *Cat's Cradle*, in which the novelist explored the consequences of ice-nine. When water freezes it normally takes the form of ice-one, but other conformations might be possible, depending on the molecular shape or template. Just imagine a crystal: the seed teaches atoms how to stack and lock, so with different seeds the crystal will take novel shape even when its molecular composition is the same. In the 1963 book, the character Dr. Breed tells the reader, "Two different crystals of the same substance can have quite different physical properties." Felix Honnecker, a Nobel laureate in physics, makes a chip of ice-nine, a new way for water atoms to freeze, with disastrous consequences. When water makes contact with this seed, it freezes as ice-nine. Papa Monzano, the dictator of the Caribbean island of San Lorenzo, swallows some and his body fluids become ice-nine. His corpse falls into the sea and the world's oceans turn into ice-nine. The pathological form of water propagates irresistibly, causing molecular apocalypse.[72]

Prusiner also was fond of quoting from Michael Crichton's novel *The Lost World*. The story derived from the recent success of *Jurassic Park*. On an island off Costa Rica, scientists were recreating dinosaurs, but then they foolishly fed the babies protein extract made from rendered sheep. Not surprisingly, the dinosaurs eventually came down with a new prion disease, called DX, which killed them all. "Prions," the book's heroic scientist informs readers, "are the simplest disease-causing entities known, even simpler than viruses. They're just protein fragments. They're so simple they can't even invade a body—they have to be passively ingested. But once eaten they cause disease: scrapie, in sheep; mad cow disease; and kuru, a brain disease in human beings."[73] Prusiner reveled in this example of the appeal of prions in popular culture. He pleaded with the producers of the film adaptation not to cut the dinosaurs' proteinaceous demise from the story. But Hollywood, far from the madding cows, was unmoved: prions did not yet sell movies.[74]

Even if its entertainment value was still dubious, the prion became by the end of the millennium a powerful and pervasive entity in science, economics, and politics. Its assiduous cultivation had largely supplanted the delicate, endangered slow virus.[75] It seems Vonnegut predicted this too. In *Cat's Cradle* he reported a conversation in the Cape Cod Room of the Del Prado Hotel:

"What is the secret of life?" I asked.
  "I forget," said Sandra.
  "Protein," the bartender declared, "They found out something about protein."
  "Yeah," said Sandra, "That's it."[76]

AFTER A LONG DAY preparing for the interview and then answering questions, John Collinge was growing testy. The leading British prion scientist was giving evidence to the committee of inquiry into the BSE epidemic early in the summer of 1998. In a thick Midlands accent, he insisted that he was not a government scientist as the committee claimed on its website. He told them about the shoddy scientific review procedures of previous government committees and the repeated failure to heed his advice. Under his thatch of dark hair and beetling brow, the scientist could look saturnine and distrusting. At the end, when asked the obligatory question about his diet, Collinge was not impressed. If he went to a friend's house he would still eat whatever was put in front of him, he sullenly revealed. But, he continued, "I have definitely reduced my consumption of processed products that are likely to contain offal."[77]

In 1998 Collinge directed the Medical Research Council's Prion Unit, located in a shabby modern building off Queen Square in London. At the beginning of his research career, some ten years earlier, the Bristol medical graduate had been interested in the genetics of schizophrenia and neurological disorders. After hearing Prusiner speak about prions, Collinge began to wonder whether physicians were missing transmissible spongiform encephalopathies simply because they did not think of looking for an infectious protein. Indeed, when he screened a small group of patients who died with dementia of unknown cause, he found a few had suffered from an inherited prion disease.[78] Awarded a grant from the Wellcome Trust, Collinge moved to St. Mary's Hospital Medical School, London, where he started to investigate prion diseases in earnest. He sought to determine the genetic basis of susceptibility to the pathogenic proteins.

It soon became apparent that a tiny segment of the PrP gene influenced the likelihood of CJD occurring in an infected individual. Within a gene, it takes a nucleotide triplet, called a codon, to specify the information needed to produce a single amino acid for the protein. Most cases of sporadic CJD affected individuals whose codon at position 129 in both copies, or alleles, of the PrP gene produced either methionine or valine amino acids—that is, those who were homozygotes. When one allele at codon 129 made methionine and the other valine, these heterozygotes seemed relatively resistant to prions. Collinge reasoned that uniform presence of a normal protein with matching structure aided propagation of the rogue form—when the sequence was identical, the protein was more readily converted on contact with the deformed template.[79] Thus a subtle inherited difference in amino acid sequence substantially influenced predisposition to prion disease. Evidently, all victims of vCJD were homozygous for methionine at codon 129.[80]

The problem of strain preservation in transmission of prion diseases also fascinated Collinge. How might a single protein code for different types of disease? For Collinge, the "molecular basis of strain specificity" was "really the key problem in understanding prion propagation."[81] By 1996, he was able to demonstrate that each strain corresponded to a distinct conformation of the prion and the pattern of sugar molecules attached to it.[82] This discovery allowed Collinge and his coworkers to differentiate biochemically the PrP^{sc} of vCJD from the PrP^{sc} of other strains of CJD. These conformational differences also helped to explain the species barrier for transmission of prion diseases. Thus, when the primary structure of the PrP protein of the species from which the prion derived differed from that of the inoculated host, the prion would propagate less efficiently. It was harder for the inoculated prion to deform the normal host protein. But the species barrier was rarely absolute. Indeed, BSE spread effectively to a wide range of species, suggesting it possessed an especially potent conformation, or strain. This particular prion shape evidently favored transformation of many other types of PrP, allowing the strain to become remarkably promiscuous.[83] Indeed, Collinge and his team were able to demonstrate that the molecular conformations of the BSE and vCJD prions were similar.[84] Such identification of molecular markers of disease strains potentially permitted prion typing in days, rather than the years it took to determine their effects in experimental animals. But mass molecular diagnosis remained in practice a distant goal.

Around 1995, Collinge decided he should try to find the prion strain that caused kuru. He wrote to Michael Alpers, asking for some kuru brain tissue, but none could be found. The paucity of prion material frustrated Collinge— for years the Edinburgh group that ran British CJD surveillance had limited his access to their autopsy samples too. Alpers suggested they might seek an autopsy on one of the few lingering kuru victims and, at the same time, conduct more genetic studies of Fore survivors. "Just don't send someone over who's going to wander around in a pith helmet," he warned the Englishman.[85] The plan interested Collinge. After securing Wellcome Trust funding, he recruited a lean and ascetic young adventurer, Jerome Whitfield, who happened to have some army nursing training and bush experience. Collinge told him to try to get a kuru autopsy within a few months. But Alpers secretly expected it would take years, ensuring the continuation of the field base among the Fore.

In November 1996, after learning how to remove brains from cadavers,

Jerome Whitfield with Geoffrey Yota (left) *and David Pako* (right), *early 2000s.*
Photograph courtesy Jerome Whitfield

Whitfield flew to Goroka and then drove down to Waisa. "I'd been out in various jungle places before, so it wasn't that new," he glibly told me at the bar of the Bird of Paradise Hotel in Goroka. "The jungle's always the same wherever you go: wet and green. But culturally very different from anywhere else I had worked before."[86] He quickly picked up Pidgin and gradually acquired some Fore words too. Like Gajdusek and Alpers, he proved to be a prodigious hiker. Fore attitudes and behavior fascinated him, but at first the people kept their distance. Initially the kuru reporters resisted going on patrol with him, preferring to deal directly with Alpers. "It took a lot of time to knock everything into shape again," Whitfield recalled. He passed the hours rereading Shirley Lindenbaum's *Kuru Sorcery* and dipping into Gajdusek's journals.

Soon after Whitfield arrived in Waisa, a steady, reputable Fore man in his late thirties approached him and offered to help. Anderson said he was the son of Puwa, a village big man who knew Alpers and had once taught the white man how to make salt. Mission educated, Anderson had worked for a while in Port Moresby with Burns Philp, the Pacific traders. Now he was back home, trying to build resources and reputation so he might win an election. Anderson was thoughtful and reserved when we spoke at Ivingoi in 2003. "I had done lots of work for the community . . . for helping folks here," he said. "So

I was the first to go help Jerome, with community work, his public relations." It struck me that, while he was speaking Pidgin, he used the English words "public relations." Anderson went on, reflecting on his role in kuru research: "We became middlemen, like I know some things of the ancestors and I know something of Western styles and such, maybe I know the names of particular illnesses, I know that there are particular cures for each . . . My own strong belief is on the side of the whites with respect to medicine . . . Other folks, they believe strongly in sorcery poison." Anderson often spent hours trying to cajole a single blood sample. "Lots of people opposed it. They said: 'You haven't made anything, all this time you haven't made one cure.' But some people like myself—like you too, doctor—you know you have to find the disease before you make the cure." I found myself nodding in agreement. "I tried to counsel them," he said, "and some of them understand, many understand. But some people are jealous of our work . . . They don't get it."[87] It was hard work for Whitfield, and sometimes dangerous for those assisting him. But Anderson felt it was worth it. "Jerome encountered many obstacles that I helped him overcome, including the people who complained and complained. Jerome began to give me recognition, as did Dr. Alpers, and John Collinge too saw what I was doing."[88] Soon Anderson himself became a big man in Waisa, though as he said, never a bigheaded man.

Even with Anderson's help, for the first six months Whitfield failed to collect any blood samples or perform an autopsy. The people were wary of investigators, especially ones they hardly knew. Alpers, still based in Goroka as director of the Institute of Medical Research, encouraged the young Englishman to try to fit into the community and not to rush his studies. Two kuru deaths occurred in the first six months, but Whitfield did not feel confident asking for autopsies. "I decided to build up a relationship with the local people first," he said. To break into the community, he set up a clinic, dispensing basic medicines for common problems like malaria and pneumonia. "The important thing," he maintained above the hotel din at the Bird, "is it's all done on an exchange basis. You have to find ways of getting into communities. You have to find a workman who's got a relative in that village, or a sister who's married into it, you can't just turn up there. People are still very suspicious and very superstitious." Usually Whitfield employed the children of Alpers's former assistants, such as Anderson, but sometimes this sparked jealousy among those rejected. Soon he was caught up in local grievances and rivalries, some of them obscure to him. "If I'm employing people and they're earning a living, they have access to more money . . . They can extend their relationships with other people, they can fulfill their social obligations, and that gives them standing, and people are immediately jealous." He tried to focus on community projects like installing water tanks and providing health services. But people demanded individual recognition and compensation. "I've been getting it in the neck for seven years," Whitfield told me in 2003. "I still get people coming in complaining about one thing or another." Eventually the disputes in Waisa became so severe and threatening that he moved to the mission at Ivingoi, near Wanitabe, a rival village. "And that worked?" I asked him. "That worked fine."[89]

As he waited for the opportunity to get a kuru brain, Whitfield began collecting blood samples and acquiring accompanying family histories for genetic studies. While Fore often disregarded his attempts to explain what he was doing, they participated avidly in exchange with him. "And really, it is an exchange. I don't think that's changed." He gave me some examples. "I've put five hundred kina in for community development. That money's been used to establish preschools or buy materials for the community schools, or it's been put in toward a project or whatever. Sometimes it's just medical equipment for clinics." Under the circumstances, concern with informed consent, predicated on individual autonomy, seemed extraneous or trivial.

Collinge started to visit the Fore regularly as he became increasingly committed to studies of genetic susceptibility to prions. Kuru seemed to offer a model for the British vCJD epidemic: genetic studies of the Fore therefore might illuminate the British future. "Professor Collinge," Whitfield assured bemused readers of the Port Moresby *Post-Courier* in 2001, "is equally happy eating taro in the bush or giving presentations to the Papua New Guinea Medical Society." The professor was visiting the Fore so he could learn about "English kuru," or the "human form of mad cow disease." Local kuru researchers, Whitfield wrote, still had much to teach the world.[90] Of course Alpers was delighted.

Soon there was both a little good news and a lot of bad news for Britain out of the Fore region. In 2003, Collinge and his colleagues reported in *Science* that kuru had exerted a strong selection pressure on the Fore, virtually eliminating all codon 129 homozygotes and favoring the more resistant heterozygotes. The blood Whitfield took from healthy elderly Fore women—all of whom had eaten human flesh—indicated that a disproportionate number were codon 129 heterozygotes, thereby producing mixed amino acids which inhibited propagation of the prion, and conferring on them a marked survival advantage. Moreover, studies in European populations suggested, though less dramatically, that codon 129 heterozygosity had in the past been a significant selective advantage for them too. The scientists claimed this finding was consistent with ancestral participation of Europeans in endocannibalism and their resultant exposure to prion disease epidemics.[91] Provocatively, the scientists speculated that the cannibalism of their forebears might be providing some Britons with limited protection against vCJD. All the same, since almost 38 percent of those living in the United Kingdom were homozygous for methionine, there was still plenty of capacity for amplification of the vCJD epidemic.

The pattern of kuru more generally appeared to confirm Collinge's pessimism about the outcome of the vCJD epidemic. He frequently warned the British public that the incubation period of their prion affliction might extend fifty years or more. The epidemic would continue to evolve over many decades. "We know from the disease kuru," Collinge wrote in the London *Times,* "that human prion disease can span decades with incubation periods that can exceed forty years, with an average of around twelve years."[92] Among the Fore, Whitfield investigated all suspected cases of kuru, hoping to determine the maximum incubation period for human prion infection and to identify genetic factors influencing disease expression in recent patients. Between 1996 and 2004 he found eleven definite kuru victims. Since all transmission through mortuary practices ceased by 1960, it was assumed they had contracted the prion earlier. Interestingly, most of those who succumbed after forty or more years harboring the pathogenic protein were heterozygous at codon 129. Being heterozygous therefore did not confer absolute protection, though evidently it delayed onset of the disease, often beyond the normal human lifespan. Presumably, sporadic or chance conversion of the prion protein of a Fore person to an infective, pathogenic formation some time in the 1920s (or earlier) had led to his death from CJD, then cannibalism propagated abnormal, fatal protein first among infected homozygotes, then eventually among some heterozygotes, with further cannibalism producing a cascade of this process among the closed population. The pattern of kuru incidence implied that "mean incubation periods of human BSE infection of thirty years or more should be regarded as possible, if not probable." Kuru was providing "an insight into the probable span of the vCJD epidemic in the UK"—and the results were chilling.[93] As the editor of the *Lancet* commented in 2006: "Any belief vCJD incidence has peaked and that we are through the worst of this sinister disease must now be treated with extreme scepticism."[94]

After years of waiting, Whitfield finally managed to arrange an autopsy on a kuru victim. Some time before, soon after arriving in the region, he befriended

Animo, a Yagareba man who as a youth had worked for the Glasses. It was hard for him to watch as Animo lay dying from kuru. But since Whitfield knew the family well, the autopsy was fairly easily secured. "We checked the patient every day," he told me. "We did everything we could for pressure sores to make him comfortable . . . We treated him with antibiotics. We did everything."[95] "We took a long time to look after that man," Anderson recalled. "We waited and waited, almost a year."[96] Arriving in a helicopter, Alpers examined Animo during the recumbent stage of the illness, confirming the diagnosis. Collinge agreed to pay the family forty thousand kina for the autopsy, an incredible amount of money in a place where one struggled to find change for a twenty kina note. The key relatives each received five thousand kina to pay for the education of their children. When Animo died, Ken Boone, a Henganofi man who worked as a surgeon in Goroka, came down to do the cutting. Later, over a beer at the Bird of Paradise, Boone told me he does five or six autopsies a year so the procedure was no big deal, though operating in the bush brought challenges.[97] They packed up samples of brain and other tissues to send to London, and then they left the family to mourn. Anderson kept his distance from the operation. "The sorcery accusations circulated quickly," he told me. "People blamed each other." He feared sorcery charges directed against him. "So I wasn't around, I stayed away. I just stayed in my own house."[98] The autopsy had a lasting effect on Whitfield. A few years later, recollecting the event, he wrote to me: "I feel sad about this at times and wish it could have been someone more distant."[99] But he had done his job and left the Fore when concerns over the autopsy diminished.

"The problem with these sort of things is jealousy," Whitfield reminded me. "If other people catch on that the family might have some money, then everyone is going to be putting pressure on them."[100] Rivals began threatening Animo's survivors. Others demanded compensation. As the situation became more menacing, Whitfield tried to protect his kuru assistants and their families. After a year or so, the importunate claims, the denunciations, and the attempts at intimidation lessened, and so the tensions around Yagareba and Waisa slowly eased.

MY HELICOPTER LANDED on the decaying basketball court at Ivingoi mission a few months after Animo's autopsy. At the time, *raskols*—the cute Pidgin word for very un-cute bandits—patrolled the only road from Goroka, so flying over them seemed the safer route. But it was an ostentatious arrival, and twenty or so old men and children stood around watching closely. I was glad

Hamlet in South Fore, near Purosa, 2003. Warwick Anderson

most adults were already in the gardens or doing other business. Jerome ran up the hill to greet me and bring me to the new kuru research project house. There I met Anderson, who told me he would look after me the following weeks and find me informants, mostly the old men who once worked with the scientists. In typical Fore fashion, he bluntly told me that if I did not like this arrangement I could "piss off." He smiled as he said it. In fact, I was pleased he would get people to talk to me from the first day, even if the version of events I heard might be filtered and distorted in some way. I simply could not wait six months to get to know the Fore as Jerome had done. Over a cup of tea, Anderson expressed disapproval of the generosity of an anthropologist who had passed through the region a few months earlier, and he warned me not to inflate the price of stories. If I planned to behave like a "white man," he said, I should find the satellite phone and call back the helicopter. For when people like me do anything "heavy," he was the one who suffered the consequences. I soon understood he was trying to clean up the big mess resulting from Animo's death.[101]

During the following weeks I settled in at the Open Bible mission, fetching water from the well, brewing copious pots of tea, chewing on thick

"Highway Beef" crackers, cooking the *kaukau* (sweet potato) I bought along the roadside, and for special occasions preparing the *tinfish* and rice I had carried in. Sometimes I chatted with the pastor, a Fore who had trained in Iowa. Anderson appeared each morning with another old man for me to interview and left for Waisa in the afternoon when the mist enclosed the house and the rain began. Each day was full of curious visitors. Everyone wanted to tell his story and to have his name attached to it. Everyone needed to explain his crucial contribution to kuru research. Everyone asked for more compensation, now a New Guinea convention.

After a few days at Ivingoi, I decided to walk to Okapa. The winding road was muddy and steep in places. Soon I acquired a retinue of a hundred or so children, who expressed surprise at seeing a white man walking and carrying a pack. Some of the younger ones appeared frightened when the adolescents teased them, telling them I was a ghost. The braver ones wanted to touch my skin. At the end of three or so hours of hard walking, I arrived at the dilapidated station. It was market day, and hundreds of Fore clamored around me. I found a grassy clearing and sat down next to a solitary, proud old man. Piles

Masasa with Warwick Anderson, Okapa Market, 2003. Photograph by Tom Strong, collection of Warwick Anderson

of fruit and vegetables accumulated in front of me. Eventually the old man leant over and told me he was Masasa, and he had come from Yagusa to talk with me. I knew him from Carleton's journals, but Anderson had been "unable" to find him. Among other things, Masasa needed to tell me that he never received adequate compensation for his research work.

The following week I trudged over the divide and down to Purosa. Some boys I met along the way, trying to kill birds with slingshots, warned me about the sorcerers at Kamira. Without Anderson again, I sought out Turi, another of Anua's sons, a scrawny, bashful young man, just as his father once was. Wearing a t-shirt emblazoned with "Hair Dressers are Creative Professionals," Turi found some of Carleton's former assistants. For a few days we talked with the old men, sweaty in their secondhand clothes, crouching or lounging on the floor of the aid post, leaning against the pastel murals. At night I stayed at an empty house on the local coffee plantation. Then, on the morning I planned to climb to Agakamatasa to find Tiu, one of the few surviving *dokta bois,* the rain poured down. Tired, sunburned, scratching my fleabites, I reluctantly decided to return to Ivingoi. I worried that tropical ulcers were developing where leeches had broken the skin, and I seemed to have acquired some lice. In any case, I heard rumors that Agakamatasa was in mourning for Tiu's sister, and no one in the village would talk with me.

Clinging to the eastern margins of the highlands, the Fore region is still a poor place, with few pigs and sparse material culture. Though promising much, coffee production has delivered little. Most plants are abandoned—and anyhow it is hard to get the beans past the *raskols.* Little cash circulates, even at the roadside stalls and trade stores. There are, however, more aid-posts than elsewhere in the highlands, a legacy of kuru.

When I was there, everyone was talking about compensation. A sort of postcolonial melancholy pervaded conversations, a sense that as individuals and as a people they were unfairly excluded from globalization and its presumed rewards. Kuru research once led development of the region: the roads were kuru roads; the fibro buildings, kuru buildings. Kuru investigators supported the schools. Kuru brought cargo. But fewer resources now came down the track from Goroka and less was happening. It seemed the white men they collected had let down the Fore. Yet a desire, a yearning, had been stimulated or amplified. White men came and went, got *bigpela* prizes and perhaps plenty of money, and left the Fore people with demands unmet and expectations dashed. Now everyone wanted more compensation, especially some of Car-

Anderson Puwa at Alpers's house at Waisa, 2003. Warwick Anderson

leton's former assistants. A few even wrote to him, but he replied that he was a poor man, living in exile from his own country. They did not believe him. Instead, they harangued Jerome and Michael, and tried to take out their frustrations on Anderson. It was not so much a demand for redress of wrongdoing—such as organ theft—as an expression of resentment and jealousy, a breakdown of trust, a plea for recognition and respect.

Around him, Anderson saw nothing but signs of disappointment and decline. "Now we see that money is changing everything, its value extends to everything," he lamented. "That kind of idea has come along and 'fouled' the ideas of many men."[102] With everyone striving for money, nothing much was left to hold the community together. Ceremonial exchanges were dwindling.

People claimed individual ownership of things that once were public goods. As he said this, I thought of the scientists in San Francisco, London, and Bethesda in their exciting new economy with their patents and biotechnology companies.

Along the track, Fore would come up to me and ask if I could find a record of their mother or sister who died of kuru. They heard that Michael kept the photographs of victims somewhere and wondered if they could get them back. I wrote down the details but nothing ever came of it. On other occasions, as I walked along, Fore talked animatedly about English kuru. Someone told them kuru was afflicting another people. But how, they wondered, had these other people learned the correct techniques of kuru sorcery? They never doubted that kuru was a thing made by man, not a natural disease.

John Collinge and Michael Alpers held a conference in 2007 at the Royal Society of London to mark "the end of kuru" and celebrate fifty years of research into this plague on the Fore people. Kuru may have stopped, but there is no real end. Humans will continue to make disease.

DÉNOUEMENT WAS A BIT DIFFICULT

THE TRANSFORMATION OF PARTS OF FORE
persons into things of scientific value made Carleton Gajdusek famous. In fashioning and circulating these valuables, he became a big man, a dream man, in modern science. His reputation preceded him wherever he went. After the Nobel Prize, his name possessed as much value as any of the kuru brains he once collected and distributed. He had become known. His name traveled. It could move and influence others.[1] "You have increasingly divorced yourself from the work of science," his colleague Paul Brown declared in 1985, "and become more and more a traveling purveyor of ideas (I stress the word 'traveling')."[2] It turned out the passage between fame and defamation on these travels was shorter than anyone assumed.

Gajdusek continued to move around the world, lecturing and hectoring audiences, supporting fellow scientists in obscure laboratories and isolated field sites, recounting the kuru story, and speculating on the causes of even rarer brain diseases. Wondering about the epidemic of amyotrophic lateral sclerosis on Guam, he visited the island and came back with a theory about trace metal imbalances in the environment.[3] More frequently, though, he was drawn to the Siberian taiga, far from the tropics of his youth, to study Viliuisk encephalomyelitis (VE), another emerging fatal disease of unknown cause. This brain degeneration continues to afflict the Sakha people, but it never achieved the international notoriety of kuru. For more than thirty years, Gajdusek and others have suspected it spreads through infection, but no agent has been identified or causal mechanism postulated. "This is serious and our new goal," the aging Nobel laureate instructed Joe Gibbs back at the National Institutes of Health (NIH) in 1991, "to solve VE. It is as good as the kuru story!!! We

Gajdusek, the
successful scientist.
AAPI

must inoculate monkeys and chimpanzees *and* other animals with fresh and frozen tissue. We can have all the sera, leucocytes, CSF and autopsy tissue we want, collated as we wish."[4] With a glint in his eye, Gajdusek foresaw the same specimen rush he once led in New Guinea; the same insistent search for blood, cerebrospinal fluid, and brains; the same deathwatches and fast autopsies. "Our real quest here," he wrote in his journal on September 3, 1992, "is material for further inoculation and further serology."[5] But so far, nothing has been found.

In the 1990s, things started to fall apart for Gajdusek. His interest in kuru and New Guinea finally waned—it had become too tidy and civilized as other white men crowded him out. In Siberia, he eventually realized he had lost the drive and energy needed to acquire and circulate specimens, and to manipulate people, as he did in younger days. Back in Bethesda, he found himself on the margins of prion research and impatient with what he regarded as its biochemical pedantry. Above all, he was tired of the demands of reciprocity in his relationships with others, whether potential research subjects or loved ones. Once so adept at securing personal advantage and commitment though exchange, he now retreated from the human visibility and attachment it entailed, attempting to freeze others out. There was even a moment when he flirted with the new, more alienated, economy of the biomedical sciences, briefly becoming the owner of a cell line. Not surprisingly, this venture into the market, away from the entanglements of gift exchange, proved uncomfortable and embarrassing. It was just part of the general breakdown of Gajdusek's personal relations in old age—far worse was to come.

IN 1996 THE PAPUA NEW GUINEA Institute of Medical Research in Goroka became a major salient in the struggle against bioprospecting and commodification of indigenous bodies. It was a trying time for Michael Alpers, its director, and for Gajdusek, who became involved against his better judgment.

Carol Jenkins, an American medical anthropologist at the institute, had worked with the Hagahai people of the western Schrader Range, in a remote corner of Madang province, since the early 1980s. Numbering fewer than three hundred, this isolated fringe highlands population of hunter-gatherers had little experience of diseases endemic elsewhere in New Guinea. As they now were exposed to mumps, influenza, hepatitis, and other infections, the population appeared to be falling. Jenkins visited them frequently between 1984 and 1986. She was mostly interested in their adaptation to the harsh, isolated environment and their beliefs about security, risk, and disease. [6] But she also took samples of blood and feces and obtained nasal swabs. In Goroka and Bethesda, the blood was examined for unusual biochemical and immunological patterns. Scientists in Gajdusek's laboratory soon found genetically distinct versions of the human T-lymphotropic virus type 1 (HTLV-1) in Hagahai specimens. In other groups, this virus sometimes causes adult T-cell lymphoma, but the Hagahai variant seemed benign and therefore perhaps useful in producing a vaccine. The scientists developed a cell line—a sample of cells adapted to laboratory cultivation and thus "immortalized"—that was infected with the Hagahai HTLV-1. Since the cell line was discontinuous with the person from whom it came, it was regarded as a biological invention and therefore patentable. In the early 1990s, it was NIH policy to seek patents on its "inventions," so Gajdusek and Jenkins felt obliged to apply to the U.S. Patent and Trademark Office to become owners of the cell line, which they kept frozen in liquid nitrogen at the American Type Culture Collection. [7]

The issue of the patent (5,397,696) on the Hagahai cell line on March 14, 1995, attracted widespread attention and considerable indignation. In particular, the Rural Advancement Foundation International (RAFI), a Canadian non-governmental organization dedicated to limiting Western commercial exploitation of plant and animal resources from the Third World, declared the patenting of indigenous genetic material an egregious act of "biopiracy" or "biocolonialism." RAFI alleged that Jenkins, Gajdusek, and their colleagues were claiming ownership of biological material essential to Hagahai physical identity: they were commodifying humanity. Scientists had supposedly uncovered the "genetic secrets" of indigenous people in order to make money. The controversy over patenting of human genetic material led to Jenkins being hauled off a plane in Port Moresby and asked to explain the claim to a government minister. She pointed out that she was merely complying with NIH policy and, in any case, she had assured the Hagahai that half of any income from the patent would be paid to them. Without patent protection, they would receive nothing in the unlikely event of commercial exploitation of the cell line.[8] Alpers also defended the intellectual property claim and "abhorred" RAFI's ideologically driven campaign.[9] But in 1997 the patent was quietly dropped—to everyone's relief.

The Hagahai patent controversy serves to indicate just how much the economy of science was changing toward the end of the twentieth century. Once Gajdusek had distributed his kuru brains and blood as gifts, expecting in return some recognition and assistance from other scientists. He had circulated these biological materials in ways that made claims on other persons. Now he was involved instead in making property claims on "biologicals" extracted from indigenous people—that is, he was helping to make "biological" a noun, not just an adjective. Once a transactor seeking to make other persons visible and related, now he was caught up, perhaps inadvertently and certainly reluctantly, in ownership of biological commodities, looking for real capital, not just symbolic capital. Strangely, the economy of science was being transformed in parallel to the monetarization of the New Guinea economies from which so much of this biological material had come. Everyone, including Gajdusek, toward the end of the century was becoming enmeshed in a different, disorienting set of human relations.[10]

THE ARREST OUTSIDE the house in Deer Spring, Maryland, was dramatic. It was an early spring day in 1996, just before Easter, and Gajdusek was returning home from a week of acclaim in Slovakia, his father's country. He had been traveling in part to put the Hagahai business behind him. As the corpulent, rumpled Nobel laureate pulled up in his car, five or six police vehicles with sirens sounding and flashing lights surrounded him. Agents from the Federal Bureau of Investigation (FBI) ordered him out and told him to put his hands on the car roof. Then he was handcuffed. "Hollywood would have loved it," Gajdusek observed bitterly.[11] They took him into the house, which was dirty and messy, but refused to let him see the four boys from the Pacific who lived there. Distressed and tearful, Gajdusek heard an FBI agent charge him with "child abuse" and "perverted practice," meaning oral sex. In jail later the same day, awaiting release on bond, Gajdusek began to despair. Paul Brown reported that the NIH had put the Nobel laureate on indefinite administrative leave. "Without books, without razor or soap or towel or sheets until now, I have largely abandoned hope, which is what such detention is all about." With a defective pen, Gajdusek was still scratching notes on a scrap of paper. "As I see the complications outside develop and new obstacles to even a minimal return to the living, I find it better to consider myself a CJD victim . . . I then entertain no false hopes and settle for the human cycle as I have done." Some friends, like Brown, Alpers, and Bob Gallo promised to stand by him, but he was not allowed to see any of the thirty or more Melanesian and Micronesian boys he had brought to live with him in the United States since the early 1960s. "I am truly free of material objects in my meditations," he wrote in jail, thinking first of Descartes and then of Dante. "A removal of contact with my own family . . . Life is truly over and only the Inferno lies ahead."[12]

For more than a month, Gajdusek had expected trouble. The FBI was interviewing his children and colleagues; it studied the more salacious passages in his journals; it removed his personal papers and files. A disaffected former colleague may have complained to the police. A boy at the house may have cooperated with their investigation. "I am really outraged by these inquiries," Gajdusek wrote in March 1996. He deplored the presence of "such persecuting and inquisitorial inspectors on my tail and on that of our whole family as a result of informants who are jealous, vindictive, and disturbed, probably psychotic." The investigation was making the children anxious and frightened. "It embarrasses them in school and it drives them to academic failure and to consternation." Gajdusek became miserable, distracted, and defensive. "I have obviously lived too long," the seventy-two-year old scientist grumbled. "Where in retirement I hoped to think and read and write in quiet, active, challenging personal research and solitude, I find myself victim of this ethical moral

Gajdusek at Frederick County Sheriff's Office after arrest. AP via AAP [John Mummert] © 1997

inquiry engendered by pathological hate and moral aberration."[13] He could not resist comparing the FBI with the witch hunters of Salem, the KGB, and the Gestapo. "Obviously, I have challenged the Gods by my defiant disregard of Hubris and I can only pray to my Pantheon of Gods to deliver me from this pack of wolves."[14]

On February 18, 1997, in the Frederick County, Maryland, Circuit Court, Gajdusek pleaded guilty to charges of sexually molesting one of his Micronesian "sons," who was sixteen at the time. In so doing, he avoided a long and embarrassing trial and a struggle against growing fears of pedophilia in the community. The plea agreement ensured he would serve less than a year in prison.[15] He retired immediately from the NIH. After his release from prison, he quickly left the United States.

For the remainder of his life as an exile, whether alone in his room in Amsterdam or traveling inexorably, Gajdusek cast himself as a tragic figure. "At this late stage of life, decades beyond where I ever planned to live and beyond the age to which my father lived, I really had no plans and no expectations for anything beyond this, so even this far I continually wonder for the last several years why I am still alive and how inappropriate it is that I am."[16] At the end of the twentieth century, while Stan Prusiner and John Collinge were successfully attaching prions to the British mad cow epidemic and garnering their rewards, Gajdusek was trying to find his way out of infamy and disgrace.

HOW HAD IT come to this?

On the evening of April 2, 1960, surrounded by his boys in Kainantu, Gajdusek was reflecting wistfully on his relations with them, especially the youngest, Mbaginta'o. "All have worked wonderfully for me and I cannot adequately repay them. How much I would like to have them in the U.S. and share my home with them!"[17] A Kukukuku boy lodging at Agakamatasa, Mbaginta'o seemed the most promising candidate for adoption. About the same age as Wolfgang when he came to live with Gajdusek's mother, the young Melanesian was lively, affectionate, and adaptable. A few days later, Gajdusek's thoughts turned again to the subject of patronage. "I only wish I could be more generous to them than I shall be able to be," he wrote in his journal. "They are faithful friends, not just employees, and as such I treasure and value each and every one of them."[18] But Wanevi and Masasa were rapidly becoming independent young men, and therefore ineligible for incorporation into the scientist's family. Little Mbaginta'o—"brash, brazen, subtle, insinuating," like most Kuk youths—was another matter.[19]

In 1963, the colonial administration approved Gajdusek's request to take Mbaginta'o to the United States. Gajdusek wanted him to live in the Maryland suburbs and attend the local school. When they heard this, the other boys became jealous and resentful, but Gajdusek ignored their protests. On their way to America, the scientist and the first of his new "sons" visited Jack and Lois Baker at Mendi in the southern highlands. Jack was "as full of puns, witticisms and urbane humor as ever." The *kiap* and his wife, Gajdusek's former research assistant—both exceptionally "worldly wise and humanistic and easygoing"—took Mbaginta'o into their own family for a time and taught him how to use the toilet and shower, and how to eat at the table and brush his teeth.[20] Traveling back to Bethesda, Gajdusek found he enjoyed the lad's company, his curiosity and excitement.

Gajdusek and local children, Solomon Islands, c. 1963. Photograph courtesy Peabody-Essex Museum, Gajdusek Collection

A few days after landing in America, Gajdusek wrote to Alpers, who was pursuing clinical activities and collecting specimens in the bush south of Okapa. "Mbaginta'o is doing excellently in Washington . . . you would think he lived here all his life!!! He works at the NIH, plays with everyone's children, takes to tutoring in arithmetic, reading and writing, and to English very quickly and well, and enjoys playing with the chimpanzees in the laboratory."[21] On the whole, Gajdusek felt it was a "very unusual and rewarding experience to have him with me."[22] At first Mbaginta'o displayed a passion to learn. He excelled in art and athletics at Georgetown Day School and took pains to spell and write English. He avoided speaking Pidgin and Fore. When Gajdusek traveled, Dick Sorenson or Joe Wegstein, the computer scientist, looked after the boy.[23] But Mbaginta'o—or Ivan Gajdusek as he began to call himself—remained unsettled and troubled in the Maryland suburbs. When Gajdusek visited New Guinea again in 1967, J. K. McCarthy, the chief of the Department of District Administration, "brought up Mbaginta'o, and I told them of his progress, his difficulty with the English language and verbal expression and learning, and his successes with other aspects of adjustments and learning." It was soon clear that the boy would not fit in back in New Guinea—Gajdusek realized that most local adolescents now spoke better English than his American son.[24]

Still, Ivan Mbaginta'o struggled on, studying for a while at Swarthmore and working in his Papa's laboratory at the NIH and then at the Salem Peabody Museum, cataloging its New Guinea collection. He was among the boys who accompanied Gajdusek to Stockholm in 1976. In the 1980s he returned to the highlands, becoming a curator at the J. K. McCarthy Museum in Goroka. Gajdusek helped him buy a house there. Suffering from asthma and heart disease, Ivan Mbaginta'o died suddenly in 2005. He had just written a letter to Gajdusek. "Papa," he typed, "last things I want to say that I love you always and I will never forget you . . . I am very proud to be your first adapted [sic] son from Papua New Guinea. I'm very thankful for every thing that you have given for me such as money, love, freedom to travel the world and countries, life to become who I am now and most of all, education. Without you being there for me, I don't think I would know that EDUCATION is ever existed in my time."[25]

Mbaginta'o was the first of many Melanesian and Micronesian children to pass through Gajdusek's house, none of them formally adopted, most on student visas. As a "father," Gajdusek could be liberal and indulgent one moment and severe the next. Tenderness alternated with admonition, playfulness with discipline. He had his favorites and he could be capricious. The boys' rowdiness and irreverence often slipped into drunkenness and led to sex with girls from the neighborhood. They mostly studied television and played sports and failed school. They kept asking for fast cars and video games. Gajdusek ruminated on their academic struggles:

> It is asking a lot for me to take kids, deracinated as mine are, and hope that from stone age cultures I can push them into a Western mold of high attainment, competitive achievement, so foreign to their own cultures and families. The kids pick up all the U.S. teenage mores, fads and mannerisms and dress styles, feeding them back to me as though they were their own.[26]

Irritated when they "fucked around" with their education, Gajdusek frequently lost patience with them. Just as often, he admired their high spirits and intractability. Sometimes he wished they were more ambitious, other times he just wanted them to live as they pleased:

> It is extremely complicated to know how to advise them and how to fulfill their aspirations and how to give them a happy adult life. And as I try not to invade and make decisions, and advise on marriage and push for success

and attainment, I finally feel that I am deserting them. And when I do these things they tell me that I am bugging them and making them anxious, filled with anxiety. There is no way out of such Catch-22s, and in daring to start such a labor of love, start such an atypical family, I knew I was asking for trouble, and that I have been trying to survive successive catastrophes has not surprised me.[27]

But even as he despaired, one of them would suddenly do something or say something that enchanted him.

Gajdusek made sure his children engaged with the many Nobel laureates, scientists, and anthropologists who dropped in for dinner. The boys cooked hamburgers for Margaret Mead and Claude Lévi-Strauss, telling them picaresque stories about coming of age in Chevy Chase. Gajdusek expected his guests to admire his philanthropic experiment. But often it merely provoked titillation and innuendo. The strange household was the subject of much whispering and winking at the NIH; scientists elsewhere raised their eyebrows and gave knowing looks when Gajdusek and his sons were mentioned.[28] But all except one of the "adapted" children remained loyal to Gajdusek, ready to protect him from his accusers and detractors. As adults, some of them sent their children to live with him—Mbaginta'o, for example, loaned him his own children.

Speculation abounded, but no one could be sure of the precise character of the bonds within this experimental family. In 1977, fleeing the fame of the Nobel Prize, Gajdusek openly expressed the ambivalence of his motives:

> While I have money, I am determined to invest it in education aimed at saving the most exotic and fragile and divergent of small primitive cultures—not "as they were," but as they may become in the modern world, retaining all possible awareness of their roots and of the possibilities open to man. The basic reason for such idealism comes down, however, to the fact that I like to have young boys around me![29]

Evidently some libidinal element inspired his collecting of "primitive" boys, as it did his attempted appropriation of Fore body parts. But this does not mean that he must have had sex with the boys, any more than his constant yearning for specimens implies he really loved the Fore. Then again, both such dénouements are possible.

Gajdusek at home with his sons. AP via AAP © 1976

RETURNING TO BETHESDA after his own expedition to the kuru region, Dick Sorenson wryly observed his friend and boss. "Carleton, as usual, as if pursued by demons, was busily engaging and disengaging, organizing and disorganizing, in his accustomed flap." In particular, the budding anthropologist noticed, "Carleton does like to monopolize the activities and allegiances of all the boys, and he jealously guards this prerogative."[30]

Gajdusek's collecting of children sheds light on how he attempted to realize value in others. His "treasuring" of the boys became another means of making persons, an involuted process of social reproduction that eventually largely substituted for reciprocity in his relations.[31] Like most Melanesians, Gajdusek remained committed to making persons visible—he understood persons as deriving from the agency of others, not as abstracted or separable packages of temperament. Try as he might, he found it difficult—as in the case of the Hagahai cell line—to treat bodies as alienated commodities or persons as alienated selves. In this sense, he was singularly unprepared to cope with the emerging capitalist economies of science and modern American life. Yet at the same time he was also shrinking from the demands of reciprocity, the duress of personification, in his transactions with others, whether scientists or "primitives." When he first adopted the boys he imagined himself repaying a social debt. But by the 1970s, the creation of a family seemed to offer instead a way to avoid such fraught and onerous exchange relations. Gajdusek came to dream of nonreciprocity, of what his guest Lévi-Strauss once called "a world in which one might keep to oneself"—an arena, nonetheless, in which value was reproduced and persons regenerated.[32]

Late in the 1960s, Gajdusek often thought about truncating his relations with others, of cutting his social networks, of asserting self-control. He worried especially about the closeness of some laboratory colleagues. "They develop a parasitic-paternal relationship with me in which I am forced to battle the adolescent search for identity and independence while accepting the burden of guidance and support that they would thrust upon me. It is this I have always to escape."[33] About this time he wrote in his journal that intense relationships, loving relationships, "begin to weigh heavily upon me . . . placing upon me continued personal demands to an extent which fully restricts my ability to plow other soil." The roots of his need for detachment were deep. "Whenever I detect anything akin to my mother's unloosening relationship with me, I flee it as I fled her . . . but never do I escape unscathed." He wished "to explore freely, without the encumbrance of long-established ties; restless, anxious to try to forge new ties, and willing to accept obligation from the old in material assistance but not supervision of my future life and actions."[34] As time wore on, he mostly abandoned material obligations too, finding they were never personally unencumbered. In trying to exercise control over his relationships, to stop actions that made claims on him as a person, he was asserting self-possession. This fantasy of withdrawal from reciprocity and obligation thus was an effort, ultimately forlorn, to condense and manage his own social identity or individuality, a naïve yet irresistible nostalgia for autonomy, always one of Gajdusek's romantic weaknesses. Like Don Quixote he was engaged in a "quest, and for a quest one must divest oneself of old loves."[35]

"I am not built to share another's ego," Gajdusek wrote in 1976, "nor do I require or enjoy such sharing of mine." He took little pleasure in the give and take of friendship and loving relationships. He went on writing in the New Guinea highlands, reflecting unhappily:

> I realize that everyone finally learns that I really 'use' people, that when people no longer want to be 'used,' they and I had best part company . . . Totally independent people I do not 'use,' nor do I strike up friendships, love relationships or collaborative ventures with them, and wisely so . . . If I ceased to 'use' them—which is one way of saying that I truly share my life with

them—our relationship would be meaningless, for I have so fused work-play into one 'life' that there is no possible separation.[36]

Recent disputes with Sorenson had left him miserable and sore:

It largely stems from his seeking to tail, trail and cling to me and my work. I respect this and encourage this, and even want it, but when it is accompanied by unhappiness, resentment, paranoia on both sides and a perpetual chafing, it is high time we dissolved our close association. The same syndrome lies now between me and my carriers and boys. After a while, the same occurs in the laboratory.[37]

Some years afterward:

The great friendships and lavish self-sacrifice, hospitality, and generosity all my colleagues are showing me will be very, very hard to reciprocate in kind. I am truly embarrassed as a result.[38]

It was a question, then, of how to "use" people without them making claims on him as a person.

Late one night in Paris, under stress during the investigation of the sexual molestation rumors, Gajdusek made a striking statement in his journal. "Never in all my dealings with Melanesians and Micronesians," he wrote, "have they ever paid a debt back to me or taken on any of the expenses for their kids I have sponsored, adopted or taken care of."[39] It was a desperate attempt to deny any entanglement with the people of the Pacific, or to admit any obligation to them. He was projecting onto others his own bad faith, thus hoping to erase social debts. It was a withdrawal from exchange relations. Instead, through hoarding and display of the boys he sought social reproduction without the claims of reciprocity:

I feel it is unwise from their point of view as well as my own to be closely involved with their lives. Distant mutual respect . . . is what I crave. I do not want to be in day-to-day confrontation or exchange, nor do I want to work myself into a role of mutual dependency. How can I achieve the distant relationship of mutual gratitude and respect that I have had with John Edsall and John Enders, with Michael Heidelberger and Max Delbrück, with Linus Pauling and Joe Smadel, with Tante Irene and William J.

Youden . . . In none of these did I ever launch a demand or construct an expectation which left me disappointed or bitter, but only that of a deep love and respect for what they have done for me. [40]

He would keep the boys to himself, possessed by him yet visible to others. In this way he could still circulate the unstable hybrid identities of wild white men and cosmopolitan primitives, of scientist sorcerers and specimen sons. He could continue to realize and reproduce value in others, to make other persons, without exposing himself as a person.

But as soon as he imagined this particular dénouement, he saw it was absurd. "As I think back on my heroes from mythology through the ancients to the moderns," he continued, "I see that they were all human, all torn by the passions of being human and the problems these present . . . The major message I can read is not to let a few failures in human relationships and the despair and depression they produce, override the joy of life and love, and passionate devotion."[41] He convinced himself he would always be implicated in the lives of others. Then came his arrest and imprisonment.

And so we leave him, alone now in his room in Amsterdam or inexorably traveling. Like Lord Jim, then, "he passes away under a cloud, inscrutable at heart, forgotten, unforgiven, and excessively romantic." Like Jim, too, "an obscure conqueror of fame."[42]

ACKNOWLEDGMENTS

Many years ago, as a medical student and then an intern on the Clinical Research Unit of the Hall Institute in Melbourne, Australia, I heard about kuru and Carleton Gajdusek from Ian Mackay and John Mathews, my teachers. Another ten years would pass before I found Gajdusek's journals in the anthropology library at the University of Pennsylvania, where I was studying for my Ph.D. in the history of science and medicine. I recklessly seized on kuru as the topic for my dissertation, but my adviser, Charles Rosenberg, wisely warned me against it, suggesting I leave it for a later book, when my skills might better match the story. In their graduate seminar on the history of field work, Rob Kohler and Riki Kuklick also sensibly urged caution. On the other hand, Bruno Latour, after reading an early essay, charitably advised me to go ahead, quickly.

Since 1989, then, I have been thinking about kuru, though biding my time. While an assistant professor at Harvard University I received encouragement from Arthur Kleinman, Mary Steedly, Anna Tsing, Allan Brandt, and Anne Harrington. Peter Galison casually told me I should look into economic anthropology—a revelation for me, if not for him. Later, at Melbourne University, Martha Macintyre introduced me to the complexities of Melanesian anthropology, inspiring me to take the kuru project more seriously. In 1999 I wrote an essay on the investigation of kuru for a seminar at the Davis Center, Princeton University, which was published the following year in *Comparative Studies in Society and History*. I am grateful to Gyan Prakash, Suman Seth, and others for their comments at Princeton, and to the editors and anonymous reviewers of the journal for their advice. Matthew Klugman was an estimable research assistant during my years at Melbourne.

I believe I owe my appointment at the University of California, San Francisco, to kuru: the topic greatly appealed to the mixed group of Bay Area anthropologists, sociologists, and historians interested in medicine. Adele Clarke, Philippe Bourgois, and Vincanne Adams insisted I write the book, and it proudly bears their imprint. It is hard to imagine a more stimulating and supportive intellectual environment for this sort of work. I thank Judith Barker, Lawrence Cohen, Matthew Gilmartin, Sharon Kaufman, Tom Laqueur,

Paul Rabinow, Nancy Rockafellar, Eunice Stephens, James Vernon, the graduate students of the medical anthropology program, and other Bay Area scholars for helping me to frame this project. From the beginning, Marilyn Strathern has guided me through the intricacies of exchange among Melanesians and scientists. Caroline Hannaway, Vicki Harden, and Buhm Soon Park encouraged me to delve deeper into NIH history.

More recently, I benefited from the comments and advice of my colleagues at the University of Wisconsin–Madison, in particular Joan Fujimura, Ken George, Linda Hogle, Maria Lepowsky, Gabriela Soto Laveaga, Rob Nixon, and Gabriele zu Rhein. A succession of dedicated research assistants— Mitch Aso, Fae Dremock, Bridget Collins, and Lucienne Loh—allowed me to keep writing despite the demands of chairing the Department of Medical History and Bioethics. Jean von Allmen, Lorraine Rondon, and Sharon Russ greatly eased the administrative load, though not the psychic burden, of the job. Carol Dizack prepared many of the images into a form suitable for publication. Funds from the Robert Turell chair assisted research for this book.

A grant from the National Science Foundation (SES-0114731) permitted me to conduct the necessary digging in the United States, Australia, and New Guinea. I also received grants from the Rockefeller Archive Center for travel to its collections. The American Council of Learned Societies awarded me a Frederick Burkhardt fellowship for 2005–6, held in the School of Social Sciences at the Institute for Advanced Study, Princeton. I was able to write the first half of the book at Princeton, helped especially by discussions with Clifford Geertz, Heinrich von Staden, Joan Scott, Caroline Bynum, Michael Peletz, Rena Lederman, and Angela Creager. For my time there (when I wasn't in New York enjoying Elena del Rivero's hospitality), the staff at the IAS allowed me to be a very productive hermit in the woods.

I have spoken about kuru at countless seminars and workshops, and the responses of the audiences shaped this narrative in manifold ways. I am particularly grateful to Chris Crenner for inviting me to give the 2006 Alice V. and Don Carlos Peete Lecture at the University of Kansas Medical School. In 2007 I was the inaugural Wellcome Visiting Senior Professor at the University of Manchester, where I delivered an Astra-Zeneca lecture: my thanks to Mick Worboys, John Pickstone, Elizabeth Toon, Emm Barnes, and others at the Centre for the History of Science, Technology, and Medicine for sharing with me their unrivalled knowledge of twentieth-century biomedical sciences. Kihueng Kim invited me to participate in his workshop on prions at the Wellcome Trust Centre for the History of Medicine at University College London, which greatly enlarged my understanding of the proteinaceous aspects of the kuru story. Thanks to Michael Alpers and John Collinge, I was able to attend their meeting on the End of Kuru at the Royal Society, London, late in 2007.

I have been extraordinarily fortunate to find a group of learned, engaged, and tough readers for this manuscript. I would like to thank Michael Alpers, Judith Farquhar, Janet Golden, Maria Lepowsky, Shirley Lindenbaum, Ian Mackay, John and Coralie Mathews, Hank Nelson, Lisa O'Sullivan, Mark Veitch, and Fiona Wilson—they read every word and offered extensive commentary. Their intellectual generosity humbles me. Geoffrey Gray read the first two chapters, and Ceridwen Spark offered advice on the conclusion. I am also grateful to Jacqueline Wehmueller and the two anonymous readers at the Johns Hopkins University Press for their guidance and patience.

I want to express my appreciation of the many librarians and archivists who have assisted me at collections in North America, Australia, and New Guinea. Kathy Creely at the Melanesian Studies Resource Center, University of California at San Diego; Charlie Greifenstein at the American Philosophical Society; and Widya Paul at the Institute for Medical Research in Goroka were especially helpful. In difficult circumstances, Sidney Berger and Christine Michelini gave me access to the Peabody-Essex Museum's extensive holdings of Carleton Gajdusek's photographs relating to kuru.

Without the assistance and persistence of Jim Hoesterey, this book would not be illustrated. A historian *manqué*, he scoured the storage rooms of the Salem Peabody, turning up some amazing images. Jim also compiled the index.

Tom Strong taught me about highland sociality and accompanied me on the first visit to the Fore, where he did most of the interpreting and translating. For the past five years he has helped to give direction to my meandering research, chiding me about the need to do justice to the kuru story. He was more a collaborator than a research assistant.

My deepest thanks go to the many scientists, anthropologists, and Fore people I have interviewed. In particular, I have been in regular and rewarding conversation for many years with Carleton Gajdusek, Michael Alpers, Shirley Lindenbaum, and John Mathews who have done their best to make sure I get the story and the science right. I could not have completed this book without the generous assistance of the Fore who took time to speak with me and explain the kuru investigations. Anderson Puwa and Jerome Whitfield arranged my research trips south of Okapa, while John Reeder and Deb Chapman provided hospitality in Goroka. My encounters with Fore showed me the tragic import of kuru, its true significance.

ABBREVIATIONS

ALS	amyotrophic lateral sclerosis
APS	American Philosophical Society
ANU	Australian National University
BSE	bovine spongiform encephalopathy
CJD	Creutzfeldt-Jakob disease
DNA	deoxyribonucleic acid
FBI	Federal Bureau of Investigation
FFI	fatal familial insomnia
GSS	Gerstmann-Sträussler-Scheinker syndrome
HGDP	Human Genome Diversity Project
IBP	International Biological Programme
IMR	(Papua New Guinea) Institute of Medical Research, Goroka
NFIP	National Foundation for Infantile Paralysis
NIH	National Institutes of Health
NINDB	National Institute of Neurological Diseases and Blindness
NINDS	National Institute of Neurological Disorders and Stroke
NLM	National Library of Medicine
PDC	parkinsonism-dementia complex (of Guam)
PNG	Papua New Guinea
PNGNA	Papua New Guinea National Archives
PrP	prion protein
RNA	ribonucleic acid
TSE	transmissible spongiform encephalopathies
UMA	University of Melbourne Archives
VE	Viliuisk encephalomyelitis

NOTES

INTRODUCTION. THE DISEASE EUROPEANS CATCH FROM KURU

1. Gajdusek, April 1, 1965, in *Melanesian Journal, January 23, 1965–April 7, 1965,* 124. Gajdusek is pronounced Guy-du-shek.

2. D. Carleton Gajdusek to J. E. Smadel, August 25, 1957, in Farquhar and Gajdusek, *Kuru,* 121.

3. Gajdusek, February 15, 1965, in *Melanesian Journal,* 53.

4. For other accounts of kuru investigation, see Denoon, Dugan, and Marshall, *Public Health in Papua New Guinea;* Nelson, "Kuru"; Rhodes, *Deadly Feasts;* and Spark, "Learning from the locals."

5. Rosenberg, *Cholera Years;* Rosenberg, "Cholera in nineteenth-century Europe"; idem, "What is an epidemic?"; idem, "Framing disease"; and idem, "Explaining epidemics."

6. Lindenbaum, *Kuru Sorcery;* and Lindenbaum, "Kuru, prions, and human affairs."

7. For an earlier version of this argument, see Anderson, "Possession of kuru." While there is other work on material cultures of scientific exchange, none draws so heavily on the economic anthropology developed in studying Melanesian societies—here applied to scientists. See Oudshoorn, "On the making of sex hormones"; Findlen, "Economy of scientific exchange in early modern Italy"; Biagioli, *Galileo Courtier;* Kohler, *Lords of the Fly;* Clarke, "Research materials and reproductive science"; Galison, *Image and Logic;* and Strasser, "Collecting and experimenting." For an overview of the macroeconomy of science, see Mirowski and Sent, *Science Bought and Sold.*

8. This point is not new: see Fox and Swazey, "Medical morality is not bioethics"; Fox and Swazey, "Examining American bioethics"; Fox, "Evolution of American bioethics"; Rosenberg, "Meanings, policies, and medicine"; Kleinman, "Moral experience and ethical reflection"; and Meskell and Pels, *Embedding Ethics.* On moral sensibilities, see Geertz, "Found in translation." On structures of feeling, see Williams, *Marxism and Literature.*

9. I have more to say about this approach in Anderson, "Postcolonial technoscience"; and Anderson and Adams, "Pramoedya's chickens." For an elaboration of some of these ideas, see McNeil, "Postcolonial technoscience." For a critique, see Abraham, "Contradictory spaces of postcolonial technoscience."

10. Gajdusek, January 26, 1967, in *South Pacific Expedition,* 3. The quotation is from Melville, *Moby Dick,* 15.

CHAPTER 1. STRANGER RELATIONS

1. The term *Fore* was used by people in Kainantu and Henganofi to refer to those who live in the south before it became formalized as a linguistic grouping.

2. G. W. Toogood, Report of Patrol Covering the South-Western Area—Kainantu, No. 1

of 1949/50, Central Highlands District, Box 4, MSS 0215: Papua New Guinea Patrol Reports, Melanesian Studies Resource Center, Mandeville Special Collections Library, University of California at San Diego. (Unless otherwise noted, all later patrol reports can be found here.) Earlier, R. I. Skinner had ventured as far south as Moke, in the north of the Fore region, before heading east: see Patrol Report No. 5 of 1949/50, Central Highlands District.

3. Souter, *New Guinea;* Griffin, Nelson, and Firth, *Papua New Guinea;* Latukefu, *Papua New Guinea;* and Moore, *New Guinea.*

4. Chinnery, "Mountain tribes"; and Leahy, "Stone-age people."

5. Flierl, *Christ in New Guinea;* Leahy, *Explorations into Highland New Guinea;* and Hides, *Papuan Wonderland.* See also Nelson, *Black, White and Gold;* Connolly and Anderson, *First Contact;* Schieffelin and Crittenden, *Like People You See in a Dream;* and Gamage, *Sky Travellers.*

6. Gamage, "Police and power." Taylor established the first administrative outpost in the highlands at Kainantu in 1932, and in 1946 he became the first district officer in charge of the Central Highlands District, containing almost half the population of New Guinea.

7. Radford, *Highlanders and Foreigners.*

8. Ian Downs claimed that all patrols into previously uncontacted areas were "tested by limited aggression," but "when all went well our gifts overcame fear and gave us time to satisfy curiosity" (*Last Mountain,* 115, 114). Downs was district officer of the Eastern Highlands District (created in 1951) between 1952 and 1956.

9. A *laplap* is a skirt-like item of clothing.

10. McCarthy on this and a later patrol established a post at Menyamya on the Tauri River in the heart of Kukukuku territory, but it was abandoned in 1935 and reopened only in 1950. See McCarthy, *Patrol into Yesterday.* In 1937 on a patrol among the Kukukuku, Downs found them "disciplined and economic fighters" and "hungry for steel" (*Last Mountain,* 3). In the interests of historical accuracy, because they have no name for themselves, I have used "Kukukuku" (pronounced cooka-cooka) to designate these people, even though it is a derogatory term applied to them by adjacent groups. The name "Anga"—a suggestion of D. Carleton Gajdusek and Dick Lloyd (from the Summer Institute of Linguistics)—was not commonly used until the 1970s.

11. McPherson, "'Wanted: young man, must like adventure.'"

12. McLaren, "In the footprints of Reo Fortune." The little that Fortune later wrote about the Kamano reveals his disgust at their violence: see "Rules of relationship behavior," and "Law and force in Papuan societies." Mead (in *Blackberry Winter*) argued that Fortune's own personality drew him to other violent people. Fortune had earlier written an ethnographic classic, *Sorcerers of Dobu.* After he left the highlands, he returned to the Arapesh and wrote a critique of Mead's claims for their nonviolence: "Arapesh warfare." Fortune eventually settled down after World War II as a lecturer in anthropology at Cambridge. See also Gray, "'Being honest to my science'"; and Roscoe, "Margaret Mead, Reo Fortune, and Mountain Arapesh warfare."

13. Blackwood, *Kukukuku of the Upper Watut.* See also Blackwood, "Life on the Upper Watut." Because of violence, the highlands were effectively closed to whites without special permission between 1935 and 1952. Blackwood had earlier written *Both Sides of the Buka Passage.* She spent the rest of her career as a beloved lecturer in anthropology at Oxford. See Knowles, "Reverse trajectories."

14. For example, an influenza epidemic swept through the highlands in 1936, and a dysentery epidemic in 1943.

15. After World War II, Australia held New Guinea as a trusteeship of the United Nations. See Hasluck, *A Time for Building;* and Downs, *Australian Trusteeship.*

16. G. Linsley, Patrol Report Kainantu Sub-District, No. 3 of 1950/51, Central Highlands District, 1. For a detailed account of early contact with the Fore, see Nelson, "Kuru."

17. Linsley, Patrol Report Kainantu Sub-District, No. 3 of 1950/51, Central Highlands District, 13.

18. Ibid., 16. The persistence of sorcery also frustrated K. I. Morgan, Patrol Report, No. 1 of 1951/52, Central Highlands District, 3.

19. W. J. Kelly, Patrol Report Kainantu Sub-District, No. 8 of 1951/52, Eastern Highlands District.

20. R. R. Havilland, Patrol Report, No. 3 of 1952/53, Eastern Highlands District.

21. J. R. McArthur, Report of a Patrol Originating at Okapa, No. 4 of 1954/55, Eastern Highlands District, 9. Mathews refers to McArthur's 1953 patrol in "Kuru: A puzzle in cultural and environmental medicine," 11.

22. John Colman, Report of a Patrol of the South Fore Census Area, No. 14 of 1954/55, Eastern Highlands District, 6, 7.

23. Linsley, Patrol Report No. 3 of 1950/51, 16.

24. Havilland, Patrol Report No. 3 of 1952/53, 11.

25. Campbell, "Anthropology and the professionalisation of colonial administration"; and Westermark, "Anthropology and administration." F. E. Williams was the government anthropologist in Papua (1922–43), and E. W. P. Chinnery his counterpart in New Guinea (1924–38). See Stocking, "Gatekeeper to the field"; and Gray, "There are many difficult problems."

26. Radcliffe-Brown, "Applied anthropology"; Firth, "Anthropology and native administration"; Elkin, "Notes on anthropology"; and R. Berndt, "Anthropology and administration." In his 1930 address, Radcliffe-Brown may have been the first to use the term "applied anthropology." See also Malinowski, "Practical anthropology."

27. Among their teachers at ASOPA were Camilla Wedgwood, H. Ian Hogbin, and Kenneth Read. Peter Lawrence, Elkin's successor at Sydney as professor of anthropology, expected the courses to "impress upon those attending them that they should regard the natives as human beings, respect their customs . . . see these customs as a distinct and organized way of life" ("Social anthropology and the training of administrative officers," 198).

28. Sinclair, *Kiap.*

29. Elkin, *Social Anthropology in Melanesia,* 15, 14, 76, 142. See also R. Berndt and C. Berndt, "A. P. Elkin—the man and the anthropologist"; Wise, *The Self-Made Anthropologist;* and Gray, *Before It's Too Late.* Although some of Elkin's statements echo those of Malinowski—especially the latter's desire to "grasp the native's point of view, his relation to life, to realize his vision of his world" (*Argonauts of the Western Pacific,* 25)—he was never a functionalist. Perhaps the most significant intellectual influence on Elkin was W. H. R. Rivers's study of the psychological causes of indigenous depopulation.

30. Nadel, *A Black Byzantium,* vi. See also Faris, "Pax Britannica and the Sudan." On British social anthropology see Kuper, *Anthropology and Anthropologists;* Kuklick, *The Savage Within;* and Stocking, *After Tylor.*

31. Read, *High Valley*. Read spent two years in the early 1950s among the Gahuku, "a highly demonstrative, extroverted, and aggressive people" (19).

32. Barnes, Nadel's successor at the ANU, wrote of the "African mirage in New Guinea" ("African models in the New Guinea highlands," 5). More generally, see Hays, "Historical background to anthropology in the New Guinea highlands"; and Jaarsma, "Conceiving New Guinea."

33. Read, *Return to the High Valley*.

34. Nadel offered them a research fellowship, but they refused. Instead they used their New Guinea research for Ph.D. theses through the London School of Economics, with Firth as supervisor. Sydney did not yet offer the Ph.D. in anthropology.

35. Gray, "'You are . . . my anthropological children.'" During the war, J. B. Cleland, the leader of the Adelaide anthropologists, had questioned Ronald's patriotism and wondered if his German ancestry might lead him to favor Nazism.

36. Leahy and Crain, *The Land That Time Forgot;* and Hides, *Through Wildest Papua.*

37. R. Berndt, *Excess and Restraint,* viii.

38. R. Berndt, "Into the unknown!" 72. See also C. Berndt, "Journey along mythic paths."

39. R. Berndt to Mr. and Mrs. Crawford, January 20, 1952, 1951–52 Correspondence—New Guinea Folder, Berndt Papers, Berndt Museum of Anthropology, University of Western Australia.

40. R. Berndt, *Excess and Restraint,* viii, ix.

41. Catherine Berndt to Mr. and Mrs. Tews, December 21, 1951, 1951–52 Correspondence—New Guinea Folder, Berndt Papers.

42. R. Berndt, *Excess and Restraint,* vii. The transference in this passage is striking.

43. Ronald Berndt to S. Woodward-Smith, February 3, 1952, 1951–52 Correspondence—New Guinea Folder, Berndt Papers.

44. Ann McLaren, Fortune's niece, favors the compensation explanation in "In the footprints of Reo Fortune." In 1953, Ronald Berndt tried to get Elkin to prevent James B. Watson from working near Kainantu, as he wanted to claim the whole district as his own anthropological territory: see Gray, "'You are . . . my anthropological children,'" 95.

45. Ronald Berndt to Professor [Elkin], [December 1951], 1951–52 Correspondence—New Guinea Folder, Berndt Papers.

46. The importance of exchange relations was recognized in later (mostly western and southern) highland ethnographies. See, for example, Ryan, "Gift exchange in the Mendi Valley"; A. Strathern, *The Rope of Moka;* Lederman, *What Gifts Engender;* and M. Strathern, *Gender of the Gift.*

47. R. Berndt, *Excess and Restraint,* ix.

48. K. E. Read to Ronald Berndt, June 28, 1951, 1951–52 Correspondence—New Guinea Folder, Berndt Papers.

49. Ronald Berndt to Mr. Smith and Mr. Nelson, [n.d.], 1951–52 Correspondence—New Guinea Folder, Berndt Papers.

50. Ronald Berndt to Professor [Elkin], January 6, 1952, 1951–52 Correspondence—New Guinea Folder, Berndt Papers. Yet in 1947 the Berndts had made the same sort of complaint against the "grasping" Aboriginal people at Yirrkala, Arnhem Land; see Gray, "'You are . . . my anthropological children,'" 93.

51. R. Berndt, *Excess and Restraint,* ix.

52. Ronald Berndt to Ralph Linton, April 3, 1953, 1952–53 Correspondence—New Guinea Folder, Berndt Papers.

53. Catherine Berndt to Phyllis Kaberry, [n.d.], 1952–53 Correspondence—New Guinea Folder, Berndt Papers.

54. Catherine Berndt to Janet Cotton, [1953], 1952–53 Correspondence—New Guinea Folder, Berndt Papers.

55. Ronald Berndt to Raymond Firth, [1953], 1952–53 Correspondence—New Guinea Folder, Berndt Papers.

56. Interview with Catherine Berndt, August 1992. Camilla Wedgwood told Catherine that she wished "you had a less revoltingly unpleasant people to work among" (C. H. Wedgwood to Catherine Berndt, September 29, 1952, 1952–53 Correspondence—New Guinea Folder, Berndt Papers). Edmund Leach later expressed interest in "your singularly unpleasant New Guinea cannibals" (E. R. Leach to Ronald Berndt, October 21, 1953, 1952–53 Correspondence—New Guinea Folder, Berndt Papers).

57. The Berndts resumed distinguished careers in Aboriginal Australian anthropology, founding the Department of Anthropology at the University of Western Australia.

58. Barnes, "African models," 6.

59. In *Excess and Restraint*, Ronald Berndt emphasized force more than exchange in the acquiring of status. Marshall D. Sahlins's influential essay on "big men" and ceremonial exchange was not published until 1963: "Poor man, rich man, big-man, chief." See Strathern and Stewart, *Arrow Talk*.

60. C. Berndt, "Journey along mythic paths."

61. A. Strathern, "Looking backward and forward," 250.

62. Interview with Catherine Berndt, August 16, 1993, Perth, Western Australia.

63. R. Berndt, *Excess and Restraint*, 209.

64. Ibid., 218–19.

65. R. Berndt, "A 'devastating disease syndrome,'" 25.

66. Catherine Berndt to Warwick Anderson, September 3, 1990, in the possession of Warwick Anderson. See also C. Berndt, "Journey along mythic paths." Ronald Berndt claimed that he had been offered some partly cooked flesh from a kuru victim but that he was too squeamish to eat it ("Into the unknown!" 93).

67. R. Berndt, *Excess and Restraint*, 231.

68. Ibid., 270, 271.

69. Ronald Berndt to [Elkin], [1952], 1951–52 Correspondence—New Guinea Folder, Berndt Papers.

70. Arens, *Man-Eating Myth*, 99. Arens argues that "the list of New Guinea cannibals and the recorders of their unseen deed is almost endless" (98).

71. R. Berndt, "Reaction to contact," 228.

72. Ibid.; and C. Berndt, "Socio-cultural change."

73. R. Berndt, "Cargo movement," 57. On guria more generally as "a culturally determined expression of a variety of excitatory themes including physical illness and inter-personal and ecological tensions," see Hoskin, Kiloh, and Cawte, "Epilepsy and guria." For a later treatment of the same topic in northern Australia, see R. Berndt, *Adjustment Movement in Arnhem Land*.

74. R. Berndt, "Reaction to contact," 226.

75. R. Berndt, "Cargo movement," 65. *Kiaps* became more active in the breaking up of cargo movements in the late 1950s.

76. Kelly, Patrol Report Kainantu Sub-District, No. 8 of 1951/52, 6, 7.

77. G. H. Waldron, Forae-Kumano Linguistic Group, August 5–20, 1952, file 21/29/300: Patrol Reports, Eastern Highlands-Kainantu, box 3120, series 149, accession 23, Papua New Guinea National Archives (PNGNA), Port Moresby, PNG.

78. This was the opinion of John Colman, Report of a Patrol of the South Forei Census Area, 1956, Eastern Highlands District, 4.

79. Hibberd, Patrol Report, No. 6 of 1953/54, 4.

80. H. W. West to District Commissioner, Eastern Highlands, July 8, 1955, attached to John Colman, Report of a Patrol of the South Fore Census Area, No. 14 of 1954/55, Eastern Highlands District.

81. H. F. Earl, South Gimi, South Forei and Moke, June 21–August 5, 1956, file 21/29/300: Patrol Reports, Eastern Highlands-Kainantu, box 3120, series 149, accession 23, PNGNA. See also John Colman, Report of a Patrol of the South Forei Census Area, 1956, Eastern Highlands District. Colman asked for a "native hospital," staffed by a European medical assistant, at Moke.

82. Denoon, Dugan, and Marshall, *Public Health in Papua New Guinea.* Denoon observes that Gunther's "abrasive manner woefully failed to conceal his compassion" (115). See also Scragg, "From medical tultul to doctor of medicine."

83. Gunther, "Post-war medical services in Papua New Guinea," 60; Gunther, "Public health problems of New Guinea," [c. 1955], file 54/6/12, box 5361, series 153, accession 23, PNGNA; and idem, "Medical services, history."

84. Scragg, "Sir John Gunther"; and Griffin, "John Gunther and medicine." Gunther became assistant administrator of Papua New Guinea in 1957 and the first vice-chancellor of the University of Papua New Guinea in 1966.

85. The description of McArthur is Zigas's: see *Laughing Death.* Gajdusek, in his foreword, fondly describes the book as "abstract expressionist ironical parody in the form of a historical and biographical novel" (vi). For an embittered recollection of Zigas by his former wife, see Chalmers, *Kundus, Cannibals and Cargo Cults.*

86. Zigas, *Laughing Death,* 163, 174.

87. De Derka, who boasted a Ph.D. in philosophy from Budapest, knew Zigas well. She and her mother lived in a *gemütlich* apartment in the hills above Ela beach (on which she sometimes held court). Gajdusek also became part of the de Derka circle and one of her "pets." At Gunther's request, she later organized the library at the University of Papua New Guinea. She retired to Rome, then Oxford.

88. French et al., "Murray Valley encephalitis."

89. Zigas, *Laughing Death,* 100.

90. Zigas to director, Public Health Department, December 25, 1956, Gunther Papers, in the possession of Hank Nelson.

91. Charles Julius to director of public health, February 28, 1957, 2, 6, 8, Gunther Papers. Julius had been Elkins's student at Sydney before World War II.

92. Ibid., 9.

93. Interview with Masasa, August 8, 2003, Okapa. Masasa's father became the first *luluai* at Yagusa.

94. M. Strathern, "The decomposition of an event." Marilyn Strathern writes that "what convinced the Hagen men that the Australians were human lay in the things they brought" (247). They had "the kind of things that would move men" (249).

95. Trouillot, "Anthropology and the savage slot." This characterization is not entirely fair, since the Berndts did try to emphasize social change and adjustment in some of their accounts of the Fore—on the other hand, they took pains to illustrate Fore "savagery" too. I am not referring here to later anthropologists and medical investigators who worked in the region, as will become clear in subsequent chapters. Also, it seems that the *kiaps* based at Okapa after 1954—McArthur, Colman, and Jack Baker—soon became sensitive and sympathetic observers of the Fore, far more so than required in any effort to control these people.

96. The quotation is from Dening, *History's Anthropology*, 99.

CHAPTER 2. PORTRAIT OF THE SCIENTIST AS A YOUNG MAN

1. Interview with D. C. Gajdusek, March 25, 2005, Amsterdam. Provine argues that "Huxley's discussion of evolution was the single most encompassing presentation of a neo-Darwinian viewpoint available in 1930," and he suggests that "the influence of *The Science of Life* on scientists and the educated public deserves careful study by historians" ("Introduction to Section 11: England," 332).

2. Keller, *A Feeling for the Organism;* and Comfort, *Tangled Field.*

3. Robin became a poet and taught at San Francisco State University.

4. "My intellectual curiosity and largely the nature of my emotional responses I attribute in great proportion to the environment which Mother has supplied. Daddy I respect, but know that his influence in directing my mental development has been small" (D. C. Gajdusek, February 3, 1944, in "Nascent physician: Harvard Medical School, Boston Children's Hospital, Babies' Hospital Columbia Presbyterian Medical Center, August 6, 1943–April 18, 1946" [typescript in the possession of Warwick Anderson], 8).

5. D. C. Gajdusek to Mahtil Gajdusek, April 24, 1943, box 15, D. Carleton Gajdusek Correspondence, ms. C565, National Library of Medicine (NLM), Bethesda, MD. (I originally consulted this collection through the Gajdusek Papers, ms. 58, American Philosophical Society, but since the NLM copies are the only ones open to scholars, I will refer to this duplicate source.)

6. D. C. Gajdusek to Mahtil Gajdusek, January 29, 1942, box 15, Gajdusek Correspondence, NLM.

7. D. C. Gajdusek to Mahtil Gajdusek, November 25, 1942, box 15, Gajdusek Correspondence, NLM.

8. D. C. Gajdusek to Mahtil Gajdusek, September 26, 1943, box 15, Gajdusek Correspondence, NLM. Wilson was professor of vital statistics at the Harvard School of Public Health, with a special interest in mathematical aspects of epidemiology. Edsall was the leading protein chemist at Harvard and the editor of the *Journal of Biological Chemistry.*

9. D. C. Gajdusek to Mahtil Gajdusek, April 11, 1943, box 15, Gajdusek Correspondence, NLM.

10. D. C. Gajdusek, November 30, 1945, in "Nascent physician," 9.

11. Qin became professor of pediatrics and a leading hematologist at Beijing Medical Col-

lege. She died in 2004. I am grateful to Marta Hanson for tracking down the Chinese elements of the story.

12. D. C. Gajdusek to Mahtil Gajdusek, September 20, 1947, box 15, Gajdusek Correspondence, NLM.

13. Pauling was awarded the Nobel Prize in Chemistry in 1954 for his research into the nature of the chemical bond and the Nobel Peace Prize in 1962. Kirkwood was a physical chemist who later worked in polymer chemistry and the mechanics of molecular transport processes. He was at Caltech between 1947 and 1951, before he became professor of chemistry at Yale.

14. Burnet, "'Smooth-rough' variation"; and Burnet and Lush, "Induced lysogenicity and mutation." Burnet received the Nobel Prize in Physiology or Medicine in 1960 for his discovery of immunological tolerance. It might be argued that H. J. Muller anticipated Burnet in recognizing that phage could reveal genetic mechanisms: see "Variation due to change in an individual gene."

15. Lederberg, "Infectious history."

16. Delbrück, "A physicist looks at biology." Delbrück believed that Bohr's complementarity principle would reveal the limits to the reductionist project, so much so that he worried that it would open the door to "wild and unreasonable speculations of a vitalistic kind" (22). See also Fleming, "Emigré physicists and the biological revolution"; Yoxen, "Schrödinger's *What is Life?*"; Roll-Hansen, "Application of complementarity to biology"; and McKaughan, "Influence of Niels Bohr on Max Delbrück." Delbrück shared the Nobel Prize in Physiology or Medicine in 1969 for discoveries concerning the replication method and genetic structure of viruses.

17. Kay, *Molecular Vision of Life*, 255. See also Olby, *Path to the Double Helix;* Judson, *Eighth Day of Creation;* Keller, "Physics and the emergence of molecular biology"; de Chadarevian, "Sequences, conformation, information"; and Morange, *History of Molecular Biology.*

18. Cairns, Stent, and Watson, *Phage and the Origins of Molecular Biology.* Watson shared the Nobel Prize in Physiology or Medicine in 1962 for the discovery in 1953 of the molecular structure of DNA: see Watson, *Double Helix.* Stent became professor of molecular biology at the University of California at Berkeley: see Stent, *Molecular Genetics;* and Stent, *Nazis, Women, and Molecular Biology.*

19. Kay, *Molecular Vision of Life*, 256.

20. D. C. Gajdusek to Mahtil Gajdusek, January 22, 1949, box 15, Gajdusek Correspondence, NLM.

21. D. C. Gajdusek to Mahtil Gajdusek, November 1, 1948, box 15, Gajdusek Correspondence, NLM.

22. Watson, "Growing up in the phage group," 243.

23. Van Helvoort, "What is a virus?"; and Creager, *Life of a Virus.*

24. Van Helvoort, "Controversy between John H. Northrop and Max Delbrück."

25. Avery, MacLeod, and McCarty, "Induction of transformation," 152.

26. Hershey and Chase, "Independent functions of viral protein and nucleic acid." See also Hershey, "Injection of DNA into cells by phage." By the late 1950s, the consensus was that a virus consists of nucleic acid surrounded by protein: see Crick and Watson, "Structure of small viruses."

27. Gajdusek to Corydon B. Dunham, June 8, 1947, box 11, Gajdusek Correspondence,

NLM. Dunham grew up to become executive vice president and general counsel for NBC and write *Fighting for the First Amendment*.

28. D. C. Gajdusek to Mahtil Gajdusek, November 29, 1950, box 15, Gajdusek Correspondence, NLM.

29. Enders shared the Nobel Prize in Physiology or Medicine in 1954 for the cultivation of the polio virus in 1948. He also worked at growing measles and mumps viruses at Boston Children's Hospital.

30. Robert S. Morison, August 25, 1955, in "Diaries," Rockefeller Foundation Archives, Rockefeller Archive Center, Tarrytown, New York, 142. Morison later describes Gajdusek as a "person of almost unbelievable energy and mental brilliance" with a "talent for turning up in various parts of the world where unusual events of medical importance are taking place." But he observed "his speedy and energetic ways of doing things tends to rub the more average type of individual the wrong way and he may develop sufficient opposition so that many of his objectives will become unobtainable" (September 16, 1958, 88, 90).

31. Smadel to Richard Masland, NINDB, January 10, 1958, in Farquhar and Gajdusek, *Kuru*, 274.

32. D. C. Gajdusek to Mahtil Gajdusek, [c. 1955], box 15, Gajdusek Correspondence, NLM. In his journal Gajdusek wrote: "I find it uncomfortable to work for him and basically his freedom-granting is spurious. I can only be grateful to Dr. Smadel, but I must get out" (April 3, 1955, in "Return to Washington DC: Army Medical Service Graduate School, Walter Reed Institute of Research, March 25, 1955–August 9, 1955" [typescript in the possession of Warwick Anderson], 3).

33. Smadel received his M.D. from Washington University, St. Louis, and then worked at the Rockefeller Institute between 1934 and 1946. In 1956 he became associate director for intramural programs at the National Institutes of Health. He received the Lasker Award in 1962 for his research into the treatment of typhus. See Harden, *Rocky Mountain Spotted Fever*.

34. D. C. Gajdusek to Mahtil Gajdusek, May 31, 1950, box 15, Gajdusek Correspondence, NLM.

35. D. C. Gajdusek to Mahtil Gajdusek, September 17, 1950, box 15, Gajdusek Correspondence, NLM.

36. D. C. Gajdusek to Mahtil Gajdusek, April 6, 1953, box 15, Gajdusek Correspondence, NLM.

37. D. C. Gajdusek to Barry Adels, November 13, 1952, box 1, Gajdusek Correspondence, NLM. Adels trained as a pediatrician and became director of medical services at Mt. Holyoke College in Massachusetts.

38. D. C. Gajdusek to Mahtil Gajdusek, October 7, 1948, box 15, Gajdusek Correspondence, NLM.

39. D. C. Gajdusek to Mahtil Gajdusek, April 1, 1953, box 15, Gajdusek Correspondence, NLM.

40. A protein chemist working with Pauling, Dunitz became a professor at the ETH-Zürich. Ruge was a peace and spiritual activist based in Mexico who advocated world government, under his own direction.

41. D. C. Gajdusek to Mahtil Gajdusek, [c. 1952], box 15, Gajdusek Correspondence, NLM. See Gajdusek and Rogers, "Specific serum antibodies."

42. See Gajdusek, "Hemorrhagic fevers in Asia."

43. Gajdusek to Drs. Woodward and Smadel, October 31, 1954, in Gajdusek, *Year in the Middle East,* 378.

44. D. C. Gajdusek to Mahtil Gajdusek, May 12, 1954, box 15, Gajdusek Correspondence, NLM.

45. Gajdusek, August 18, 1954, in *Year in the Middle East,* 164.

46. Gajdusek, August 20, 1954, ibid., 166.

47. Gajdusek, August 27, 1954, ibid., 175.

48. Gajdusek to Smadel, August 23, 1954, box 31, Gajdusek Correspondence, NLM.

49. Ibid.

50. Robert Traub to Gajdusek, June 28, 1954, in *Year in the Middle East,* 339. Theodore E. Woodward, a professor of medicine at the University of Maryland wrote: "Undoubtedly there is no such valuable collection available anywhere at the present time. Indeed, the sera which you have sent will undoubtedly keep the Medical School busy for quite some time" (November 11, 1954, in *Year in the Middle East,* 390).

51. Gajdusek, November 17, 1954, in *Year in the Middle East,* 230, 231.

52. Gajdusek to Adels, July 30, 1954, box 1, Gajdusek Correspondence, NLM.

53. D. C. Gajdusek to Mahtil, March 1954, box 15, Gajdusek Correspondence, NLM.

54. Gajdusek to Smadel, July 30, 1954, box 31, Gajdusek Correspondence, NLM.

55. Gajdusek, November 14, 1954, in *Journal 1955–1957,* 14.

56. Burnet, *The Walter and Eliza Hall Institute.* Among the visitors was Joshua Lederberg, a professor of medical genetics at the University of Wisconsin, who visited the Hall Institute in 1957 on a Fulbright scholarship. He received a Nobel Prize in Physiology or Medicine in 1958 for his studies of genetic recombination in bacteria and went on to become professor of genetics at Stanford and president of Rockefeller University.

57. Burnet, *Virus as Organism, 3.* On Burnet, see Burnet, *Changing Patterns;* Sexton, *Seeds of Time;* and Anderson, "Natural histories of infectious disease," and "Frank Macfarlane Burnet." Ian R. Mackay reports that some of Burnet's colleagues thought that "Darwinism was inscribed on his spectacles" (Interview, January 13, 2005, Melbourne). Morison referred to his "strongest point—his outstanding biological intuition" (R. S. Morison, "Diaries" [June 3, 4, and 5, 1958], 61).

58. Burnet, "Modification of Jerne's theory," 68; and Burnet, *Clonal Selection Theory.* For an earlier version of the theory, see Burnet and Fenner, *Production of Antibodies,* esp. 103. See also Lederberg, "Ontogeny of the clonal selection theory." Lederberg observes that "Burnet's uncanny biological intuition was not matched by his resonance with molecular biology or a detailed familiarity with its classical precepts" (179).

59. D. C. Gajdusek to Mahtil, May 12, 1956, box 15, Gajdusek Correspondence, NLM.

60. Gajdusek, February 22, 1957, in *Journal 1955–1957,* 72.

61. Burnet to Gunther, April 1957, Gunther Papers (also in Farquhar and Gajdusek, *Kuru,* 41).

62. D. C. Gajdusek to Mahtil, December 23, 1955, box 15, Gajdusek Correspondence, NLM.

63. D. C. Gajdusek to Mahtil, December 6, 1955, box 15, Gajdusek Correspondence, NLM.

64. The previous year the Christesens had appeared before the Royal Commission on Espionage (Petrov Commission) defending exchanges with the Soviet Union.

65. D. C. Gajdusek to Mahtil, December 23, 1955.

66. Gajdusek to Adels, May 24, 1956, box 1, Gajdusek Correspondence, NLM.

67. D. C. Gajdusek to Mahtil, December 23, 1955. Fazekas de St. Groth became professor of microbiology at the ANU.

68. C. A. Valentine to Gajdusek, n.d., box 33, Gajdusek Correspondence, NLM.

69. O'Leary, *North and Aloft.*

70. D. C. Gajdusek to Mahtil, June 29, 1956, box 15, Gajdusek Correspondence, NLM.

71. Simmons, Graydon, and Gajdusek, "Blood group genetical survey"; and Simmons, Gajdusek, and Larkin, "Blood group genetical survey in New Britain." The measles-related sera were transferred to Smadel for tests for complement fixation and neutralization antibodies by tissue culture techniques.

72. D. C. Gajdusek to Mahtil, June 29, 1956.

73. Gajdusek, June 30, 1956, in *Journal 1955–1957,* 45.

74. D. C. Gajdusek to Mahtil, June 30, 1956, box 15, Gajdusek Correspondence, NLM.

75. McCarthy, *Patrol into Yesterday.*

76. Cilento, "Story of a massacre," *Sydney Morning Herald* (December 30, 1928). The "massacre" refers to the killing of the prospectors, not the Nakanai. See McPherson, "'Wanted: Young man, must like adventure,'" 82–110.

77. Val and Edie to Jess and Charlie, November 6, [1956], Correspondence Folder, Charles A. Valentine Papers, University of Pennsylvania Museum Archives, Philadelphia. The Valentines later referred to him fondly as "our mad genius doctor friend" (Val to Mother, March 7, 1960, Correspondence Folder, Valentine Papers).

78. Val to Mother and Daddy, May 28, [1955], Correspondence Folder, Valentine Papers. This is a description from their earlier field work.

79. Val and Edie to Jess and Charlie, September 13, 1956, Correspondence Folder, Valentine Papers.

80. C. A. Valentine to Gajdusek, August 18, 1956, box 15, Gajdusek Correspondence, NLM. Val told Carleton about various beatings since his departure and warned him his letters to the Nakanai might be intercepted.

81. Val to Mother and Daddy, November 4, 1957, Correspondence Folder, Valentine Papers.

82. Val to Mother and Daddy, October 24, 1956, Correspondence Folder, Valentine Papers. Back in the United States, Val and Edie divorced and he became more deeply involved in civil rights activism, largely abandoning anthropological work. See Valentine, *Culture and Poverty.* In the 1960s Carleton tried to persuade Val to study kuru among the Fore, but he was not interested. On the West New Britain work, see Valentine, "Introduction to the history of changing ways of life"; and Jebens, "'Vali did that too.'"

83. Gajdusek to C. A. Valentine, n.d., box 33, Gajdusek Correspondence, NLM.

84. Gajdusek to Vuaroa, August 22, 1956, box 33, Gajdusek Correspondence, NLM (filed with Valentine folder).

85. Gajdusek to C. A. and E. Valentine, February 26, 1960, box 33, Gajdusek Correspondence, NLM. See also Gajdusek, *Solomon Islands.*

86. Gajdusek, August 5, 1956, in *Journal 1955–1957,* 53.

87. Gajdusek to Adels, [August 1956], box 1, Gajdusek Correspondence, NLM.

88. D. C. Gajdusek to Mahtil, August 19, 1956, box 15, Gajdusek Correspondence, NLM.

89. Gajdusek to Adels, November 25, 1956, box 1, Gajdusek Correspondence, NLM.

90. D. C. Gajdusek to Mahtil, November 21, 1956, box 15, Gajdusek Correspondence, NLM.

91. Gajdusek to Adels, December 12, 1956, box 1, Gajdusek Correspondence, NLM.

92. Gajdusek to Smadel, December 29, 1956, box 31, Gajdusek Correspondence, NLM. He went on: "I am sole author of the work . . . although a dozen people here, including Sir Mac, are standing ready to pounce on the problem with my departure."

93. Gajdusek, "Autoimmune reaction against human tissue antigens in certain chronic diseases"; Gajdusek, "Autoimmune reaction against human tissue antigens in certain acute and chronic diseases"; and Gajdusek and Mackay, "Autoimmune reaction against human tissue antigens." See also Mackay and Burnet, *Autoimmune Diseases*.

94. Interview with Ian R. Mackay, January 13, 2005, Melbourne.

95. Interview with S. Gray Anderson, September 28, 2005, London. Anderson conducted research primarily on Murray Valley and rubella viruses. See Anderson et al., "Murray Valley encephalitis"; and Anderson et al., "Murray Valley encephalitis in Papua and New Guinea: II."

96. Frank Fenner to Burnet, October 31, 1955, Folder 97, File 1 (Correspondence), Burnet Papers, University of Melbourne Archives, Melbourne, Victoria.

97. Burnet to Gunther, July 4, 1956, Folder 2, File 10 (Kuru), Burnet Papers.

98. Burnet to Paul Hasluck, federal minister for territories, April 12, 1956, Folder 2, File 10 (Kuru), Burnet Papers.

99. Morison to Burnet, May 8, 1956, Folder 2, File 10 (Kuru), Burnet Papers. Morison had worked with Smadel and kept him informed of these developments. Burnet wrote to Hale to tell him that in Australia there was "a great deal of interest in the medical 'exploration' of New Guinea," urging him to move to Indonesia (Burnet to Hale, May 9, 1956, file 12/16/10, box 1902, series 114, accession 23, PNGNA).

100. Gunther to Burnet, May 7, 1956, Folder 2, File 10 (Kuru), Burnet Papers. Original emphasis.

101. Interview with S. Gray Anderson, September 28, 2005, London. Burnet's increasing commitment to immunology had made Anderson uncomfortable at the Hall Institute, so in 1961 he moved to London.

102. Strickland, *Politics, Science and Dread Disease;* Harden, *Inventing the NIH;* and Park, "Development of the intramural research program."

103. Smadel to Gajdusek, July 3, 1956, box 31, Gajdusek Correspondence, NLM.

104. Gajdusek to Smadel, September 28, 1956, box 31, Gajdusek Correspondence, NLM.

105. D. C. Gajdusek to Bobby, April 15, 1957, box 17, Gajdusek Correspondence, NLM.

CHAPTER 3. A CONTEMPTUOUS TENDERNESS

1. "Aoga," Kuru Case Records, Cabinet 3.3, Alpers Papers, Curtin University, Perth, Western Australia. The first notes are dated March 15, 1957.

2. Gajdusek, March 8, 1957, in *Journal 1955–1957,* 77. It is certain, however, that the commitment of both Burnet and Anderson to New Guinea investigations predated Gajdusek's arrival at the Hall Institute—see Chapter 2. Gajdusek must have known at least about the visits of Burnet and Anderson to the territories, even if he was unaware of a specific in-

terest in kuru. Burnet remained convinced that Gajdusek first heard about kuru at the Hall Institute.

3. Ibid., 78. Original emphasis.

4. Anderson to Gunther, January 17, 1957, Gunther Papers.

5. Burnet to Gunther, February 12, 1957; and Gunther to Burnet, February 15, 1957, folder 2, series 10, Burnet Papers.

6. Anderson to Gunther, February 12, 1957, Gunther Papers.

7. Burnet to Gunther, March 11, 1957, Gunther Papers.

8. Gajdusek, March 9, 1957, in *1955–1957 Journal and Field Notes,* 78.

9. Gajdusek, March 1957, in *1955–1957 Journal and Field Notes,* 81, 80.

10. Mother to D. C. Gajdusek, March 1957, box 15, Gajdusek Correspondence, NLM.

11. Gajdusek to Smadel, March 15, 1957 (Moke), box 31, Gajdusek Correspondence, NLM (in Gajudsek, *Correspondence on the Discovery,* 50). Smadel tried to find intramural support for Gajdusek but "we cannot put a man on a fellowship or on a staff appointment without having a warm body and a signature" (Smadel to Marion C. Morris, National Foundation for Infantile Paralysis [NFIP], March 29, 1957, box 31, Gajdusek Correspondence, NLM). The NFIP director came up with some discretionary money: see Thomas M. Rivers to Gajdusek, May 21, 1957, folder 2, series 10, Burnet Papers. Known for its yearly fundraiser, the March of Dimes, the NFIP granted millions of dollars to virus researchers in the 1940s and 1950s, including John Enders. See Benison, *Tom Rivers;* and Kevles and Geison, "Experimental life sciences."

12. Gajdusek to Burnet, Anderson, and Wood, March 15, 1957, folder 2, series 10, Burnet Papers, and Gunther Papers (in Gajdusek *Correspondence on the Discovery,* 53). Ian Wood was the director of the Clinical Research Unit. For more discussion of the possibility of taking a Fore child out of the region, see Gajdusek to Scragg, March 18, 1957, Gunther Papers.

13. Gajdusek *Correspondence on the Discovery,* 52.

14. Ibid. To Scragg he also suggested Wilson's disease, a disorder of copper metabolism (Gajdusek to Scragg, March 18, 1957, Gunther Papers).

15. Gajdusek to Burnet, Wood, and Anderson, March 20, 1957, folder 2, series 10, Burnet Papers, and Gunther Papers (in Gajdusek, *Correspondence on the Discovery,* 57).

16. "Taranto," "Yo'iea," "Tasiko," Kuru Case Records, Cabinet 3.3, Alpers Papers.

17. Interview with Tarubi, August 10 and 11, 2003, Purosa.

18. David Ikabala interview with Tiu Pekiyeva, July 31, 2003, available at Institute for Medical Research (IMR), Goroka, PNG.

19. Interview with Paudamba, August 11, 2003, Purosa.

20. Interview with Pako, August 6, 2003, Waisa.

21. Scragg to Anderson, March 5, 1957, file 54/6/12, box 5361, series 153, accession 23, PNGNA.

22. Gunther to Gajdusek, April 16, 1957, Gunther Papers.

23. Sydney Sunderland to Scragg, November 12, 1957, file 12/16/21, box 1902, series 114, accession 23, PNGNA.

24. Interview with Donald Simpson, December 22, 2005, Adelaide. See Simpson, Lander, and Robson, "Observations of kuru: II."

25. Gajdusek to Smadel, April 3, 1957, box 31, Gajdusek Correspondence, NLM (in Farquhar and Gajdusek, *Kuru,* 29).

26. Gajdusek to Smadel, May 28, 1957, box 31, Gajdusek Correspondence, NLM (in Farquhar and Gajdusek, *Kuru,* 65).

27. Interview with Jack and Lois Baker, January 3, 2006, Bribie Island, Queensland.

28. J. C. Baker, Patrol Report, No. 6 of 1956/57, Okapa, Eastern Highlands District.

29. J. C. Baker, Patrol Report, No. 6 of 1957/58, Kainantu, Eastern Highlands District.

30. Gajdusek to Burnet and Anderson, November 16, 1957, folder 2, series 10, Burnet Papers.

31. Gajdusek, October 5, 1957, in *Kuru Epidemiological Patrols.*

32. Gajdusek, September 28, 1957, in ibid., 59.

33. Gajdusek, October 14, 1957, in ibid., 117.

34. Gajdusek, April 3, 1960, in *Solomon Islands,* 113.

35. Gajdusek, October 8, 1957, in *Kuru Epidemiological Patrols,* 95. His feelings only deepened in later years: "No more 'bastardly' people in the world could be found to compare with them—but such charming, 'darling,' bastards!!! I really like them! They carry my eulogy of individuality to a mocking extreme" (December 24, 1961, in Gajdusek, *New Guinea Journal,* 156–57).

36. Gajdusek, September 3, 1957, in *Kuru Epidemiological Patrols,* 1, 2. Later, Gajdusek reflected that he could "never repay them for the richness they have given to my life as they now give while they sit leaning against me, or surrounding me as I work" (August 24, 1959, in *Journal of Continued Quest for the Etiology of Kuru,* 112).

37. Gajdusek, September 6, 1957, in *Kuru Epidemiological Patrols,* 8.

38. *Haus Kapa* acquired his odd nickname when he first saw a modern house and was overcome with emotion.

39. Gajdusek, October 12, 1957, in *Kuru Epidemiological Patrols,* 107, 108.

40. Gajdusek, November 10, 1957, in ibid., 180–81. A few years later, among the south Fore, Gajdusek noted in his journal: "An evening of finishing Proust's *Within a Budding Grove* at long last and finding too close parallels between myself and the narrator in his pursuit of the 'little band' to read perceptively or comfortably" (April 16, 1960, in *Solomon Islands,* 136).

41. Interview with Masasa, August 8, 2003, Okapa.

42. Interview with Tarubi, August 10, 2003, Purosa.

43. Interview with Masasa, August 8, 2003, Okapa.

44. David Ikabala interview with Koiye Tasa, [c. 2003], [n.p.], available at IMR. Employed as a carrier, Paudamba recalled thinking, "We've come to a distant place, some place our parents have never seen. And we thought of our parents" (interview with Paudamba, August 11, 2003, Purosa).

45. Interview with Masasa, August 8, 2003, Okapa.

46. Interview with Tarubi, August 11, 2003, Purosa.

47. Sena Anua interview with Ai clan, Ketebe village, [c. 2003], Purosa, available at IMR.

48. Gajdusek, "Introduction," in Farquhar and Gajdusek, *Kuru,* xxiii.

49. Gajdusek to Scragg, March 20, 1957, Gunther Papers (in Farquhar and Gajdusek, *Kuru,* 22).

50. Gajdusek to Smadel, April 3, 1957, box 31, Gajdusek Correspondence, NLM (in Farquhar and Gajdusek, *Kuru,* 29). Gajdusek and Zigas had a bed in the corner of the *kiap*'s hut.

51. Gajdusek to Smadel, July 8, 1957, box 31, Gajdusek Correspondence, NLM (in Farquhar and Gajdusek, *Kuru*, 87).

52. Gajdusek to Smadel, August 6, 1957, Gajdusek Correspondence, NLM (in Gajdusek, *Correspondence on the Discovery*, 172).

53. Gajdusek, to Burnet, Anderson, and Ian Wood, August 6, 1957, folder 2, series 10, Burnet Papers.

54. Gajdusek to Cyril Curtain, November 13, 1957, box 9, Gajdusek Correspondence, NLM.

55. Gajdusek to Smadel, July 10, 1957, box 31, Gajdusek Correspondence, NLM (in Farquhar and Gajdusek, *Kuru*, 91).

56. Gajdusek to Scragg, [mid-March 1957], Gunther Papers.

57. Gajdusek to Scragg, April 6, 1957, Gunther Papers (the month is typed as "March" but Scragg has written "April" alongside it).

58. Jack Baker, Report from Okapa Patrol Post for the Quarter Ending 30 November 1957, Accession 2003/042, Box 001, File: Medical Patrol Reports 1956–57, IMR Archives, PNG.

59. Robson was professor of medicine at Adelaide, and later became vice-chancellor of the University of Sheffield and then the University of Edinburgh. Lander, a physician interested in heavy metal poisoning, later became dean of the medical school in Fiji and then professor of international health in the University of Hawaii School of Public Health.

60. Rheinberger, *Toward a History of Epistemic Things*, 3.

61. Knorr-Cetina, "Ethnographic study of scientific work"; and Knorr-Cetina, *Epistemic Cultures.* As Knorr-Cetina writes elsewhere, "The laboratory subjects natural conditions to a social overhaul and derives epistemic effects from the new situation" ("The couch, the cathedral, and the laboratory," 117). See also Fleck, *Genesis and Development;* and Latour and Woolgar, *Laboratory Life.*

62. Gajdusek to Scragg, November 3, 1958, file 32/35/2, box 3467, series 114, accession 23, PNGNA.

63. Weber, *Protestant Ethic*, 182.

64. Lewis, *Arrowsmith*, 365.

65. Nietzsche, *Genealogy of Morals*, 174.

66. Ibid., 286–87, 290.

67. Gajdusek, March 10, 1965, in *Melanesian Journal, January 23, 1965–April 7, 1965*, 92. Gajdusek was rereading *The Genealogy of Morals* among the Kukukuku.

68. Gajdusek, February 22, 1957, in *1955–1957 Journal and Field Notes*, 71.

69. Gajdusek, June 25, 1959, in *New Guinea Journal, June 10, 1959–August 15, 1959*, 11.

70. Gajdusek and Zigas, "Degenerative disease," 975. Gajdusek and Zigas initially emphasized similarities with *paralysis agitans* or Parkinsonism, but Simpson's later clinical studies led them to shift focus to cerebellar dysfunction. The observation of emotional lability became rare in later years. To the annoyance of Burnet and Gunther, Gajdusek and Zigas announced their "discovery" in an Australian journal *after* the U.S. publication: see Zigas and Gajdusek, "Kuru." Even so, Gajdusek regarded publication of the *MJA* paper as premature and swore never again to publish in an Australian journal (Gajdusek to Cyril Curtin, November 13, 1957, Gajdusek Correspondence, NLM).

71. Gajdusek and Zigas, "Degenerative disease," 978.

72. Klatzo, Gajdusek, and Zigas, "Pathology of kuru." See also Fowler and Robertson, "Observations on kuru III."

73. Klatzo to Gajdusek, September 13, 1957, box 22, Gajdusek Correspondence, NLM (in Farquhar and Gajdusek, *Kuru,* 155–56). The pathology report on the brain of Yabaiotu is attached to this letter. CJD was first described by German neurologists Hans Gerhard Creutzfeldt and Alfons Maria Jakob.

74. R. Berndt, "Devastating disease syndrome," 21.

75. Ibid., 11, 27. Berndt discusses Cannon, "'Voodoo' death." See Dror, "'Voodoo death.'" In 1957, Gajdusek wrote to Ronald Berndt: "We cannot agree that kuru is a psychosis or hysterical phenomenon, although we have seen the entity of kuru with recovery which currently we attribute to hysterical mimicry of true kuru" (box 5, Gajdusek Correspondence, NLM).

76. Gajdusek to Curtin, November 13, 1957, Gajdusek Correspondence, NLM.

77. Gajdusek to Burnet and Anderson, November 16, 1957, folder 2, series 10, Burnet Papers.

78. Gajdusek to Smadel, November 10, 1957, in Farquhar and Gajdusek, *Kuru,* 239.

79. Anonymous, "The laughing death," *Time* (November 11, 1957): 55–56: here, 55.

80. Kathleen Teltsch, "Rare disease and strange cult disturb New Guinea territory," *New York Times* (June 26, 1959): 3.

81. Burnet to Scragg, March 26, 1957, folder 2, series 10, Burnet Papers.

82. Burnet to Gajdusek, March 26, 1957, folder 2, series 10, Burnet Papers. Burnet later told Gajdusek that his actions were "indefensible" (Burnet to Gajdusek, April 9, 1957, folder 2, series 10, Burnet Papers). But he told T. M. Rivers at the NFIP that despite his annoyance, he retained "a sort of exasperated affection for Gajdusek and a great admiration for his drive, courage and capacity for hard work" (Burnet to Rivers, May 27, 1957, folder 2, series 10, Burnet Papers).

83. Anderson to Scragg, March 22, 1957, folder 2, series 10, Burnet Papers.

84. Burnet to Scragg, April 5, 1957, folder 2, series 10, Burnet Papers.

85. Gajdusek to Scragg, April 6, 1957, Gunther Papers. Scragg had previously let slip: "With your presence in the field, it will hardly be necessary for Dr. Anderson to visit the area" (Scragg to Gajdusek, March 28, 1957, file 12/16/21, box 1902, series 114, accession 23, PNGNA).

86. Gajdusek to Burnet, April 20, 1957, file 12/16/21, box 1902, series 114, accession 23, PNGNA.

87. Gunther to Gajdusek, April 9, 1957, Gunther Papers.

88. Gunther to Burnet, April 9, 1957, folder 2, series 10, Burnet Papers.

89. Gajdusek to Gunther, April 20, 1957, Gunther Papers.

90. Gunther to Burnet, April 24, 1957, folder 2, series 10, Burnet Papers.

91. Gunther to Gajdusek, May 3, 1957, Gunther Papers.

92. Interview with S. Gray Anderson, September 28, 2005, London.

93. S. G. Anderson, "Report on cases of kuru Seen at Okapa, May 16, 1957," Gunther Papers.

94. Interview with Ian R. Mackay, January 13, 2005, Melbourne.

95. Interview with Donald Simpson, December 22, 2005, Adelaide.

96. D. M. Cleland, administrator, to secretary, Department of Territories, May 15, 1957, Gunther Papers.

97. Burnet to Scragg, September 25, 1957, Gunther Papers. Gajdusek felt the Australian pathology results were slower to arrive and less accurate than those he could obtain in the U.S.

98. Goodenough, "Moral outrage"; and Goffman, *Relations in Public.*

99. Burnet to Gunther, January 6, 1959, Gunther Papers.

100. Robert S. Morison, September 16, 1958, in "Diaries."

101. Ibid.

102. Burnet to Gunther, January 6, 1959, Gunther Papers.

103. Gunther to J. H. Bennett, November 30, 1959, Gunther Papers.

104. J. K. McCarthy to Gunther, December 13, 1957, file 32/35/2, box 3467, series 114, accession 23, PNGNA.

105. J. T. Gunther, "Aide-Memoire (Kuru)," February 2, 1959, Gunther Papers. Gajdusek noted that Zigas "has been effectively 'blacklisted' by the Public Health Department and administration and everyone knows it" (March 11, 1960, in *Solomon Islands,* 43).

106. H. N. Robson, S. Sunderland, and J. C. Eccles, "Report on the Kuru Disease," November 1957, file 32/35/2, box 3467, series 114, accession 23, PNGNA.

107. "Adelaide University Study of Kuru, December 1957–January 1958: Preliminary Report," file 32/35/2, box 3467, series 114, accession 23, PNGNA. The members of this team were Robson (on his second visit), Harry Lander, D. A. Simpson (on his second visit), and the aging pathologist J. B. Cleland (who developed prostate trouble and was flown back to Adelaide soon after arriving in New Guinea). F. A. Rhodes from the Public Health Department accompanied them.

108. I describe the Adelaide investigations more extensively in Chapter 5.

109. J. T. Gunther notes on margins of letter from D. S. Barnes to the director, Department of Public Health, November 14, 1958, file 32/35/2, box 3467, series 114, accession 23, PNGNA.

110. Scragg to Harry White, December 24, 1958, file 32/35/2, box 3467, series 114, accession 23, PNGNA.

111. Gajdusek, February 19, 1959, in *Journal of Continued Quest,* 35.

112. Gajdusek, March 12, 1960, in *Solomon Islands,* 47.

113. Interview with Tarubi, August 11, 2003, Purosa; interview with Masasa, August 8, 2003, Okapa. Agakamatasa had been founded relatively recently, probably in the 1940s, by refugees from Purosa fleeing warfare and disease.

114. Interview with Paudamba, August 11, 2003, Purosa.

115. Gajdusek, August 17, 1959, in *Journal of Continued Quest,* 77.

116. D. C. Gajdusek to Mahtil Gajdusek, March 6, 1962, box 15, Gajdusek Correspondence, NLM.

117. D. C. Gajdusek to Mahtil Gajdusek, April 12, 1962, box 15, Gajdusek Correspondence, NLM.

118. Conrad, *Lord Jim,* 204, 226.

CHAPTER 4. THE SCIENTIST AND HIS MAGIC

1. Sena Anua interview with Andemba Katago, February 2004, Purosa, available at IMR.

2. Fore sorcery is brilliantly described in Lindenbaum, *Kuru Sorcery.* Lindenbaum points out that the typical sorcerer was supposed to be a marginal person, a "rubbish" man, with a grudge. See also her "Sorcery and structure."

3. Gajdusek to Smadel, [late May 1957], box 31, Gajdusek Correspondence, NLM (in Gaj-dusek, *Correspondence on the Discovery,* 95).

4. Rumsey, "White man as cannibal."

5. Interview with Masasa, August 8, 2003, Okapa.

6. David Ikabala interview with Tiu Pekiyeva, July 31, 2003, Agakamatasa, available at IMR.

7. On the gift, see Mauss, *The Gift;* Sahlins, *Stone Age Economics;* Gregory, *Gifts and Com-modities;* Lederman, *What Gifts Engender;* M. Strathern, *Gender of the Gift;* Thomas, *Entan-gled Objects;* Weiner, *Inalienable Possessions;* and Godelier, *Enigma of the Gift.* On strategy and calculation in gift exchange see Bourdieu, *Outline of a Theory of Practice.*

8. By "barter" I mean the determining of an exchange ratio or substitutability during bargaining, where the relationship between transactors becomes one of "reciprocal indepen-dence," rather than the "reciprocal dependence" of gift exchange. See Humphery and Hugh-Jones, "Introduction: Barter, exchange and value," 11. See also Gregory, *Gifts and Commodi-ties,* esp. 42.

9. Parry, "On the moral perils of exchange."

10. Malinowski, *Argonauts of the Western Pacific,* 25.

11. M. Strathern, "Qualified value." Strathern notes that in gift exchange, "people must com-pel others to enter into debt: an object in the regard of one actor must be made to become an object in the regard of another" (177).

12. Gregory observes that in a commodity economy, things and persons assume the social form of things, while in a gift economy they assume the social form of persons (*Gifts and Com-modities,* 42).

13. As Mauss points out, in gift exchange "objects are never completely separated from the people who give them" (*Gift,* 31). See also Weiner, *Inalienable Possessions.*

14. On these ceremonies, see Lindenbaum, "A wife is the hand of man," esp. 57; and R. Berndt, *Excess and Restraint,* 94–97, 104.

15. Gajdusek to Smadel, May 28, 1957, box 31, Gajdusek Correspondence, NLM (in *Cor-respondence on the Discovery,* 90).

16. Interview with Pako, August 6, 2003, Waisa.

17. Interview with Tarubi, August 11, 2003, Purosa.

18. Interview with Tarubi, August 11, 2003, Purosa.

19. Interview with Pako, August 6, 2003, Waisa.

20. Lindenbaum, *Kuru Sorcery.*

21. It was not until the 1990s that Fore began to pay for Western medical services, on the rare occasions they could obtain them.

22. Interview with Inamba, August 4, 2003, Ivingoi.

23. Gajdusek to Smadel, [May 1957], box 31, Gajdusek Correspondence, NLM (in Gajdusek, *Correspondence on the Discovery,* 88).

24. On tournaments of value, see Appadurai, "Introduction: Commodities."

25. Gajdusek, December 13, 1961, in *New Guinea Journal, Part I,* 124.

26. Gajdusek to Smadel, [late May 1957], box 31, Gajdusek Correspondence, NLM (in Gaj-dusek, *Correspondence on the Discovery,* 93).

27. Gajdusek to Burnet and Anderson, May 19, 1957, folder 2, series 10, Burnet Papers (in Farquhar and Gajdusek, *Kuru,* 57).

28. Gajdusek, April 16, 1960, in *Solomon Islands,* 137.

29. Interview with Jack and Lois Baker, January 3, 2006, Bribie Island, Queensland. "Goodies," Jack's wife Lois Larkin Baker interjected.

30. This description is based largely on my interviews with Michael Alpers, one of the scientists later involved in these procedures (see Chapter 5), July 24, 2002, and August 4, 2004, Fremantle.

31. Or made this clear during the procedure: Vin Zigas described an autopsy where he managed to get a brain and some spinal cord, but "couldn't get viscera as relatives were very impatient and became quite cranky" (Zigas to Gajdusek, [January 1959], box 36, Gajdusek Correspondence, NLM).

32. Lindenbaum, *Kuru Sorcery,* 21.

33. Michael Alpers told me that in his experience this was usually not such a problem, since a master or mistress of ceremonies would normally intervene to make sure gifts were properly distributed.

34. Interview with Masasa, August 8, 2003, Okapa.

35. Interview with Inamba, August 5, 2003, Ivingoi.

36. Interview with Tarubi, August 10, 2003, Purosa. Further conversation with Tarubi established that although he used the word "germ" he meant *liklik binatang,* or small insect. Like many others, he would sometimes use the terms of biomedical science without its explanatory apparatus.

37. Gajdusek to Smadel, [late May 1957], box 31, Gajdusek Correspondence, NLM (in Gajdusek, *Correspondence on the Discovery,* 92).

38. Gajdusek to Smadel, November 24, 1957, box 31, Gajdusek Correspondence, NLM (in Gajdusek, *Correspondence on the Discovery,* 309–10).

39. D. M. Cleland to secretary, Australian Department of Territories, December 9, 1958, file 32/35/2, box 3467, series 114, accession 23, PNGNA.

40. Scragg to assistant director, Department of Public Health, November 26, 1958, file 32/35/2, box 3467, series 114, accession 23, PNGNA.

41. Scragg to administrator, September 28, 1959, Gunther Papers. Yet Baker noted: "Voluntary attendance at the hospital has never been better, which tends to discount any idea that loss of prestige due to what has been termed the 'ostentatious failure' of the kuru project has adversely affected the general medical work at this post" (Report from the Okapa Patrol Post for the Quarter Ending February 28, 1958, file: Medical Patrol Reports, 1957–1959, box 001, series 2003/042, IMR Archives).

42. Gajdusek to Smadel, November 24, 1957, box 31, Gajdusek Correspondence, NLM (in Gajdusek, *Correspondence on the Discovery,* 310). My emphasis.

43. Gajdusek to Smadel, March 15, 1957, box 31, Gajdusek Correspondence, NLM (in Gajdusek, *Correspondence on the Discovery,* 51). Smadel was concerned that Gajdusek might be eaten: "What will happen to the records, the material, and the information you carry in your head if the plane comes down in the jungle or if one of the indigenes decides to revert to cannibalism?" (Smadel to Gajdusek, August 16, 1957, box 31, Gajdusek Correspondence, NLM, and Gajdusek, *Correspondence on the Discovery,* 177).

44. Gajdusek to Smadel, September 27, 1957, box 31, Gajdusek Correspondence, NLM (in Gajdusek, *Correspondence on the Discovery,* 234).

45. Gajdusek to Smadel, [late May 1957], box 31, Gajdusek Correspondence, NLM (in Gajdusek, *Correspondence on the Discovery,* 95). On the continuing appeal of the cannibal metaphor in medicine and science, see Arens, "Rethinking anthropophagy."

46. Lindenbaum, *Kuru Sorcery.* See also Sanday, *Divine Hunger;* and Brown and Tuzin, *Ethnography of Cannibalism.*

47. On this problem of social reproduction see Weiner, "Sexuality among the anthropologists"; and Rosaldo, "Skulls and causality."

48. Thus "cannibal appetite is its own impossible desire" (Bartolovich, "Consumerism," 234).

49. Janet Hoskins argues that "heads are taken—in imagination as in traditional practice—to seize an emblem of power, to terrify one's opponents, and to transfer life from one group to another" ("Introduction: Head hunting," 38). On exchange models for understanding the cultural logic of head hunting, see George, "Head hunting, history and exchange"; and Pannell, "Travelling in other worlds."

50. M. Strathern, "Cutting the network."

51. Perhaps one should add the autopsy theater to the bush lab as a site of transformation, but the former is functionally an instantiation of the latter, though differently bounded. The process of the "cultural transformation" of medieval saints' relics, in which they acquired new status and meaning in a different context, is described in Geary, "Sacred commodities," 181. Even if originally gifts, these things often acquired more value if represented as thefts. Moreover, on "translation," these relics were subject to tests of authenticity (186).

52. Magic being "based on man's confidence that he can dominate nature directly, if only he knows the laws which govern it magically, is in this akin to science" (Malinowski, "Magic, science, and religion," 3). See also Tambiah, *Magic, Science, Religion;* and Verrips, "Dr. Jekyll and Mr. Hyde."

53. On "boundary work" in science see Gieryn, "Boundary-work." On the creation of "boundary objects," see Star and Griesemer, "Institutional ecology." According to Waldby and Mitchell, "Establishment of intellectual property in human tissues requires the prior dispossession of the donor" (*Tissue Economies,* 86).

54. Lederman, *What Gifts Engender.* As Malinowski points out, magic is at the same time the "institutionalization of human optimism" ("Ethnographic theory of the magical world," 239). Sigmund Freud, of course, describes it as a symptom of "omnipotence of thought" (*Totem and Taboo,* 120).

55. An "aura" in the sense of Benjamin, "Work of art." For Gajdusek's involvement in global scientific exchange, see Chapter 6.

56. As Marilyn Strathern argues with respect to the failure of men in ceremonial exchange to recognize women's labor in the production of pigs: "Whatever work went into the production of the pigs, others cannot appropriate it as work; they acquire the gift as a debt to be repaid" (*Gender of the Gift,* 156). But Melanesian male agency in ceremonial exchange—perhaps somewhat like Gajdusek's agency in exchange with other scientists—is achieved only "through a set of systematic separations that render the sphere in which they claim prestige apart from domestic activity, where they are, quite emphatically, not the singular proprietors of their persons" (158).

57. Gajdusek to Burnet, March 13, 1957, Burnet Papers (in Farquhar and Gajdusek, *Kuru,*

6). Wood was the director of the Clinical Research Unit of Burnet's Hall Institute—Gajdusek's former collaborator Ian Mackay was the associate director.

58. Macfarlane Burnet thought it would be worthwhile removing a typical patient from the "tribal environment" to see if there was any beneficial effect and to allow expert neurological assessment. Unlike Gajdusek, he wanted "in the event of death of the patient to have a fully adequate post-mortem examination made" (Burnet to Scragg, March 27, 1957, Gunther Papers). In the early 1960s, Burnet tried to take kuru orphans to Melbourne for further investigation and to remove them from potential environmental toxins, but his efforts also were rejected.

59. Interview with Masasa, August 8, 2003, Okapa.

60. Malinowski, *Argonauts of the Western Pacific*, 25.

61. Pratt, *Imperial Eyes*, 7. Roy MacLeod urges abandonment of center-periphery models in the history of science and proposes instead a study of the traffic of ideas and institutions, a recognition of reciprocity using "perspectives colored by the complexities of contact" ("Introduction," 6).

62. Galison, "Computer simulations," 119. See also Galison, "Material culture." The bush laboratory and the autopsy hut—already reconfigured as cargo houses—thus resemble also the cultural beaches that Greg Denning describes in the Pacific, strands that must be crossed to enter or leave different social worlds and to make or change social worlds—even though we know such a metaphoric beach in the New Guinea highlands is a geographical anomaly (*Islands and Beaches*).

63. Gajdusek, January 26, 1962, in *New Guinea Journal, Part I*, 269.

64. Gajdusek, April 20, 1962, in *New Guinea Journal, Part II*, 86, 87.

65. I am drawing together here the idea of incommensurability between scientific practices and theories and anthropological understandings of incommensurability: on the former see Kuhn, *Structure of Scientific Revolutions;* and Hoyningen-Huene and Sankey, *Incommensurability and Related Matters*. On the latter, see Povinelli, "Radical worlds."

CHAPTER 5. HEARTS OF DARKNESS

1. Zigas to Gajdusek, February 23, 1958, box 36, Gajdusek Correspondence, NLM.

2. "Aoga," Kuru Case Records, Cabinet 3.3, Alpers Papers.

3. Zigas to Gajdusek, February 23, 1958, box 36, Gajdusek Correspondence, NLM. Zigas later extolled Robson: "I felt he had it all: good looks, the common touch, and a sense of humor when needed. What's more, he injected a marvelous dose of Scottish blood and common sense into the Adelaide team" (*Laughing Death*, 268).

4. Zigas to Gajdusek, February 23, 1958, box 36, Gajdusek Correspondence, NLM.

5. Zigas to Gajdusek, June 14, 1958, box 36, Gajdusek Correspondence, NLM. Around this time, Zigas's marriage was breaking up. "I was sick and tired of trying to cope with the medical people coming and going," his wife Gloria recalled (Chalmers, *Kundus, Cannibals and Cargo Cults*, 110).

6. Zigas to Gajdusek, May 15, 1958, box 36, Gajdusek Correspondence, NLM.

7. Zigas to Gajdusek, February 23, 1958, box 36, Gajdusek Correspondence, NLM.

8. Gajdusek to Baker, May 8, 1958, box 3, Gajdusek Correspondence, NLM.

9. Gajdusek to Baker, August 25, 1958, box 3, Gajdusek Correspondence, NLM.

10. Baker to Gajdusek, September 3, 1958, box 3, Gajdusek Correspondence, NLM.

11. Interview with R. F. R. Scragg, November 22, 2006, Glenelg, South Australia. Scragg later became professor of social medicine at the University of Papua New Guinea.

12. Gajdusek to Zigas, December 31, 1958, box 36, Gajdusek Correspondence, NLM (in Farquhar and Gajdusek, *Kuru*, 38).

13. Burnet to Gajdusek, April 9, 1957, folder 3, series 10, Burnet Papers (in Farquhar and Gajdusek, *Kuru*, 117).

14. Gajdusek to Scragg, September 24, 1957, Gunther Papers (in Farquhar and Gajdusek, *Kuru*, 166). Leonard Kurland had early discounted any genetic hypothesis: see Kurland to Gajdusek, August 20, 1957, box 22, Gajdusek Correspondence, NLM.

15. Gunther to Bennett, January 3, 1959, Gunther Papers.

16. See Scragg to Zigas, February 16, 1959, Kuru research papers, box 001, 2003/042, IMR Archives, and in Gunther Papers. For background detail from the Adelaide perspective, see Bennett to Brigadier Cleland, administrator of PNG, January 15, 1959, Gunther Papers. In this letter, Bennett complains that Gajdusek and Zigas were too "lavish . . . in rewarding of Fore natives for information and specimens collected from them." According to Zigas, Auricht complained to him that he and Gajdusek were "spoiling kanakas" (Zigas to Gajdusek, February 1, 1959, box 36, Gajdusek Correspondence, NLM).

17. Gabb later completed his Ph.D. with Bennett, becoming the first medical geneticist trained in Australia. He went on to teach at Adelaide University.

18. Robson to Bennett, February 28, 1959, Gunther Papers. Morison regarded Robson as "a first class clinician and good administrator, in whom one could have every confidence," R. S. Morison, August 19, 1958, in "Diaries." Richmond K. Anderson, an associate director at the foundation, later visited Adelaide and New Guinea and offered further support for the Australians.

19. Robson to Gunther, March 10, 1959, Gunther Papers.

20. Robson to Bennett, March 9, 1959, Gunther Papers. See also Smadel to Scragg, March 12, 1959, folder 27, box 3, series 3, record group 1.2, Rockefeller Foundation Archives, Rockefeller Archive Center.

21. Scragg to regional medical officer, Goroka, June 30, 1959, Gunther Papers.

22. Gajdusek to Schofield, August 13, 1959, in Gajdusek, *New Guinea Journal, 1959,* 149. Gajdusek had intended to return to New Guinea in March 1959, but the Public Health Department managed to defer his arrival until June.

23. Bennett and Robson to administrator, August 14, 1959, Gunther Papers. Bennett also urged Burnet to protest against Gajdusek and any "other odd types who will be attracted to the kuru region" (Bennett to Burnet, September 11, 1959, folder 3, series 10, Burnet Papers).

24. Fisher to Burnet, November 17, 1959, Gajdusek Correspondence, NLM. See Fisher, *Statistical Methods for Research Workers;* Fisher, *Genetical Theory of Natural Selection;* and Bennett, *Collected Papers of R. A. Fisher.*

25. Gajdusek notes in Fisher folder, c. 2000, box 13, Gajdusek Correspondence, NLM.

26. Baker to Gajdusek, April 4, 1959, box 3, Gajdusek Correspondence, NLM. Zigas also initially regarded Gray as a "very nice chap" (Zigas to Gajdusek, March 12, 1959, box 36, Gajdusek Correspondence, NLM).

27. Gray to Bennett, October 18, 1959, Gunther Papers.

28. Zigas to Scragg, September 14, 1959, Gunther Papers.

29. Gray to Scragg, December 22, 1959, Gunther Papers.

30. H. N. White to A. J. Gray, January 4, 1960, Gunther Papers.

31. Gajdusek, December 13, 1961, in *New Guinea Journal, Part I*, 127.

32. Robson to Scragg, December 15, 1959, Gunther Papers.

33. Simpson to Zigas, February 1959, box 36, Gajdusek Correspondence, NLM. Like Gajdusek, Zigas got on well with Simpson, regarding him as a "most sensible and honest gentleman" (Zigas to Gajdusek, February 1, 1959, box 36, Gajdusek Correspondence, NLM).

34. Zigas to Gajdusek, September 10, 1959, box 36, Gajdusek Correspondence, NLM.

35. Scragg to Zigas, September 29, 1959, Gunther Papers.

36. Conrad, *Youth.*

37. Read to Wentworth, November 13, 1959, Gunther Papers. See Read's *High Valley.*

38. Read to Wentworth, March 17, 1960, Gunther Papers. In any case, Read did not return to New Guinea for many more years, conducting instead an ethnography of a Seattle gay bar (*Other Voices*).

39. Bennett, Rhodes, and Robson, "Observations on kuru. I"; and Bennett, Gray, and Auricht, "Genetical study of kuru."

40. For a description of Gray's fieldwork, see Gray to Scragg, October 20, 1959, kuru research papers file, box 2003/042, IMR Archives. By October 1959, Gray had conducted twenty autopsies. Malcolm Fowler, an Adelaide pathologist, examined most of their brains. See Fowler and E. Robertson, "Observations on kuru. III." (This paper is based on material sent earlier by Gajdusek.)

41. Bennett to Brigadier Cleland, administrator of PNG, January 15, 1959, Gunther Papers.

42. Bennett and Robson to administrator, September 3, 1959, Gunther Papers.

43. Zigas and Gajdusek, "Kuru."

44. Sir John Gunther, "Australia, kuru and a Nobel Prize," n.d., typescript, Gunther Papers.

45. Interview with J. Henry Bennett, December 21 and 22, 2005, Adelaide. Bennett had trained in mathematics under Thomas Cherry at the University of Melbourne before completing his Ph.D. with Fisher at Cambridge. Bennett initially struck Zigas as "quite a young chap, English conservative type . . . a good listener" (Zigas to Gajdusek, February 1, 1959, box 36, Gajdusek Correspondence, NLM).

46. "Report of Committee Appointed by the Administrator to Consider the Administrative and Public Health Aspects of Kuru," December 21, 1959, file 54/2/5, box 5360, series 153, accession 23, PNGNA. Other members of the committee were Edward Ford (director of the School of Public Health and Tropical Medicine, University of Sydney), Hugh K. Ward (professor of bacteriology, University of Sydney), Robert J. Walsh (a geneticist and director of the New South Wales Division of the Australian Red Cross Blood Transfusion Service), Charles Julius, and Scragg, with Fisher as an adviser. Bennett's submission to the committee and Burnet's handwritten notes are in the Burnet Papers.

47. "Report of Committee Appointed by the Administrator to Consider the Administrative and Public Health Aspects of Kuru," December 21, 1959, file 54/2/5, box 5360, series 153, accession 23, PNGNA.

48. Burnet to Scragg, January 5, 1960, folder 3, series 10, Burnet Papers. Burnet also

reported that his son Ian had said that patrol officers much preferred Gajdusek to the "academics."

49. Smadel to Burnet, February 10, 1960, folder 3, series 10, Burnet Papers.

50. Roy to John, December 23, 1959, Gunther Papers.

51. F. D. Schofield, "The future of medical research in the territory of Papua and New Guinea," November 9, 1959, typescript, file 54/10/2, box 5364, series 153, accession 23, PNGNA.

52. Gajdusek, August 13, 1959, in *New Guinea Journal, 1959*, 146. Schofield came to despair of ever reconciling "two groups of scientists who do not collaborate with each other and have exhibited not only personal but even national rivalries" (Schofield to administrator, PNG, March 22, 1962, file 54/10/2, box 5364, series 153, accession 23, PNGNA).

53. Interview with Frank D. Schofield and Robert McLennan, January 2, 2006, Brisbane, Queensland. Schofield left New Guinea in 1964 to take a job with the WHO in Addis Ababa, Ethiopia. While in New Guinea he conducted important studies of tetanus of the newborn. See Schofield and Parkinson, "Social medicine in New Guinea."

54. Bennett to Brigadier Cleland, administrator of PNG, January 15, 1959, Gunther Papers.

55. "Report of Committee Appointed by the Administrator to Consider the Administrative and Public Health Aspects of Kuru," December 21, 1959, file 54/2/5, box 5360, series 153, accession 23, PNGNA. Burnet was vigorously opposed to Fore isolation.

56. Roy to John, December 23, 1959, Gunther Papers. Robson had told Scragg directly that his and Bennett's opinions "do not exactly coincide" (Robson to Scragg, December 15, 1959, Gunther Papers). Gajdusek earlier observed the discrepancy between Robson's "complete assurance of total cooperation" and Bennett's "instructions for NO cooperation, no sharing of data or experience, and as little contact as possible" (Gajdusek to Masland, Smadel, and Bailey, June 30, 1959, box 26, Gajdusek Correspondence, NLM).

57. Bennett to administrator, February 24, 1960, Gunther Papers.

58. Gunther, "Australia, Kuru and a Nobel Prize," Gunther Papers, 33. Gunther claimed it was a justified eugenic measure, and if enforced, further euthenic (educational and economic) interventions would have accompanied it. For instructions to limit movement out of the region, see J. K. McCarthy, acting director, Department of Native Affairs, to district officer, Goroka, May 16, 1960, cabinet 1.4, Alpers Papers. See also John A. Osmundsen, "Tribe in New Guinea is isolated to curb rare inherited disease," *New York Times* (July 18, 1960): 45.

59. Gajdusek and Zigas, "Studies in kuru."

60. Dobzhansky, "Eugenics in New Guinea," 77.

61. "Report of Meeting of Ad-hoc Committee on Medical Research, Adelaide, May 19–20, 1962," file 54/2/5, box 5360, series 153, accession 23, PNGNA.

62. Gajdusek to Zigas, March 6, 1959, box 36, Gajdusek Correspondence, NLM. This research program is reiterated in Gajdusek to Burnet, December 3, 1959, folder 3, series 10, Burnet Papers. Zigas continued to work as a district medical officer until retirement and never went on to develop the research career he wanted.

63. Gajdusek to Masland, Smadel, and Bailey, June 30, 1959, box 26, Gajdusek Correspondence, NLM.

64. Gajdusek to Masland, July 7, 1959, box 26, Gajdusek Correspondence, NLM.

65. Gajdusek to Masland and Smadel, January 30, 1960, box 26, Gajdusek Correspondence, NLM. Kurland's visit, however, started badly. "There was some delay in Dr. Kurland getting

through Customs' facilities," H. N. White, the assistant director of public health, reported to the assistant administrator, "as he had brought a small automatic pistol, apparently on the advice of Dr. Zigas and Dr. Gajdusek" (February 22, 1960, file 54/2/11, box 8743, and file 54/10/2, box 5364, series 153, accession 23, PNGNA).

66. Schofield to Charles B. Wisseman, July 11, 1962, box 31, Gajdusek Correspondence, NLM. Wisseman, a microbiology professor at the University of Maryland, frequently collaborated with the NINDB team.

67. Burnet to Terry Abbott, September 17, 1964, file 54/2/5, box 5360, series 153, accession 23, PNGNA.

68. Abbott to Burnet, September 17, 1964, file 54/2/5, box 5360, series 153, accession 23, PNGNA.

69. Bennett to Robson, April 7, 1959, Gunther Papers.

70. Gajdusek to Fenner, December 28, 1965, file 21/5, box 39, Fenner Papers, ms. 143, Basser Library, Australian Academy of Science, Canberra.

71. Fenner to Gajdusek, January 23, 1966, Fenner Papers. Fenner was professor of microbiology at the ANU and a former protégé of Burnet. Michael Alpers recalls talking with him about Gajdusek, soon realizing "they were obviously worried about Carleton's homosexuality as well as their inability to control his scientific activities" (Alpers to Warwick Anderson, January 31, 2007, in the possession of Warwick Anderson).

72. Paul Siple to Burnet, January 21, 1966, Fenner Papers; and file 54/2/5, box 5360, series 153, accession 23, PNGNA. As an eagle scout, Siple had accompanied Richard E. Byrd to Antarctica in 1928, and later returned there on the 1933 and 1939 expeditions. In the 1940s he coined the term "wind-chill factor."

73. Siple to Scragg, January 21, 1966, Fenner Papers; and file 54/2/5, box 5360, series 153, accession 23, PNGNA.

74. Burnet to Robson, May 31, 1965, in Burnet file, box 6, Gajdusek Correspondence, NLM. Burnet felt Gajdusek was not malicious—it was just the result of "his manic necessity to be always doing, understanding and talking."

75. Hadlow to Gajdusek, July 21, 1959, box 19, Gajdusek Correspondence, NLM; also in Prusiner et al., *Prion Diseases of Humans and Animals,* 47–48. See also Hadlow, "Scrapie and kuru." Hadlow concluded that it might be "profitable" to "examine the possibility of the experimental induction of kuru in a laboratory primate" (290). Hadlow was based at the Rocky Mountain laboratories of the NIH in Montana. For his later reflections on his involvement in kuru research see Hadlow, "The scrapie-kuru connection"; and Hadlow, "Neuropathology and the scrapie-kuru connection." See also Boggs, "Biographical sketch of Dr. William Hadlow and his career at Rocky Mountain Labs, in Hamilton, Montana" (Office of NIH History, 2002, typescript in the possession of Warwick Anderson). Hadlow's general observation of the similar pathological pictures of scrapie and kuru was confirmed by Beck, Daniel, and Parry, "Degeneration." See also Beck and Daniel, "Prion diseases."

76. The first report of the transmissibility of scrapie was Cuillé and Chelle, "Pathologie animale." "Slow infections" of sheep were first described, and the criteria for their diagnosis defined, in the laboratory of Björn Sigurdsson in Iceland in the 1950s. See Sigurdsson, "Observations on three slow infections of sheep." For an overview, see Pattison, "Sideways look at the scrapie saga."

77. Gajdusek to H. K. Ward, December 3, 1959, folder 3, series 10, Burnet Papers.

78. Gajdusek to Masland and Smadel, December 15, 1959, box 26, Gajdusek Correspondence, NLM. Even in 1964 he was doubtful. He wrote to Carl Eklund that "we must face the fact that [inoculation of primates] may be a complete bust as far as human disease is concerned" (February 25, 1964, box 12, Gajdusek Correspondence, NLM).

CHAPTER 6. SPECIMEN DAYS

1. I use "fetish" here in the Marxist sense of the commodity form obscuring social relations. See Pietz, "Problem of the fetish I." Pietz claims that the fetish is a concept-thing that comes about in the moment of contact of two cultures.

2. For a functionalist account of gift giving among scientists, see Hagstrom, *Scientific Community*. Bruno Latour and Steve Woolgar take issue with the normative aspects of Hagstrom's argument in *Laboratory Life,* esp. 203–4. Recently, Robert E. Kohler has described the "moral economy" of laboratory researchers in *Lords of the Fly.* My own approach follows Pierre Bourdieu's account of the strategic use of reciprocity among the Kabyle (*Outline of a Theory of Practice*). Similarly, Mario Biagioli acknowledges a debt to Bourdieu in *Galileo Courtier.*

3. Sahlins, *Stone Age Economics,* chap. 5.

4. Sahlins, "Poor man, rich man, big-man, chief."

5. M. Strathern, *Gender of the Gift,* 164.

6. I do not want to suggest a homogeneous unity in either economy—the heterogeneity of the Fore economy is evident in earlier chapters, and the emergence of market forces in science becomes clearer later in the book. I use the term "gift economy" to point to the predominant form, not the singular mode, of exchange among Fore and scientists in this particular time and place.

7. Gajdusek to Burnet, June 27, 1958, folder 3, series 10, Burnet Papers.

8. Stetten Jr., and Carrigan, *NIH;* Harden, *Inventing the NIH;* Fox, "Politics of the NIH"; and Park, "Development of the intramural research program."

9. On cold war science, see Kevles, "Principles and politics"; Kevles, "Cold war and hot physics"; Leslie, *Cold War and American Science;* and Rasmussen, "Mid-century biophysics bubble." The earlier research of Smadel on hemorrhagic fevers and Shannon on malaria reinforced the international orientation of these scientists.

10. Kurland and Mulder, "Epidemiologic investigations of amyotrophic lateral sclerosis." Gajdusek also became involved in research on Guam ALS. See Chen, *Mysterious Diseases of Guam.*

11. Rowland, *NINDS at 50;* and Farreras, Hannaway, and Harden, *Mind, Brain, Body and Behavior.*

12. The basic science argument also could be deployed successfully in New Guinea, of course. Difficult as the Australians found it to involve the NIH in kuru research, the intrusion of the U.S. Centers for Disease Control would have been an unimaginable insult.

13. Gajdusek to Adels, March 3, 1959, box 1, Gajdusek Correspondence, NLM.

14. Gajdusek to Adels, April 17, 1959, box 1, Gajdusek Correspondence, NLM.

15. Adels to Gajdusek, July 1, 1959, box 1, Gajdusek Correspondence, NLM. In 1962 Adels returned to the NINDS as a biologist.

16. Gajdusek to Masland, July 7, 1959, box 26, Gajdusek Correspondence, NLM. Jack and

Lois married in June 1959 and arrived in Bethesda in September, just as Adels was leaving for the first time.

17. Interview with Jack and Lois Baker, January 3, 2006, Bribie Island.

18. Early in 1957, S. Gray Anderson at the Hall Institute had inoculated chick embryos and mice with kuru brain material, but he did not monitor the experiment for more than a few weeks, the standard period for acute infections.

19. The U.S. Department of Agriculture had banned the import of scrapie material since 1947 and was attempting to eradicate the disease, which broke out in 1947 (Michigan), 1952 (Colorado), and 1954 (Ohio). Uncomfortable now with his own hypothesis, Bennett briefly explored possible virological studies of scrapie and kuru in Adelaide, but it was beyond his expertise so he soon abandoned the idea. See W. H. Howarth to Bennett, March 22, 1962; Derek Rowley to Bennett, April 3, 1962; and H. B. Parry to Bennett, April 19, 1962; all in the possession of Warwick Anderson.

20. Gajdusek to Hadlow, August 23, 1962, box 30, Gajdusek Correspondence, NLM. See Morris and Gajdusek, "Encephalopathy of mice." In England, Richard L. Chandler had beaten them to it: "Encephalopathy in mice produced with scrapie brain material."

21. Clarence J. Gibbs Jr., "Spongiform encephalopathies: Slow, latent, and temperate virus infections: In retrospect," c. 1992, typescript, box 18, Gajdusek Correspondence, NLM, 3. This essay was revised and reprinted in Prusiner et al., *Prion Diseases of Humans and Animals.*

22. Gajdusek to Gibbs, January 18, 1962, box 18, Gajdusek Correspondence, NLM.

23. Gajdusek to Gibbs, January 18, 1962, box 18, Gajdusek Correspondence, NLM. By "temperate" he means those viruses that exist dormant in the bacterial cell (as "pro-viruses") only to be activated into a reproductive state by external factors, leading to lysis, or breakup, of the cell. It was this analogy to phage that gave rise to the category of "slow, latent, and temperate" viruses.

24. Clarence J. Gibbs Jr., "Spongiform encephalopathies: Slow, latent, and temperate virus infections: In retrospect," c. 1992, typescript, box 18, Gajdusek Correspondence, NLM, 45. Gibbs remarked that "I have developed a friendship with Carleton Gajdusek that, despite the differences in our approach to problems, has provided for a unique form of scientific accomplishments—a relationship that has been great to experience and to treasure but one which is not definable" (20).

25. Gajdusek to Schofield, August 30, 1962, box 31, Gajdusek Correspondence, NLM.

26. Gajdusek to Schofield, September 24, 1962, box 31, Gajdusek Correspondence, NLM.

27. Thirty years later this monkey was alive and well.

28. Gajdusek, "Kuru in New Guinea," 4–5.

29. Clarence J. Gibbs Jr., "Spongiform encephalopathies: Slow, latent, and temperate virus infections: In retrospect," c. 1992, typescript, box 18, Gajdusek Correspondence, NLM.

30. Eklund, Hadlow, and Kennedy, "Properties of the scrapie agent."

31. Gajdusek and Gibbs Jr., "Preface," in *Slow, Latent, and Temperate Virus Infections,* ix. For an extensive report on the conference, see John A. Osmundsen, "New viruses tied to chronic ills," *New York Times* (December 13, 1964): 81. The cause of visna turned out to be a retrovirus.

32. Gajdusek, "Kuru in New Guinea," 11.

33. Sorenson arrived in the Fore region in September 1963 to initiate a study of child behavior and development, and he conducted a further four trips there between 1963 and 1970.

Sorenson received his Ph.D. degree in anthropology from Stanford in 1971 and went to work at the National Anthropological Film Center at the Smithsonian. See Sorenson and Gajdusek, "Investigation of nonrecurring phenomena"; Sorenson and Gajdusek, "Study of child behavior"; Sorenson, "Socio-cultural change"; and Sorenson, *Edge of the Forest*. In 1964, Gajdusek also encouraged the first New Guinea field work of Jared Diamond, the son of one of his teachers at Boston Children's Hospital. Diamond analyzed Fore classifications of nature: see "Zoological classification system."

34. Whitbread, "Stalking a new kind of killer," 84. Gajdusek's relaxed "management style" perhaps reflects widespread concern in the 1960s about the rise of hierarchical and industrialized "big science" in the United States. He certainly sought to avoid becoming what Norbert Wiener called a "morally irresponsible stooge in a science-factory" ("Rebellious scientist," 338). See Price, *Little Science, Big Science;* and Zilsel, "Mass production of knowledge."

35. Judith Farquhar, from her recollections presented at the End of Kuru meeting, Royal Society, London, October 11, 2007. Farquhar became a professor of anthropology at the University of Chicago and studied health and medicine in China. Richard Benfante and Ralph Garruto also were among the anthropologists who passed through the Bethesda lab.

36. D. C. Gajdusek to Mahtil Gajdusek, March 31, 1962, box 15, Gajdusek Correspondence, NLM.

37. Gajdusek to Adels, March 18, 1966, box 1, Gajdusek Correspondence, NLM.

38. Sorenson to Gajdusek, February 23, 1965, box 32, Gajdusek Correspondence, NLM.

39. Sorenson to Gajdusek, March 12, 1966, box 32, Gajdusek Correspondence, NLM.

40. Gajdusek to Adels, May 20, 1972, box 1, Gajdusek Correspondence, NLM.

41. Gajdusek to Adels, March 11, 1967, box 1, Gajdusek Correspondence, NLM.

42. Gajdusek to Frank [Schofield], April 16, 1962, file 54/2/5, box 5360, series 153, accession 23, PNGNA.

43. Gajdusek to Scragg, August 30, 1962, file 54/2/5, box 5360, series 153, accession 23, PNGNA.

44. Gajdusek, March 8, 1962, in *New Guinea Journal, Part II,* 9.

45. Robson to H. N. White, acting director of public health, July 19, 1960, file P2605, box 4732, accession 22, Papua New Guinea Department of Health Archives, Port Moresby, PNG.

46. Alpers to Scragg, February 21, 1961, file P2605, box 4732, accession 22, Papua New Guinea Department of Health Archives. As a disaffected medical student, Alpers had begun reading widely in anthropology, partly through the influence of C. Stanton Hicks, who studied Aboriginal physiology and adaptation and was a family friend. Alpers planned to work in Aboriginal communities until he saw reports about kuru in the Adelaide press and approached Robson.

47. Alpers to Scragg, December 16, 1961, file P2605, box 4732, accession 22, Papua New Guinea Department of Health Archives.

48. Interview with Michael P. Alpers, July 24, 2002, Fremantle.

49. Gajdusek, December 13, 1961, in *New Guinea Journal, Part I,* 125.

50. Gajdusek, February 6, 1962, in *New Guinea Journal, Part I,* 292. Zigas recalled Alpers as "a supremely confident yet self-effacing man, gracious in manner, polite in speech, but implacably stubborn" (Zigas, *Laughing Death,* 299).

51. Alpers, "Reflections and highlights," 68.

52. Interview with Michael P. Alpers, July 24, 2002, Fremantle. Alpers continued with his clinical and epidemiological studies: see his "Kuru: A clinical study" (U.S. Department of Health, Education, and Welfare, typescript in the possession of Warwick Anderson, c. 1963), and the discussion of Alpers's epidemiological research in Chapter 7.

53. Alpers still supports the local preschools and helps pay school fees for children from Waisa and other Fore communities. "Michael" and "Alpers" are now popular Fore names, as are "Carleton" and "Dickson," an abbreviation of Dick Sorenson.

54. On the importance of age-mates and their reciprocal obligations, see Lindenbaum and Glasse, "Fore age mates."

55. Interview with Pako, August 6, 2003, Waisa.

56. Interview with Michael P. Alpers, July 24, 2002, Fremantle.

57. Eiro case record, cabinet 3.3, Alpers Papers.

58. Kigea case record, cabinet 3.3, Alpers Papers.

59. Interview with Michael P. Alpers, July 24, 2002, Fremantle. In my interview with him on August 4, 2004, Alpers told me that doing an autopsy was "very emotional, the whole thing, because all of these patients were people I knew extremely well. They were *close* to me, and having to do the autopsy was quite a task."

60. Kigea case record, cabinet 3.3, Alpers Papers.

61. Gajdusek to Schofield, January 8, 1963, box 31, Gajdusek Correspondence, NLM.

62. Gajdusek to Schofield, September 24, 1962, box 31, Gajdusek Correspondence, NLM. This included Eiro's brain. See also Gajdusek to Alpers, September 21, 1962, box 4, Gajdusek Correspondence, NLM—this letter makes it clear that Alpers's reward will be a position at the NIH. Transport of specimens remained erratic and uncertain, however. In 1967 Gajdusek wrote to Adels: "All of our shipments are arriving incorrectly and many are not delivered . . . this is the same problem—EXACTLY the same—that has faced us years and years over. IT MUST BE SOLVED" (February 15, 1967, box 1, Gajdusek Correspondence, NLM).

63. Malinowski, *Argonauts of the Western Pacific;* Fortune, *Sorcerers of Dobu;* and A. Strathern, "Kula in comparative perspective." As Malinowski describes it, the *kula* unites with social bonds a vast area, making participants "follow minute rules and observations in a concerted manner" (510).

64. Schofield to Gajdusek, October 8, 1963, box 31, Gajdusek Correspondence, NLM.

65. Paul Brown, November 6, 1963, in "New Guinea" (manuscript in the possession of Warwick Anderson), 4. Brown completed his M.D. at Hopkins and thought he might fulfill his selective service requirement through working at the NIH. He started in Gajdusek's lab in spring 1963, and by August was visiting the Caroline Islands, studying the durability of measles vaccine in an isolated population in which there was a presumed absence of inapparent circulating "boosters." Bown told me that "the idea of traveling to exotic places in the western Pacific was a little intimidating, a little anxiety producing, but also fairly exciting" (interview with Paul Brown, January 25, 2006, Chevy Chase, MD).

66. Brown, November 7, 1963, in "New Guinea," 5.

67. Sorenson, November 6, 1963, in *Journal Account*, 73. Sorenson had earlier made available to the kin three blankets and several *laplaps*.

68. Brown, November 14, 1963, in "New Guinea," 21–22.

69. Interview with Paul Brown, January 25, 2006, Chevy Chase.

70. Sorenson, November 23, 1963, in *Journal Account of the Expedition,* 81.

71. Brown, November 22, 1963, in "New Guinea," 29. On the problems shipping Igi'erak-aba's brain see Brown to Gajdusek, January 2, 1964, box 6, Gajdusek Correspondence, NLM. For her case record, see Brown to Gajdusek, November 17, 1963, cabinet 1.3, Alpers Papers.

72. Sorenson, December 2, 1963, in *Journal Account of the Expedition,* 98.

73. Brown, December 3, 1963, in "New Guinea," 44. As Brown later reflected, "little did we know that I could have put [the brain tissue] in my pocket, and come around the world on a slow freighter, and it would still have been infectious when I got back" (interview with Paul Brown, January 25, 2006, Chevy Chase).

74. Brown, December 10, 1963, in "New Guinea," 50.

75. Interview with Michael P. Alpers, July 24, 2002, Fremantle.

76. Interview with Paul Brown, January 25, 2006, Chevy Chase.

77. Gajdusek, December 13, 1961, in *New Guinea Journal, Part I,* 125. "In effect, we are two of a kind," Gajdusek reflected, "both rather outspoken, both proud, both egocentric and one-minded, knowing our own minds" (125). Brightwell was respected in the Department of Native Affairs for his integrity and toughness. In 1961 he rigorously investigated a controversial incident at Wonenara after D. K. McCarthy refused. See Sinclair, *Kiap.* As D. K. Burfoot, the district officer for the Eastern Highlands, put it: "With Mr. Brightwell in charge at Okapa I am quite certain that medical researchers will no longer be allowed to ride roughshod over individual rights" (to director, Department of Native Affairs, November 9, 1961, with Okapa Patrol Report No. 3, 1960/61, ms 53, Mandeville Special Collections Library, University of California–San Diego, CA).

78. Brightwell to Schofield, October 4, 1962, file 54/2/5, box 5360, series 153, accession 23, PNGNA.

79. Brightwell to Gajdusek, June 15, 1964, box 6, Gajdusek Correspondence, NLM.

80. Brown, November 7, 1963, in "New Guinea," 7. Inamba remembered Brightwell saying, "call me *fatwel,* I have a big stomach, call me *fatwel*" (interview with Inamba, August 4, 2003, Ivingoi). "But his real name was Meka," Inamba continued.

81. Sorenson, October 23, 1963, in *Journal Account of the Expedition,* 60. But he agreed that "Mert is always looking for things which show Americans as bunglers" (November 30, 1963, in *Journal Account of the Expedition,* 93).

82. Interview with Michael P. Alpers, July 24, 2002, Fremantle.

83. Schofield to Scragg, October 12, 1962, file 54/2/5, box 5360, series 153, accession 23, PNGNA. Brown remembers Brightwell pulling up in his Land Rover and gleefully telling him President John F. Kennedy had been shot. "He seemed to think this was a huge joke on the American public, and since I was the only member of the American public there, he would stick it to me—with a smile, with a smirk, with a laugh. Tells you something about Brightwell's mentality" (interview with Paul Brown, January 25, 2006, Chevy Chase).

84. Gajdusek to R. J. Walsh, December 2, 1959, Gunther Papers.

85. Pietz, "Spirit of civilization"; and Pietz, "Death of the deodand."

86. Munn, *Fame of Gawa.*

87. Frow, "Invidious distinctions."

88. Gajdusek to Scragg, November 3, 1959, folder 27, box 3, series 3, RG 410A, Rockefeller Foundation Archives 1.2, Rockefeller Archive Center.

89. A. Strathern, *Rope of Moka,* 227.

90. E. Graeme Robertson to J. Godwin Greenfield, NINDB, October 31, 1957, in Farquhar and Gajdusek, *Kuru,* 230.

91. Gajdusek to Robertson, November 13, 1957, in Farquhar and Gajdusek, *Kuru,* 246.

92. Gajdusek to Smadel, December 7, 1957, in Farquhar and Gajdusek, *Kuru,* 266.

93. Gajdusek to Smadel, December 24, 1957, in Farquhar and Gajdusek, *Kuru,* 270. For Sunderland's reports, see Kuru neuropathology, brain of Ereio, folder A0122, Sydney Sunderland Papers, acc. 96/35, UMA. I am grateful to Ross Jones for drawing my attention to this report.

94. Gajdusek to Fenner, February 4, 1966, Fenner Papers.

95. Gajdusek, March 22, 1959, in *Journal of the Continued Quest,* 39–40.

96. M. Strathern, "Divisions of interest"; and M. Strathern, "Potential property."

97. Gajdusek to Alpers, August 28, 1965, box 4, Gajdusek Correspondence, NLM.

98. Alpers, "Reflections and highlights," 71.

99. Gibbs and Gajdusek first reported the development of this cerebellar degeneration in the inoculated chimps in an addendum to their essay, "Attempts to demonstrate a transmissible agent," 46. They also expressed a temporary reservation: "Whether these syndromes are spontaneous or related to the inoculation remains to be determined" (46).

100. Lynch, "Sacrifice and transformation."

101. Beck and Daniel, "Prion diseases," 64.

102. Gajdusek, Gibbs Jr., and Alpers, "Experimental transmission." Gajdusek was sending out off-prints of this article in August 1966. See also Beck et al., "Experimental 'kuru' in chimpanzees."

103. Gajdusek to Fenner, December 28, 1965, file 21/5, box 39, Fenner Papers.

104. Gunther to Schofield, March 17, 1966, Gunther Papers. Schofield was by then professor of public health in Addis Ababa, Ethiopia.

105. F. Macfarlane Burnet, "The pathogenesis of kuru: Speculations based on new observational material," c. 1964, typescript, folder 7, series 10, Burnet Papers, 17. See also the minutes of the Meeting to Discuss the Pathogenesis of Kuru, Ciba Foundation, London, June 15, 1964, typescript, folder 7, series 10, Burnet Papers. Burnet became increasingly interested in kuru in the mid-sixties, courageously declaring that a "proper elucidation of the etiology and pathogenesis of kuru would do more for medicine than any other specific research project available today" ("Kuru—the present position," 3).

106. Burnet, *Natural History of Infectious Disease;* and Anderson, "Natural histories of infectious disease." See also Keller, *Century of the Gene;* and Lindee, *Moments of Truth.*

107. Burkitt, "A 'tumour safari'"; and Epstein, Achong, and Barr, "Virus particles." See Emm Barnes, "Networks, microbes, and pills: Safaris, missionaries, and mosquitoes" (typescript in the possession of Warwick Anderson, 2007). I am grateful to Emm Barnes for sharing this with me.

108. Bayer, Blumberg, and Werner, "Particles associated with Australia antigen." See also Stanton, "Blood brotherhood"; and Blumberg, *Hepatitis B.*

109. Gajdusek, Gibbs Jr., and Alpers, "Transmission and passage"; and Gajdusek et al., "Transmission of experimental kuru."

110. Gibbs Jr. et al., "Creutzfeldt-Jakob disease."

111. Masters, Gajdusek, and Gibbs Jr., "Creutzfeldt-Jakob disease virus isolations." On the identification of fatal familial insomnia as a slow virus disease, see Max, *Family That Couldn't Sleep*.

112. Brown, Preece, and Will, "'Friendly fire' in medicine"; and Brown, "Environmental causes of human spongiform encephalopathy."

113. Gibbs Jr., "Spongiform encephalopathies."

CHAPTER 7. WE WERE THEIR PEOPLE

1. Leonard B. Glick, a Ph.D. student in anthropology at the University of Pennsylvania, was also conducting pioneering ethnographic studies of health beliefs among the neighboring Gimi in this period. See his "Medicine as an ethnographic category."

2. Zigas to Gajdusek, January 2, 1959, box 36, Gajdusek Correspondence, NLM.

3. Zigas, *Laughing Death*, 282. Fortune was among the Fore from September through December 1958.

4. Fortune, "Statistics of kuru." Bennett told Gunther that Fortune did not know what he was talking about (June 3, 1960, in Gunther Papers).

5. Fortune to Gajdusek, April 20, 1961, box 13, Gajdusek Correspondence, NLM.

6. W. R. Geddes to Bennett, April 26, 1960, in Gunther Papers.

7. Bennett to Gunther, May 26, 1960, in Gunther Papers.

8. J. A. Barnes to Bennett, February 7, 1961, in Gunther Papers. Robert Glasse did his field work at Tari in 1955–56 and 1959–60. First Derek Freeman and then Barnes became his thesis supervisors. See Glasse, "Huli descent system"; and Glasse, "Revenge and redress." Glasse's work with the Huli revealed to him the "optative, contingent character of group membership" among highlanders ("Encounters with the Huli," 245). While at Sydney, Shirley Glasse was entranced by Mervyn Meggitt's tales of field work in New Guinea.

9. Bennett to Gunther, February 28, 1961, in Gunther Papers. Ken Inglis, who taught history at Adelaide University during this period, later became vice-chancellor of the University of Papua New Guinea.

10. Interview with Shirley Lindenbaum, March 17, 2005, New York City. The Glasses stayed at Wanitabe from July 1961 until March 1962, then again from July 1962 until May 1963: see Lindenbaum, *Kuru Sorcery*. See also Glasse and Lindenbaum, "Kuru at Wanitabe"; and Glasse and Lindenbaum, "Fieldwork in the South Fore."

11. Interview with Shirley Lindenbaum, March 17, 2005, New York City.

12. Glasse and Lindenbaum, "Fieldwork in the South Fore," 80.

13. Interview with Shirley Lindenbaum, March 17, 2005, New York City.

14. Interview with Shirley Lindenbaum, March 17, 2005, New York City. Later, Shirley told me: "It would be better to say that I sat under the trees a lot, while they sometimes joined me, but mostly worked hard" (Lindenbaum to Warwick Anderson, February 15, 2007, in the possession of Warwick Anderson).

15. Interview with Inamba, August 4, 2003, Ivingoi.

16. Interview with Kivengi, August 5, 2003, Ivingoi. Shirley returned in 1973, 1993, 1996, and 1999, to revise genealogies collected in the 1960s. She recognized Kivengi on each occasion.

17. Interview with Shirley Lindenbaum, March 17, 2005, New York City.

18. R. M. Glasse, "South Fore society: A preliminary report" (typescript in the possession of Warwick Anderson, c. June 1962, courtesy of Henry Bennett).

19. R. M. Glasse, "The spread of Kuru among the Fore" (typescript in the possession of Warwick Anderson, c. June 1962, courtesy of Henry Bennett), 2. Gajdusek and Zigas also heard that kuru was a new problem, but they discounted the possibility: see "Kuru," 461. See also Gajdusek's journal entry for February 17, 1962, in *New Guinea Journal, Part I,* 341.

20. R. M. Glasse, "Spread of kuru among the Fore," c. June 1962, 3–4.

21. Lindenbaum to Warwick Anderson, February 15, 2007, in the possession of Warwick Anderson.

22. R. Glasse, "South Fore cannibalism and kuru" (typescript in the possession of Warwick Anderson, c. 1962, courtesy of Henry Bennett), 4. The relatively recent uptake of cannibalism is discussed in R. Glasse, "Cannibalism in the kuru region" (typescript in the possession of Warwick Anderson, c. 1964, courtesy of Henry Bennett), esp. 14. See also R. M. Glasse, "Cannibalism in the kuru region of New Guinea"; and Lindenbaum, "Thinking about cannibalism."

23. Burnet, "Kuru." Burnet was admitting his mistake. Even at the time, the Glasses' research impressed him. "Tropical medicine generally," he wrote, "could benefit from these techniques" ("Minutes of the Papua and New Guinea Medical Research Advisory Committee, Canberra, October 25, 1964," in kuru historical file, cabinet 1.2, Alpers Papers).

24. Bennett to H. B. Parry, Nuffield Institute for Medical Research, Oxford, April 30, 1962, in the possession of Warwick Anderson, courtesy of Henry Bennett.

25. H. B. Parry to Bennett, November 19, 1962, in the possession of Warwick Anderson, courtesy of Henry Bennett.

26. For example, Evans-Pritchard, "Zande cannibalism"; R. Berndt, *Excess and Restraint;* Sagan, *Cannibalism;* and Sahlins, "Cannibalism." For the 1970s, see Kidd, "Scholarly excess."

27. Gajdusek, "Kuru," 162.

28. See also Gajdusek, "Kuru in New Guinea," 3. In the reported discussion at the 1964 conference, Gajdusek had said: "I do not believe cannibalism is the answer . . . the suggestion is more in the nature of an embellishment to other etiological hypotheses" (81–82).

29. A. Fischer and J. L. Fischer, "Aetiology of kuru"; A. Fischer and J. L. Fischer, "Culture and epidemiology." See also Williams et al., "Evaluation of the kuru genetic hypothesis."

30. A. J. Gray, "Report for the Period February 1959–October 1960," in patrol reports file, cabinet 1.4, Alpers Papers.

31. Interview with Frank Schofield, January 2, 2006, Brisbane, Queensland.

32. Shirley Glasse, "The social effects of kuru" (typescript in the possession of Warwick Anderson, June 1962, courtesy of Henry Bennett).

33. Ibid., 16. In this paper, Shirley Glasse relied on Evans-Pritchard, *Witchcraft, Oracles and Magic.*

34. Lindenbaum, *Kuru Sorcery,* viii, 146. See also Lindenbaum, "Sorcery and structure."

35. Interview with Shirley Lindenbaum, March 17, 2005, New York City. The phrase echoes Charles E. Rosenberg's argument for viewing epidemic disease as a social "sampling technique" (*Cholera Years,* 4). After their New Guinea experiences the Glasses worked for three years on cholera in East Pakistan (Bangladesh), where they separated. Bob taught at Queens College, New York, between 1965 and 1990, and died in 1993. Shirley spent the rest of her career at the New School for Social Research, New York, and the CUNY Graduate Center, and became the

second president of the Society for Medical Anthropology. In the 1980s and 1990s she continued to work on social aspects of kuru and began to study another emergent disease, HIV/AIDS. See Lindenbaum and Lock, *Knowledge, Power, and Practice*.

36. Alpers to Gajdusek, November 27, 1962, box 1, Gajdusek Correspondence, NLM. Gajdusek admitted having discounted claims of the recent origin of kuru, but argued that since he was proceeding to inoculate chimps anyhow as though the disease resulted from infection, the issue was not decisive (Gajdusek to Sorenson, February 26, 1962, box 32, Gajdusek Correspondence, NLM).

37. Interview with Shirley Lindenbaum, March 17, 2005, New York City. For a contemporary plea for engagement between the social sciences and epidemiology, see Cassel, "Social science theory." Yet even in the 1980s it was still commonplace to deplore the lack of connection between infectious diseases research and social anthropology: see James, Stall, and Gifford, *Anthropology and Epidemiology*.

38. Interview with Michael Alpers, July 24, 2002, Fremantle.

39. Gajdusek, Zigas, and Baker, "Studies on kuru. III"; and Bennett, "Population studies."

40. Gajdusek and Zigas, "Studies on kuru. I." See also Gajdusek, "Kuru," 155.

41. McArthur, "Age incidence of kuru." The demographer stayed with the Glasses when she briefly visited the region. McArthur's major work was *Island Populations of the Pacific*. She later retrained as a Pacific archeologist at the ANU.

42. Alpers and Gajdusek, "Changing patterns of kuru," 878; and Alpers, "Epidemiological changes in kuru," in Gajdusek, Gibbs Jr., and Alpers, *Slow, Latent, and Temperate Virus Infections*. Bennett had hinted that the disease was becoming rare in younger Fore in "Population studies in the kuru region." He and his colleagues discussed this "striking" change in Bennett, Gabb, and Oertel, "Further changes in the pattern of kuru."

43. This trend was confirmed in Alpers, "Kuru in New Guinea"; and Alpers, "Epidemiology and ecology of kuru." The last child less than ten years old to die of kuru died in 1967; but the disease may not yet, even in 2008, have disappeared completely.

44. Alpers, "Reflections and highlights," 72. As William Arens put it, the "anthropological fixation on cannibalism in the field [had become] more compatible with laboratory experiments" (*Man-Eating Myth*, 109).

45. Alpers presented this theory at the 1967 meeting of the International Academy of Pathology. See Alpers, "Kuru: Implications."

46. Gajdusek to Scragg, August 30, 1962, file 54/2/5, box 5360, series 153, accession 23, PNGNA.

47. Interview with Michael Alpers, August 4, 2004, Fremantle. Hornabrook got on well with Brightwell, though clashing repeatedly with John F. Stephens, the medical officer assisting him. Zigas accused Hornabrook of "professional pride and jealousy" and of "doing all he could to tie Gajdusek and Alpers' hands, preventing their access to any data he could keep from them" (*Laughing Death*, 306).

48. Gajdusek, November 5, 1965, in *Melanesian and Micronesian Journal*, 165.

49. Gajdusek, February 15, 1965, in *Melanesian Journal, January 23, 1965–April 7, 1965*, 46, 47.

50. Hornabrook, "Kuru—some misconceptions and their explanation"; Hornabrook, "Kuru—a subacute cerebellar degeneration"; and idem, "Kuru: The disease."

51. Hornabrook to W. D. Symes, Medical Research Division, Port Moresby Hospital, December 28, 1965, file 54/2/15, box 16095, series 153, accession 23, PNGNA.

52. Hornabrook to E. J. Wright, assistant director (medical research), Department of Public Health, PNG, September 22, 1966, file 54/2/15, box 16095, series 153, accession 23, PNGNA.

53. Mathews to Scragg, January 21, 1965, J. D. Mathews file, P4536, box 4779, accession 22, Papua New Guinea Department of Public Health Archives.

54. Interview with Ian R. Mackay, January 13, 2005, Melbourne. In 1966, Mackay visited Mathews at Okapa. "The appearance of people with kuru was just mind boggling," he told me. There seemed no doubt at the time that cannibalism was involved.

55. Mathews to Symes, acting director of public health, PNG, September 11, 1967, file 4/7/13, box 15908, series 106, accession 23, PNGNA.

56. Interview with John D. Mathews, July 11, 2001, Canberra, ACT. Mathews also sent brains to neuropathologists in Melbourne and London.

57. J. D. Mathews, "Report to Medical Advisory Committee of Papua and New Guinea," July 13, 1967, file 4/7/13, box 15908, series 106, accession 23, PNGNA. Mathews managed ten autopsies over two years.

58. In 1966, Henry Beecher exposed numerous abuses in human experimentation in "Ethics and clinical research." At the NIH this led to the development of ethics guidelines covering all federally funded human experimentation and the establishment of institutional review boards. See Rothman, *Strangers at the Bedside*.

59. Mathews, "Kuru: A puzzle," appendix X, 2, 3, 7, 8. This thesis includes twenty fascinating kuru case studies. Later, Mathews took every opportunity to promote the use of "scientific method" to solve community health problems—otherwise, "there is a very real danger that we may be substituting one form of witchcraft for another" ("Scientific method in epidemiology").

60. Gajdusek to Wright, June 21, 1967, file 54/2/5, box 16095, series 153, accession 23, PNGNA.

61. Gajdusek, February 23, 1967, in *South Pacific Expedition,* 45.

62. Interview with John D. Mathews, July 11, 2001, Canberra.

63. Interview with John D. Mathews, July 11, 2001, Canberra. After losing at chess, Gajdusek noted grumpily, "I have more important things to do toward ego bolstering than competing and marking myself off against my contemporaries" (February 23, 1967, in *South Pacific Expedition,* 50).

64. "Administration and Management of Kuru Research Centre," Okapa, 1965–71, file 4/7/13, box 15908, series 106, accession 23, PNGNA. See also Mathews, "Statutory Declaration," file 4/7/13, box 15908, series 106, accession 23, PNGNA.

65. Mathews to Scragg, September 14, 1966, J. D. Mathews file, P4536, box 4779, accession 22, Papua New Guinea Department of Public Health Archives.

66. Scragg to Mathews, October 12, 1966, J.D. Mathews file, P4536, box 4779, accession 22, Papua New Guinea Department of Public Health Archives.

67. Mathews to Ian Riley, September 12, 1966, file 54/2/5, box 8743, series 153, accession 23, PNGNA.

68. Mathews to Scragg, May 16, 1967, J. D. Mathews file, P4536, box 4779, accession 22, PNG Papua New Guinea Department of Public Health Archives.

69. R. J. Walsh to Scragg, May 18, 1967, J. D. Mathews file, P4536, box 4779, accession 22, PNG Papua New Guinea Department of Public Health Archives.

70. Interview with John D. Mathews, July 11, 2001, Canberra. Mathews later worked on the Clinical Research Unit of the Hall Institute. He became director of the Menzies School of Health Research in Darwin, then director of the Australian Centre for Disease Control, Canberra. He is now a professorial fellow of the School of Population Health, University of Melbourne.

71. Mathews, "The changing face of kuru," 1140. See also Mathews, "Transmission model for kuru."

72. Interview with John D. Mathews, July 11, 2001, Canberra.

73. Mathews, R. Glasse, and Lindenbaum, "Kuru and cannibalism," 451. Interestingly, the authors do not cite Alpers's recent speculations on cannibalism as the route of transmission. See also Mathews, "Kuru as an epidemic disease." On the theory that a sporadic case of CJD, cannibalized after death, was the cause of the kuru epidemic, see Alpers and Rail, "Kuru and Creutzfeldt-Jakob disease."

74. For experimental proof of oral transmission, see Gibbs Jr. et al., "Oral transmission of kuru." The primates involved were squirrel monkeys: intriguingly, chimpanzees did not manifest these diseases when fed infected tissue. Debate continued about the importance of non-oral routes of transmission, including exposure of damaged mucosal membranes and skin to infected tissue, especially in mortuary rituals. For example, Steadman and Merbs, "Kuru and cannibalism?" In the early 1980s, Robert Klitzman established the long incubation period of kuru, relating clusters of the disease to specific cannibal feasts. See Klitzman, Alpers, and Gajdusek, "Natural incubation period." Klitzman evokes the frustrations of kuru field work in *Trembling Mountain.*

75. Alphonse Kutne to director, Department of Public Health, February 1, 1969, file 54/2/5, box 8743, series 153, accession 23, PNGNA.

76. Interview with Inamba, August 5, 2003, Ivingoi.

77. File 54/10/2, box 5364, series 153, accession 23, PNGNA. Although Burnet was an evolutionary ecologist, the IBP was dominated by an ecosystem approach: see Collins and Weiner, *Human Adaptability;* and Kwa, "Representations of nature."

78. Hornabrook to Susan Serjeantson, August 6, 1975, box 002, IMRA/2004/011, IMR Archives. Serjeantson was a population geneticist based at the IMR. On Cavalli-Sforza, see Stone and Lurquin, *Genetic and Cultural Odyssey.* On Neel, see his *Physician to the Gene Pool;* and Lindee, *Suffering Made Real.* Patrick Tierney's claims about Neel's activities among the Yanomami should be treated with caution (*Darkness in El Dorado*).

79. Hornabrook to Selwyn J. Baker, August 1, 1975, box 002, IMRA/2004/011, IMR Archives.

80. Hornabrook to John A. Bergin, November 11, 1975, box 002, IMRA/2004/011, IMR Archives. At the same time, he dreaded returning to New Zealand, where the "socialists" had taken over.

81. Hornabrook to C. J. Hackett, November 27, 1975, box 002, IMRA/2004/011, IMR Archives.

82. Alpers, "Past and present research."

83. Cavalli-Sforza to Kuldeep Bhatia, IMR, December 12, 1983, Alpers correspondence file, box 003, IMRA/2003/025, IMR Archives. Although not personally involved in kuru research,

Cavalli did use the disease as an example of the genetic consequences of cultural "errors" such as cannibalism, to support his theory of gene-culture co-evolution (Cavalli-Sforza and Feldman, *Cultural Transmission and Evolution*).

84. Cavalli-Sforza et al., "Call for a world-wide survey." Jonathan Marks has identified flaws in the sampling procedures of the project (*What It Means to Be 98% Chimpanzee*). For other critiques see Lock, "Interrogating the human genome diversity project"; Cunningham, "Colonial encounters"; and Reardon, "Human genome diversity project." For a reflection on the failure of the project see Greely, "Human genome diversity."

85. Gajdusek to Cavalli-Sforza, September 4, 1969, in box 7, Gajdusek correspondence, NLM. In the 1970s, Gajdusek complained that Jim Neel was "entering the area that he knows I have worked most intensely in and [been] most dedicated to" (January 7, 1976, in *Stumbling along the Tortuous Road*, 104). He was concerned that Neel and Maurice Godelier were poaching his cherished Anga (Kukukuku). On the cultural front, he felt a similar rivalry with Gilbert Herdt.

86. Alpers to Jonathan Friedländer, Temple University, August 26, 1987, Alpers correspondence file, box 006, IMRA/2003/025, IMR Archives.

87. Alpers to Robert Attenborough, May 22, 1995, Alpers correspondence file, box 001, IMRA/2003/025, IMR Archives.

88. Alpers became professor of international health at Curtin University of Technology, Perth, Western Australia. John Reeder succeeded Alpers at the IMR and expanded laboratory research in molecular biology while maintaining field investigations.

CHAPTER 8. STUMBLING ALONG THE TORTUOUS ROAD

1. Gajdusek to Barry Adels, May 20, 1972, box 1, Gajdusek Correspondence, NLM.

2. Interview with Michael Alpers, July 24, 2002, Fremantle. In a letter to Frank Fenner, Alpers wrote that Gajdusek "wouldn't buy the cannibalism idea at that time and so to get it out I had to 'go it alone.' It took a while to win Carleton around the idea" (February 24, 1994, Alpers correspondence file, box 006, IMRA/2003/025, IMR Archives).

3. Paul Brown to Marion Poms, May 5, 1971, box 6, Gajdusek Correspondence, NLM.

4. Gajdusek, October 14, 1976, in *Stumbling along the Tortuous Road*, 447, 448.

5. See Blumberg, *Hepatitis B*. Blumberg was searching for biochemical and immunological variations in human populations and initially thought the Australia antigen was a genetic product, but it turned out to be part of an infectious agent. Australia antigen became known as the hepatitis B surface antigen. Before finding the remainder of the virus, which contains nucleic acid, Blumberg had "wild ideas that we had found a new class of infectious agent that could produce protein from protein without the presence of nucleic acid" (109).

6. Gajdusek, October 16, 1976, in *Stumbling along the Tortuous Road*, 450.

7. Gajdusek, "Unconventional viruses and the origin and disappearance of kuru," in *Nobel Lectures*, 167.

8. D. C. Gajdusek, "A kuru research laboratory at the Awande Kuru Center," December 1, 1967, typescript, Gunther Papers.

9. Wilson, Anderson, and Smith, "Studies in scrapie"; and Pattison, "Resistance of the scrapie agent to formalin." For a detailed history of scrapie investigations see Yam, *Pathological Protein;* Seguin, *Infectious Processes;* and Kim, *Social Construction of Disease.*

10. Epitomized in Crick and Watson, "Structure of small viruses"; and Lwoff, "Concept of virus."

11. Alper, Haig, and Clarke, "Exceptionally small size of the scrapie agent." See Alper, "Photo- and radiobiology of the scrapie agent."

12. Alper et al., "Does the agent of scrapie replicate without nucleic acid?"; and Alper, "Nature of the scrapie agent." See also Pattison and Jones, "Possible nature of the transmissible agent of scrapie." (Iain Pattison was a veterinary pathologist at Compton; see his "Sideways look at the scrapie saga" and "Fifty years with scrapie.") For a contemporary explanation of how a protein might be able to reproduce itself, see Griffith, "Self-replication and scrapie." The UV inactivation results did not conclusively show that no nucleic acid is present in the agent as the smallest RNA viruses, called viroids, have similar resistance to inactivation.

13. Crick, "Central dogma of molecular biology," 563. Another way of putting the "central dogma" is that once information has passed to a protein it cannot get out again. For an argument reconciling autocatalytic proteins with this assertion, see Keyes, "Prion challenge."

14. Thomas, *Late Night Thoughts,* 59, 126.

15. Gajdusek, August 4, 1976, in *Stumbling along the Tortuous Road,* 413.

16. Gajdusek, September 20, 1976, in *Stumbling along the Tortuous Road,* 436. For example, see Siakotos et al., "Partial purification of the scrapie agent"; and Gibbs Jr., Gajdusek, and Latarget, "Unusual resistance."

17. Gajdusek, September 25, 1976, in *Stumbling along the Tortuous Road,* 440. See also Gajdusek, "Subacute spongiform virus encephalopathies, 493, 496.

18. Stanley B. Prusiner, "Autobiography," *Nobel e-Museum,* at www.nobel.se/medicine/laureate/1997/prusiner-autobio.html. See also Prusiner and McCarty, "Discovering DNA."

19. Kerr, *Uses of the University.* See also Soo and Carson, "Managing the research university."

20. Jong, "Organizational structures in science." More generally, Steve Sturdy describes the growth of a "molecular economy" in "Reflections."

21. See Hughes, "Making dollars out of DNA." The decision of the U.S. Supreme Court in 1980 (*Diamond v. Chakrabarty*) to allow the patenting of life forms was pivotal in directing attention to the potential to profit from human tissue research, leading scientists and institutions to claim ownership of human cell lines and genes. The same year, the passage of the Bayh-Dole Act and the Stevenson-Wydler Act allowed commercial development of research funded by federal government institutions, such as the NIH. See Andrews and Nelkin, *Body Bazaar.*

22. Prusiner, "Approach to analysis."

23. Taubes, "The game of the name is fame."

24. British researchers had developed the test but used it sparingly, doubting the validity of its results. Richard Marsh had transmitted scrapie to hamsters in the mid-1970s at the University of Wisconsin–Madison: Marsh and Kimberlin, "Comparison of scrapie."

25. Interview with Paul Brown, January 25, 2006, Chevy Chase.

26. Stanley B. Prusiner, "Neurological examinations of patients with kuru," August 1980, typescript in Alpers Papers. See Prusiner, Gajdusek, and Alpers, "Kuru with incubation periods." According to Alpers, Prusiner "wasn't an intruder: I was keen to have trained independent clinicians examine kuru patients to fill out the clinical description . . . though it would

have been better if he had stayed longer and been physically fitter" (Alpers to Warwick Anderson, July 25, 2007, in the possession of Warwick Anderson).

27. Stanley B. Prusiner, recollections at the End of Kuru meeting, October 11, 2007, Royal Society, London.

28. Rhodes, *Deadly Feasts,* 161. Interview with Inamba, August 5, 2003, Ivingoi.

29. Prusiner, "Novel proteinaceous infectious particles."

30. Prusiner quoted in Taubes, "The game of the name is fame," 28.

31. Ibid., 33.

32. Masiarz quoted in Taubes, "The game of the name is fame," 36.

33. Interview with Michael P. Alpers, July 24, 2002, Fremantle.

34. For example, Dickinson, "Host-pathogen interactions."

35. Dickinson and Outram, "The scrapie replication-site hypothesis."

36. Kimberlin, "Scrapie agent: prions or virinos?" Prusiner replied that sorting out the chemical structure was the priority ("Research on scrapie").

37. Bolton, McKinley, and Prusiner, "Identification of a protein"; McKinley, Bolton, and Prusiner, "Protease-resistant protein"; and Prusiner et al., "Purification and structural studies."

38. Kim, *Social Construction of Disease;* and Yam, *Pathological Protein.*

39. Oesch et al., "A cellular gene encodes scrapie"; Basler et al., "Scrapie and cellular PrP isoforms"; Prusiner, "Prion hypothesis"; and Bolton and Bendheim, "Modified host protein." Patricia Merz had identified scrapie-associated fibrils in 1978; see Merz et al., "Abnormal fibrils from scrapie-infected brain." On their redefinition as prion rods, see McKinley et al., "Molecular characteristics of prion rods."

40. As noted in Chapter 7, Colin Masters and others in Gajdusek's laboratory had transmitted the disease to monkeys inoculated with brain tissue of GSS victims: Masters, Gajdusek, and Gibbs Jr., "Creutzfeldt-Jakob disease virus isolation."

41. Hsiao et al., "Linkage of the prion"; and Hsiao and Prusiner, "Inherited human prion diseases." Eventually other inherited spongiform encephalopathies were associated with other versions of the prion protein gene: Medori et al., "Fatal familial insomnia"; and Lee et al., "Ancestral origins and worldwide distribution."

42. Hsiao et al., "Spontaneous neurodegeneration."

43. Weissmann et al., "Susceptibility to scrapie in mice." See also Weissmann, "Molecular biology of prion diseases"; and Weissman, "Some thoughts on the pursuit of success in science."

44. Interview with Michael P. Alpers, July 24, 2002, Fremantle.

45. Ferry, "Mad brains and the prion heresy," 35.

46. Gajdusek, "Subacute spongiform encephalopathies," 2289.

47. Mestel, "Putting prions to the test," 189.

48. Hunter, "Natural transmission." Robert Rohwer, a former member of Gajdusek's group, repeated Tikvah Alper's experiments and determined that the presence of RNA in the scrapie agent was still possible: Rohwer, "The scrapie agent."

49. Mestel, "Putting prions to the test." Prusiner and others suggested a chaperone molecule, called protein X, might be required for the conversion: Telling et al., "Prion propagation in mice." To Laura Manuelidis at Yale, "protein X" sounded just like a virus: "Force of prions."

50. Kimberlin, Walker, and Fraser, "Genomic identity"; and Bruce et al., "Transmission of bovine spongiform encephalopathy."

51. Robert Rohwer and Stanley Prusiner quoted in Gina Kolata, "Viruses or prions: an old medical debate still rages," *New York Times* (October 4, 1994), C12.

52. De Chadarevian and Kamminga, *Molecularizing Biology and Medicine;* Wright, *Molecular Politics;* Fujimura, *Crafting Science;* and Gaudillière and Rheinberger, *From Molecular Genetics to Genomics.* Kim emphasizes the process of molecularization in *Social Construction of Disease.*

53. www.inpro.com (accessed July 9, 2007). Set up in 2000, InPro stands for "intellectual property."

54. Kim, *Social Construction of Disease.* See also Andrews and Nelkin, *Body Bazaar;* and Kleinman, *Impure Cultures.*

55. There may well have been earlier cases undetected during the previous decade: *Report of the BSE Inquiry* (London: Her Majesty's Stationery Office, 2000), www.bseinquiry.gov.uk/pdf/volume3/chapter1.pdf (accessed February 17, 2007). See also Rhodes, *Deadly Feasts;* Cooke, *Cannibals, Cows and the CJD Catastrophe;* McCalman, Penny, and Cook, *Mad Cows and Modernity;* Ridley and Baker, *Fatal Protein;* Ratzan, *Mad Cow Crisis;* Pennington, *When Food Kills;* Yam, *Pathological Protein;* Schwartz, *How the Cows Turned Mad;* Seguin, "UK BSE crisis"; and Seguin, "BSE saga."

56. A new prion mutation in cattle is now generally regarded as the more plausible explanation for the origin of the epidemic. Changes in the rendering of cattle in the late 1970s prevented inactivation of the agent in meat and bone meal. See Wilesmith, Ryan, and Atkinson, "Bovine spongiform encephalopathy."

57. By the end of 2002, there were a total of 180,000 BSE cases, though the incidence was dropping to about one thousand annually. During this period the British government spent three billion pounds slaughtering cattle. In 1996, it ordered the killing of all cattle in the national herd over thirty months old, some 4.7 million beasts.

58. See Wyatt et al., "Naturally occurring scrapie-like spongiform encephalopathy." By September 1994, fifty-seven cases of FSE were reported.

59. Will et al., "New variant of Creutzfeldt-Jakob disease"; and Will et al., "Deaths from variant Creutzfeldt-Jakob disease."

60. Bruce et al., "Transmissions to mice."

61. See Ritvo, "Roast beef of old England."

62. Thomas Stuttaford, "Eating people is wrong," *The Times* (March 11, 1993).

63. Peter Martin, "Mad, bad and dangerous to eat," *Mail on Sunday* (March 6, 1994).

64. Sue Arnold, "Sick cannibals, mad cows, and Englishmen," *Observer* (May 15, 1994), 28.

65. Tim Radford, "Mad cow mystery persists," *Guardian* (December 7, 1995). Others wrote of kuru and vCJD agents "punching holes in brain tissue" (Robin McKie et al., "A conspiracy to drive us all mad," *Observer* [March 24, 1996]), or "eating it away and turning it into sponge" (Charles Arthur, "The killer protein," *Independent* [March 27, 1996]).

66. Anon., "Mad cows and Englishmen," *Economist* (March 30, 1996).

67. Geoffrey Cowley, "Cannibals to cows: the path of a deadly disease," *Newsweek* (March 12, 2001).

68. http://nobelprize.org/nobel_prizes/medicine/laureates/1997/press.html (accessed May 10, 2007). Transmissible mink encephalopathy and chronic wasting disease of deer had been known since the 1940s: see Marsh, "Transmissible mink encephalopathy"; and Williams and Young, "Chronic wasting disease of captive mule deer." On the recent identification of fatal familial insomnia as a prion disease, see Max, *Family That Couldn't Sleep.*

69. http://nobelprize.org/nobel_prizes/medicine/laureates/1997/prusiner-speech.html (accessed May 10, 2007).

70. Lawrence K. Altman, "US scientist wins Nobel for controversial work," *New York Times* (October 7, 1997).

71. Testimony of Dr. Stanley Prusiner, May 6, 1998, in *BSE Inquiry,* available at http://www.bseinquiry.gov.uk/files/tr/tab27.pdf (accessed February 17, 2007), 62.

72. Vonnegut, *Cat's Cradle,* 45.

73. Crichton, *The Lost World,* 428. See "Transcript of a lecture by Dr. Stanley Prusiner," Columbia University, New York City, October 1997, statement no. 66, in *BSE Inquiry,* available at www.bseinquiry.gov.uk/evidence/ws (accessed February 17, 2007).

74. Steven Spielberg directed the movie version of *The Lost World* in 1997, produced at Universal Pictures.

75. For a similar story, see Latour, *Pasteurization of France.*

76. Vonnegut, *Cat's Cradle,* 25.

77. Testimony of Professor John Collinge, June 4, 1998, in *BSE Inquiry,* available at www.bseinquiry.gov.uk/files/tr/tab25.pdf (accessed February 17, 2007), 144.

78. Collinge et al., "Diagnosis of Gerstmann-Sträussler syndrome." Interview with John Collinge, September 27, 2006, London.

79. Collinge, Palmer, and Dryden, "Genetic predisposition"; and Palmer et al., "Homozygous prion protein genotype."

80. John Collinge et al., "Prion protein gene analysis." About 38 percent of the British population is methionine homozygous at codon 129 and therefore especially susceptible to vCJD. (It is expected that some valine homozygotes and heterozygotes will eventually manifest vCJD after longer incubation periods.) Idiopathic CJD, in contrast, seems to favor valine homozygotes.

81. Testimony of Professor John Collinge, June 4, 1998, in *BSE Inquiry,* available at www.bseinquiry.gov.uk/files/tr/tab25.pdf (accessed February 17, 2007), 124.

82. Collinge et al., "Molecular analysis." Prusiner argued that different molecular conformations gave different sizes of protease-resistant fragments of prion, and that the glycosylation of the normal protein influenced its affinity for the rogue form: see DeArmond et al., "Selective neuronal targeting."

83. Collinge, "Prion diseases of humans and animals." See also Prusiner, "Prion diseases."

84. Hill et al., "Same prion strain causes vCJD and BSE." Collinge and his colleagues also conducted a series of experiments with transgenic mice expressing human PrP (unfortunately valine homozygous at codon 129). Much cheaper than primates, the genetically modified rodents lacked any species barrier to human prion diseases. The animal models proved relatively resistant to BSE, implying that the species barrier did offer some protection to humans, though at longer incubation periods some transmissions did occur.

85. Michael Alpers, recollections at the End of Kuru meeting, October 12, 2007, Royal Society, London.

86. Interview with Jerome Whitfield, August 16, 2003, Goroka, PNG. Previously he worked in Pakistan, Burma, and Namibia.

87. Interview with Anderson Puwa, August 14, 2003, Ivingoi, PNG.

88. Interview with Anderson Puwa, August 7, 2003, Ivingoi.

89. Interview with Jerome Whitfield, August 16, 2003, Goroka.

90. Jerome Whitfield, "Kuru: Europe learns a lesson from PNG," *Post-Courier* (June 28, 2001).

91. Mead et al., "Balancing selection."

92. Collinge, "Remember CJD?"

93. Collinge et al., "Kuru in the 21st century," 2072. For the previous period, see Alpers, "Epidemiology of kuru."

94. Edit., "Lessons from kuru."

95. Interview with Jerome Whitfield, August 16, 2003, Goroka.

96. Interview with Anderson Puwa, August 14, 2003, Ivingoi.

97. Interview with Ken E. Boone, July 27, 2004, Goroka, PNG. Boone is a medical graduate from the University of Papua New Guinea who did further surgical training in Adelaide. His father had patrolled with Jim Taylor.

98. Interview with Anderson Puwa, August 14, 2003, Ivingoi.

99. Jerome Whitfield to Warwick Anderson, January 9, 2006, in the possession of Warwick Anderson.

100. Interview with Jerome Whitfield, August 16, 2003, Goroka.

101. For another account of the legacy of the kuru investigations among the Fore, see Beasley, "Kuru truths."

102. Interview with Anderson Puwa, August 7, 2003, Ivingoi. On the meanings of "compensation," see Strathern and Stewart, *Arrow Talk.*

CONCLUSION. DÉNOUEMENT WAS A BIT DIFFICULT

1. Munn, *Fame of Gawa.*

2. Brown to Gajdusek, March 27, 1985, box 6, Gajdusek Correspondence, NLM.

3. Gajdusek, *Journal of Expeditions to the Soviet Union;* Gajdusek, "Foci of motor neuron diseases," and idem, "Interference with axonal transport."

4. Gajdusek to Gibbs, August 10, 1991, box 18, Gajdusek Correspondence, NLM. Original emphasis.

5. Gajdusek, *Viliuisk Encephalomyelitis,* 177. See also Goldfarb and Gajdusek, "Viliuisk encephalomyelitis."

6. Jenkins, "Medical anthropology"; and Boyd, "Tale of 'first contact.'" Jenkins completed her Ph.D. in anthropology at the University of Tennessee and taught at Illinois State before taking up her position at the IMR. Later she worked on AIDS awareness and conducted studies of highland sex workers. In 1997 she moved to Bangladesh to continue working as an anthropologist. She died there in 2007.

7. Bhat, "NIH and the PNG cell line"; and Carol Jenkins, "Background on Hagahai patent case," n.d., typescript, Carol Jenkins file, IMRA/2003/026, IMR archives.

8. Jenkins, "Background on Hagahai patent case."

9. Alpers, "Perspectives from New Guinea." The outrage at indigenous "gene patenting" spilled over as opposition to the human genome diversity project, even though there was no intention to claim ownership of indigenous genomes: see Ventura Santos, "Indigenous peoples." On the first patenting of a human cell line (in 1984) see Wald, "What's in a cell?"

10. See Andrews and Nelkin, *Body Bazaar;* Krimsky, "Profit of scientific discovery"; Sharp, "Commodification of the body"; Waldby and Mitchell, *Tissue Economies;* Rajan, *Biocapital;* and Kirsh, "Property effects."

11. D. Carleton Gajdusek, April 5, 1996, in "Ascent to the zenith was magnificent; Dénouement was a bit difficult. Journal: January 1, 1996–December 31, 1996," n.d., typescript in Tozzer Library, Harvard University, 239. See also Justin Gillis and Philip P. Pan, "NIH scientist's journals describe child sexuality," *Washington Post* (April 6, 1996): A1, A10; Claudia Winkler, "Ignoble Nobelman: the saga of Carleton Gajdusek, a brilliant scientist—and an accused pedophile," *Weekly Standard* (October 7, 1996); and Robert Draper, "The genius who loved boys," *GQ* (November 1999): 313–30.

12. Gajdusek, April 5, 1996, in "Ascent to the zenith," 240.

13. Gajdusek, 9:30 P.M., March 19, 1996, in "Ascent to the zenith," 175.

14. Gajdusek, 11 P.M., March 19, 1996, in "Ascent to the zenith," 176.

15. J. Gillis, "Nobel winner guilty of abusing boy. Under plea deal, former NIH scientist will spend up to a year in jail," *Washington Post* (February 19, 1997).

16. Gajdusek, March 31, 1996, in "Ascent to the zenith," 193.

17. Gajdusek, April 2, 1960, in *Solomon Islands,* 108. Gajdusek later realized that the boy's name was in fact Mbaginta, with the suffix meaning "it is me."

18. Gajdusek, April 21, 1960, in *Solomon Islands,* 160.

19. Gajdusek, May 2, 1960, in *Solomon Islands,* 179.

20. Gajdusek, June 15, 1963, in *Melanesian Journal, February 22, 1963–July 23, 1963,* 176. Dick Sorenson refers to him as a "bush Kuk manki"(January 17, 1964, in *Journal Account,* 164).

21. Gajdusek to Alpers, July 6, 1963, box 2, Gajdusek Correspondence, NLM.

22. Gajdusek, July 23, 1963, in *Melanesian Journal, February 22, 1963–July 23, 1963,* 181.

23. Peggy Thomson, "Ivan Mbagintao: from tribal tradition to the 20th century a New Guinea youth adapts, amazes with will to learn," *Washington Post* (February 14, 1965). Wegstein was chief of the Office of Information Processing Standards at the National Bureau of Standards and did pioneering work on digital image processing, especially on fingerprint comparison for the FBI.

24. Gajdusek, February 14, 1967, in *South Pacific Expedition,* 27.

25. Ivan Mbaginta'o to Gajdusek, January 17, 2005, box 27, Gajdusek Correspondence, NLM.

26. Gajdusek, March 19, 1996, in "Ascent to the zenith," 173–74.

27. Ibid., 174.

28. Ceridwen Spark has embarked on a study of Gajdusek's family: see her "Rainbow children," *The Age* [Melbourne] (April 1, 2006).

29. Gajdusek, February 25, 1977, in *Journal of a Flight from Nobility,* 15.

30. Sorenson, January 26 and January 20, 1964, in *Journal Account of the Expedition,* 178, 169.

31. Weiner, "Reproduction."

32. Lévi-Strauss, *Elementary Forms of Kinship and Marriage,* 497. See also Bloch and Parry, "Introduction: Death and the regeneration of life"; and Foster, "Dangerous circulation."

33. Gajdusek, March 5, 1967, in *South Pacific Expedition,* 76.

34. Gajdusek, March 10, 1967, in *South Pacific Expedition,* 91. His mother had died the previous year.

35. Ibid.

36. Gajdusek, March 7, 1976, in *Stumbling along the Tortuous Road,* 208.

37. Gajdusek, April 10, 1967, in *South Pacific Expedition,* 183.

38. Gajdusek, August 10, 1979, in *Viliuisk Encephalomyelitis,* 45.

39. Gajdusek, February 28, 1996, in "Ascent to the zenith," 119.

40. Gajdusek, 11:40 P.M., March 19, 1996, in ibid., 187–88.

41. Ibid., 188.

42. Conrad, *Lord Jim,* 351.

BIBLIOGRAPHY

INTERVIEWS

Conducted by Warwick Anderson

Michael Alpers, July 24, 2002 and August 4, 2004, Fremantle, Western Australia.

S. Gray Anderson, September 28, 2005, London, England.

Jack and Lois Baker, January 3, 2006, Bribie Island, Queensland.

J. Henry Bennett, December 21 and 22, 2005, Adelaide, South Australia.

Catherine Berndt, August 3, 1993, Perth, Western Australia.

Ken E. Boone, July 27, 2004, Goroka, Papua New Guinea.

Paul Brown, January 25, 2006, Chevy Chase, Maryland.

John Collinge, September 27, 2006, London, England.

Frank Fenner, June 30, 2002, Canberra, Australia.

D. Carleton Gajdusek, March 25, 2005 and August 21–22, 2006, Amsterdam, Netherlands.

Gilbert Herdt, April 11, 2003, San Francisco, California.

Shirley Lindenbaum, March 17, 2005 and February 15, 2007, New York, New York.

Ian R. Mackay, January 13, 2005, Melbourne, Victoria.

Colin Masters, January 14, 2005, Melbourne, Victoria.

John D. Mathews, July 11, 2001, Canberra, ACT.

Frank D. Schofield and Robert McLennan, January 2, 2006, Brisbane, Queensland.

Donald Simpson, December 22, 2005, Adelaide, South Australia.

Jerome Whitfield, August 16, 2003, Goroka, Papua New Guinea.

Conducted by Warwick Anderson and Thomas P. Strong

Inamba, August 4 and 5, 2003, Ivingoi, Papua New Guinea.

Kivengi, August 5, 2003, Ivingoi, Papua New Guinea.

Masasa, August 8, 2003, Okapa, Papua New Guinea.

Pako, August 6, 2003, Waisa, Papua New Guinea.

Paudamba, August 11, 2003, Purosa, Papua New Guinea.

Anderson Puwa, August 7 and 14, 2003, Ivingoi, Papua New Guinea.

Tarubi, August 10 and 11, 2003, Purosa, Papua New Guinea.

Conducted by Sena Anua (Papua New Guinea Institute for Medical Research, Goroka)

Ai clan, Ketebe village, c. 2003, Purosa, Papua New Guinea.

Andemba Katago, February 2004, Purosa, Papua New Guinea.

Conducted by David Ikabala (Papua New Guinea Institute for Medical Research, Goroka)

Tiu Pekiyeva, July 31, 2003, Agakamatasa, Papua New Guinea.
Koiye Tasa, c. 2003, n.p.

ARCHIVES

Australia

Michael P. Alpers Papers, Curtin University, Perth, Western Australia.
Ronald and Catherine Berndt Papers, Berndt Museum of Anthropology, University of Western Australia, Perth, Western Australia.
Frank Fenner Papers, ms. 14, Basser Library, Australian Academy of Science, Canberra, ACT.
John Gunther Correspondence, in the possession of Hank Nelson, Australian National University, Canberra, ACT.
Frank Macfarlane Burnet Papers, University of Melbourne Archives, Melbourne, Victoria.

Papua New Guinea

Papua New Guinea Department of Health Archives, Port Moresby.
Papua New Guinea Institute of Medical Research Archives, Goroka.
Papua New Guinea National Archives, Port Moresby.

United States of America

Awande Hospital Reports, Wartburg Theological Seminary Archives, Dubuque, Iowa.
D. Carleton Gajdusek Correspondence, ms. C565, National Library of Medicine, Bethesda, Maryland.
D. Carleton Gajdusek Papers, ms. 58, American Philosophical Society, Philadelphia, Pennsylvania.
Robert S. Morison Diaries and Record Group 410A, Rockefeller Foundation, Rockefeller Archive Center, Tarrytown, New York.
Patrol Reports, Melanesian Studies Resource Center, Mandeville Special Collections Library, University of California, San Diego, California.
Charles A. Valentine Papers, Anthropology Museum Archives, University of Pennsylvania, Philadelphia, Pennsylvania.

PUBLISHED PRIMARY SOURCES

Alper, Tikvah. "The nature of the scrapie agent." *Journal of Clinical Pathology* 6 (Supplement) (1972): 154–55.
———. "Photo- and radiobiology of the scrapie agent." In *Prion Diseases of Humans and Animals,* edited by Stanley Prusiner, John Collinge, John Powell, and Brian Anderton, 30–39. New York: Ellis Horwood, 1992.
Alper, Tikvah, W. A. Cramp, David A. Haig, and Michael C. Clarke. "Does the agent of scrapie replicate without nucleic acid?" *Nature* 214 (20 May 1967): 764–66.
Alper, Tikvah, David A. Haig, and Michael C. Clarke. "The exceptionally small size of the scrapie agent." *Biochemical and Biophysical Research Communications* 22 (1966): 278–84.

Alpers, Michael P. "Epidemiological changes in kuru, 1957–1963." In *Slow, Latent, and Temperate Virus Infections,* edited by D. Carleton Gajdusek, Clarence J. Gibbs Jr., and Michael Alpers, 65–82. Washington, DC: U.S. Department of Health, Education, and Welfare, 1965.

———. "Epidemiology and ecology of kuru." In *Slow, Latent, and Temperate Virus Infections,* edited by D. Carleton Gajdusek, Clarence J. Gibbs Jr., and Michael Alpers, 67–90. Washington, DC: U.S. Department of Health, Education, and Welfare, 1965.

———. "The epidemiology of kuru in the period 1987 to 1995." *CDI* 29 (2005): 391–99.

———. "Kuru: Implications of its transmissibility for the interpretation of its changing epidemiological pattern." In *The Central Nervous System: Some Experimental Models of Neurological Diseases,* edited by O. T. Bailey and D. E. Smith, 234–51. Baltimore: Williams and Wilkins, 1968.

———. "Kuru in New Guinea: Its changing pattern and etiological elucidation." *American Journal of Tropical Medicine and Hygiene* 19 (1970): 133–37.

———. "Past and present research activities of the Papua New Guinea Institute of Medical Research." *Papua New Guinea Medical Journal* 42 (1999): 32–51.

———. "Perspectives from New Guinea." *Cultural Survival Quarterly* 20 (1996): 32–34.

———. "Reflections and highlights: A life with kuru." In *Prion Diseases of Humans and Animals,* edited by Stanley Prusiner, John Collinge, John Powell, and Brian Anderton, 66–95. New York: Ellis Horwood, 1992.

Alpers, Michael P., and D. Carleton Gajdusek. "Changing patterns of kuru: Epidemiological changes in the period of increasing contact of the Fore people with western civilization." *American Journal of Tropical Medicine and Hygiene* 14 (1965): 852–79.

Alpers, Michael P., and L. Rail. "Kuru and Creutzfeldt-Jakob disease: Clinical and aetiological aspects." *Proceedings of the Australian Association of Neurology* 8 (1971): 7–15.

Anderson, S. G., M. Donnelly, W. J. Stevenson, N. J. Caldwell, and M. Eagle. "Murray Valley encephalitis: Surveys of human and animal sera." *Medical Journal of Australia* i (1952): 110–14.

Anderson, S. G., A. V. Price, Nanadai-Koia, and K. Slater. "Murray Valley encephalitis in Papua and New Guinea. II: Serological survey, 1956–57." *Medical Journal of Australia* ii (1960): 410–13.

Avery, Oswald T., Colin M. MacLeod, and Maclyn McCarty. "Induction of transformation by a desoxyribonucleic acid fraction isolated from pneumococcus type III." *Journal of Experimental Medicine* 79 (1944): 137–58.

Barnes, John A. "African models in the New Guinea highlands." *Man* 62 (1962): 5–9.

Basler, K., B. Oesch, M. Scott, et al. "Scrapie and cellular PrP isoforms are encoded by the same chromosomal gene." *Cell* 46 (1986): 417–28.

Bayer, M. E., B. S. Blumberg, and B. Werner. "Particles associated with Australia antigen in the sera of patients with leukemia, Down's syndrome and hepatitis." *Nature* 218 (1968): 1057–59.

Beck, Elisabeth, and P. M. Daniel. "Neuropathological changes in kuru compared and contrasted with those of some other neurological diseases." In *Essays on Kuru,* edited by R. W. Hornabrook, 117–24. Faringdon, England: E. W. Classey, 1976.

———. "Prion diseases from a neuropathologist's perspective." In *Prion Diseases of Humans and Animals,* edited by Stanley Prusiner, John Collinge, John Powell, and Brian Anderton, 63–65. London and New York: Ellis Horwood, 1992.

Beck, Elisabeth, P. M. Daniel, D. Carleton Gajdusek, and Clarence J. Gibbs Jr. "Experimental 'kuru' in chimpanzees. A pathological report." *Lancet* ii (1966): 1056–59.

Beck, Elisabeth, P. M. Daniel, and H. B. Parry. "Degeneration of the cerebellar and hypothalamo-neurohypophysial systems in sheep with scrapie and its relationship to human system degenerations." *Brain* 87 (1964): 153–76.

Beecher, Henry. "Ethics and clinical research." *New England Journal of Medicine* 74 (1966): 1354–60.

Bennett, J. Henry. "Population studies in the kuru region of New Guinea." *Oceania* 33 (1962): 24–46.

——, ed. *The Collected Papers of R. A. Fisher.* Adelaide: University of Adelaide, 1971.

Bennett, J. Henry, B. W. Gabb, and Carolyn B. Oertel. "Further changes in the pattern of kuru." *Medical Journal of Australia* i (1966): 379–86.

Bennett, J. Henry, A. J. Gray, and C. O. Auricht. "The genetical study of kuru." *Medical Journal of Australia* ii (1959): 505–8.

Bennett, J. Henry, F. A. Rhodes, and H. R. Robson. "Observations on kuru. I: A possible genetic basis." *Australasian Annals of Medicine* 7 (1958): 269–75.

Berndt, Catherine H. "Journey along mythic paths." In *Ethnographic Presents: Pioneering Anthropologists in the Papua New Guinea Highlands,* edited by Terence E. Hays, 98–136. Berkeley: University of California Press, 1992.

——. "Sociocultural change in the eastern highlands of New Guinea." *Southwestern Journal of Anthropology* 9 (1953): 112–38.

Berndt, Ronald M. *An Adjustment Movement in Arnhem Land, Northern Territory of Australia.* Paris: Mouton, 1962.

——. "Anthropology and administration." *South Pacific* 9 (1958): 611–19.

——. "A cargo movement in the eastern central highlands of New Guinea." *Oceania* 23 (1952): 42–65, 137–58, 202–34.

——. "A 'devastating disease syndrome': Kuru sorcery in the eastern central highlands of New Guinea." *Sociologus* 8 (1959): 4–28.

——. *Excess and Restraint: Social Control among a New Guinea Mountain People.* Chicago: University of Chicago Press, 1962.

——. "Into the unknown!" In *Ethnographic Presents: Pioneering Anthropologists in the Papua New Guinea Highlands,* edited by Terence E. Hays, 68–97. Berkeley: University of California Press, 1992.

——. "Reaction to contact in the eastern highlands of New Guinea." *Oceania* 24 (1953): 190–228.

Berndt, Ronald M., and Catherine H. Berndt. "A. P. Elkin—the man and the anthropologist." In *Aboriginal Man in Australia: Essays in Honour of Emeritus Professor A. P. Elkin,* edited by Ronald M. Berndt and Catherine H. Berndt, 1–26. Sydney: Angus and Robertson, 1965.

Bhat, Amar. "The NIH and the PNG cell line." *Cultural Survival Quarterly* 20 (1996): 29–31.

Blackwood, Beatrice. *Both Sides of the Buka Passage.* Oxford: Oxford University Press, 1935.

——. *The Kukukuku of the Upper Watut,* edited by C. R. Hallpike. Oxford: Pitt-Rivers Museum, 1978.

——. "Life on the Upper Watut." *Geographical Journal* 94 (1939): 11–24.

Bolton, D. C., and P. E. Bendheim. "A modified host protein model of scrapie." *Ciba Foundation Symposium* 135 (1988): 164–81.

Bolton, D. C., M. P. McKinley, and S. B. Prusiner. "Identification of a protein that purifies with the scrapie protein." *Science* 218 (1982): 1309–11.

Brown, Paul. "Environmental causes of human spongiform encephalopathy." In *Methods in Molecular Medicine: Prion Diseases,* edited by H. Baker and R.M. Ridley, 139–54. Totawa, NJ: Humana Press, 1996.

Brown, Paul, Lev G. Goldfarb, and D. Carleton Gajdusek. "The new biology of spongiform encephalopathy: Infectious amyloidoses with a genetic twist." *Lancet* 337 (1991): 1019–22.

Brown, Paul, M. A. Preece, and R. G. Will. "'Friendly fire' in medicine: Hormones, homografts and Creutzfeldt-Jakob disease." *Lancet* 340 (1992): 24–27.

Bruce, M. E., A. Chree, I. McConnell, J. Foster, G. Pearson, and H. Fraser. "Transmission of bovine spongiform encephalopathy and scrapie to mice: Strain variation and the species barrier." *Philosophical Transactions of the Royal Society of London, Series B* 343 (1994): 405–11.

Bruce, M. E., R. Will, J. Ironside, et al. "Transmissions to mice indicate that 'new variant' CJD is caused by the BSE agent." *Nature* 389 (1997): 498–501.

Burkitt, Denis Parsons. "A 'tumour safari' in East and Central Africa." *British Journal of Cancer* 16 (1962): 379–86.

Burnet, F. Macfarlane. *Changing Patterns: An Atypical Autobiography.* Melbourne: Heinemann, 1968.

——. *The Clonal Selection Theory of Acquired Immunity.* Nashville: Vanderbilt University Press, 1959.

——. "Kuru." In *Encyclopaedia of Papua New Guinea,* vol. 1, edited by Peter Ryan, 586–88. Carlton: Melbourne University Press and the University of Papua New Guinea, 1972.

——. "Kuru—the present position." *Papua and New Guinea Medical Journal* 8 (1965): 3–7.

——. "A modification of Jerne's theory of antibody production using the concept of clonal selection." *Australian Journal of Science* 20 (1957): 67–69.

——. "'Smooth-rough' variation in bacteria in its relation to bacteriophage." *Journal of Pathology and Bacteriology* 32 (1929): 15–42.

——. *Virus as Organism: Evolutionary and Ecological Aspects of Some Human Virus Diseases.* Cambridge, MA: Harvard University Press, 1945.

——. *The Walter and Eliza Hall Institute, 1915–1965.* Carlton: Melbourne University Press, 1971.

Burnet, F. Macfarlane, and Frank Fenner. *The Production of Antibodies,* 2nd ed. Melbourne: Macmillan, 1949.

Burnet, F. Macfarlane, and Dora Lush. "Induced lysogenicity and mutation of bacteriophage within lysogenic bacteria." *Australian Journal of Experimental Biology and Medical Science* 14 (1936): 27–38.

Burnet, F. Macfarlane, and David O. White. *The Natural History of Infectious Disease,* 3rd ed. Cambridge: Cambridge University Press, 1962.

Cairns, John, Gunther S. Stent, and James D. Watson, eds. *Phage and the Origins of Molecular Biology*, exp. ed. New York: Cold Spring Harbor Laboratory Press, 1992.

Cannon, Walter B. "'Voodoo' death." *American Anthropologist* 44 (1942): 169–81.

Cavalli-Sforza, L. Luca, and Marcus W. Feldman. *Cultural Transmission and Evolution: A Quantitative Approach*. Princeton, NJ: Princeton University Press, 1981.

Cavalli-Sforza, L. Luca, A. C. Wilson, C. R. Cantor, R. M. Cook-Deegan, and M. C. King. "Call for a world-wide survey of human genetic diversity: A vanishing opportunity for the human genome diversity project." *Genomics* 11 (1991): 490–91.

Chalmers, Gloria. *Kundus, Cannibals and Cargo Cults: Papua New Guinea in the 1950s*. Watsons Bay, New South Wales: Books and Writers Network, 2006.

Chandler, Richard L. "Encephalopathy in mice produced with scrapie brain material." *Lancet* i (June 24, 1961): 1378–79.

Chinnery, E. W. P. "Mountain tribes of the mandated territory of New Guinea from Mt. Chapman to Mt. Hagen." *Man* 34 (1934): 113–21.

Collinge, John. "Prion diseases of humans and animals: Their causes and molecular basis." *Annual Review of Neuroscience* 24 (2001): 519–50.

———. "Remember CJD? It hasn't gone away." *The (London) Times,* January 3, 2004.

Collinge, John, J. Beck, T. A. Campbell, et al. "Prion protein gene analysis in new variant cases of Creutzfeldt-Jakob disease." *Lancet* 348 (1996): 56.

Collinge, John, A. E. Harding, F. Owen, et al. "Diagnosis of Gerstmann-Sträussler syndrome in familial dementia with prion protein gene analysis." *Lancet* ii (1989): 15–17.

Collinge, John, M. S. Palmer, and A. J. Dryden. "Genetic predisposition to iatrogenic Creutzfeldt-Jakob disease." *Lancet* 337 (1991): 1351–54.

Collinge, John, K. C. L. Sidle, J. Meads, et al. "Molecular analysis of prion strain variation and the aetiology of 'new variant' CJD." *Nature* 383 (1996): 685–90.

Collinge, John, Jerome Whitfield, Edward McKintosh, et al. "Kuru in the 21st century—an acquired human prion disease with very long incubation periods." *Lancet* 367 (2006): 2068–74.

Conrad, Joseph. *Lord Jim.* 1900. London: Penguin Books, 1986.

———. *Youth, a Narrative, and Two Other Stories.* Edinburgh: W. Blackwood and Sons, 1902.

Crichton, Michael. *The Lost World.* New York: Ballantine Books, 1995.

Crick, Francis H. C. "Central dogma of molecular biology." *Nature* 227 (August 8, 1970): 561–63.

Crick, Francis H. C., and James D. Watson. "Structure of small viruses." *Nature* 177 (1956): 473–75.

Cuillé, Jean, and Paul-Louis Chelle. "Pathologie animale. La maladie dite tremblante du mouton, est-elle inoculable?" *Comptes rendus de l'academie des sciences* 203 (1936): 1552–54.

DeArmond, S. J., H. Sánchez, F. Yehiely, et al. "Selective neuronal targeting in prion disease." *Neuron* 19 (1997): 1337–48.

Delbrück, Max. "A physicist looks at biology." 1949. In *Phage and the Origins of Molecular Biology*, exp. ed., edited by John Cairns, Gunther S. Stent, and James D. Watson, 9–22. New York: Cold Spring Harbor Laboratory Press, 1992.

Diamond, Jared. "Zoological classification system of a primitive people." *Science* 151 (1966): 1102–4.

Dickinson, A. G. "Host-pathogen interactions in scrapie." *Genetics* 79 (Supplement) (1975): 387–95.

Dickinson, A. G., and G. W. Outram. "The scrapie replication-site hypothesis and its implications for pathogenesis." In *Slow Transmissible Diseases of the Nervous System 2,* edited by Stanley B. Prusiner and W. J. Hadlow, 13–31. London: Academic Press, 1979.

Dobzhansky, Theodosius. "Eugenics in New Guinea." *Science* 132 (July 8, 1960): 77.

Downs, Ian. *The Last Mountain: A Life in Papua New Guinea.* St. Lucia: University of Queensland Press, 1986.

Dunham, Corydon B. *Fighting for the First Amendment: Stanton of CBS vs. Congress and the Nixon White House.* New York: Praeger, 1997.

Edit., "Lessons from kuru." *Lancet* 367 (2006): 2034.

Eklund, C., W. J. Hadlow, and R. C. Kennedy. "Some properties of the scrapie agent and its behavior in mice." *Proceedings of the Society for Experimental Biology and Medicine* 112 (1963): 974–79.

Ekman, Paul. *Emotions Revealed: Recognizing Faces and Feelings to Improve Communication and Emotional Life.* New York: Holt, 2004.

Elkin, A. P. "Notes on anthropology and the future of Australian territories." *Oceania* 15 (1944): 85–88.

———. *Social Anthropology in Melanesia: A Review of Research.* London: Oxford University Press, 1953.

Epstein, M. A., B. G. Achong, and Y. M. Barr. "Virus particles in cultured lymphoblasts from Burkitt's lymphoma." *Lancet* i (1964): 702–3.

Evans-Pritchard, E. E. *Witchcraft, Oracles, and Magic among the Azande.* Oxford: Clarendon Press, 1937.

———. "Zande cannibalism." *Journal of the Royal Anthropological Institute of Great Britain and Ireland* 90 (1960): 238–58.

Farquhar, Judith, and D. Carleton Gajdusek, eds. *Kuru: Early Letters and Fieldnotes from the Collection of D. Carleton Gajdusek.* New York: Raven Press, 1981.

Fenner, Frank. "Epilogue: The broader significance of kuru." In *Essays on Kuru,* edited by R. W. Hornabrook, 146–47. Faringdon, England: E. W. Classey, 1976.

Ferry, Georgina. "Mad brains and the prion heresy." *New Scientist* (28 May 1994): 32–37.

Firth, Raymond. "Anthropology and native administration." *Oceania* 2 (1931): 1–8.

Fischer, Ann, and J. L. Fischer. "Aetiology of kuru." *Lancet* i (1960): 1417–18.

———. "Culture and epidemiology: A theoretical investigation of kuru." *Journal of Health and Human Behavior* 2 (1961): 16–25.

Fisher, R. A. *The Genetical Theory of Natural Selection.* Oxford: Clarendon Press, 1930.

———. *Statistical Methods for Research Workers.* Edinburgh: Oliver and Boyd, 1925.

Flierl, Johann. *Christ in New Guinea: Former Cannibals Become Evangelists by the Marvelous Grace of God: A Short History of Missionwork Done by the Native Helpers and Teachers in the Lutheran Mission in New Guinea.* Tanunda, South Australia: Auricht Publishing Co., 1932.

Fortune, Reo. "Arapesh warfare." *American Anthropologist* 41 (1939): 22–41.

———. "Law and force in Papuan societies." *American Anthropologist* 49 (1947): 244–59.

———. "The rules of relationship behavior in one variety of primitive warfare." *Man* 47 (1947): 108–10.

———. *The Sorcerers of Dobu: The Social Anthropology of the Dobu Islanders of the Western Pacific.* London: Routledge and Kegan Paul, 1932.

———. "Statistics of kuru," *Medical Journal of Australia* i (1960): 764–65.

Fowler, Malcolm, and E. Graeme Robertson. "Observations on kuru. III: Pathological features in five cases." *Australasian Annals of Medicine* 8 (1959): 16–26.

French, E. L., S. G. Anderson, A. V. G. Price, and F. A. Rhodes. "Murray Valley encephalitis in New Guinea." *American Journal of Tropical Medicine and Hygiene* 6 (1957): 827–34.

Gajdusek, D. Carleton. "An autoimmune reaction against human tissue antigens in certain acute and chronic diseases: Serological investigations." *Archives of Internal Medicine* 101 (1958): 9–29.

———. "An autoimmune reaction against human tissue antigens in certain chronic diseases." *Nature* 179 (March 30, 1957): 666–68.

———. "Foci of motor neuron diseases in high incidence in isolated populations of East Asia and Western Pacific." In *Human Motor Neuron Disease,* edited by L. P. Rowland, 363–93. London: Pitman Books, 1982.

———. "Foreword." In *Laughing Death: The Untold Story of Kuru,* by Vincent Zigas, v–vii. Clifton, NJ: Humana Press, 1990.

———. "Hemorrhagic fevers in Asia: A problem in medical ecology." *Geographical Review* 46 (1955): 20–42.

———. "Interference with axonal transport of neurofilaments as the common etiology and pathogenesis of neurofibrillary tangles, ALS, PDC, and many other degenerations of the central nervous system: A series of hypotheses." *New England Journal of Medicine* 312 (1985): 714–19.

———. *Journal 1955–1957: Australia and New Guinea.* Bethesda, MD: Laboratory of Central Nervous System Studies, National Institutes of Health, 1996.

———. *Journal of Continued Quest for the Etiology of Kuru with Return to New Guinea, January 1, 1959–December 31, 1959.* Gif-sur-Yvette, France: Institut de Neurobiologie Alfred Fessard, 2002.

———. *Journal of Expeditions to the Soviet Union, Africa, the Islands of Madagascar, la Réunion and Mauritius, Indonesia, and to East and West New Guinea, Australia and Guam . . . June 1, 1969–March 3, 1970.* Bethesda, MD: National Institute of Neurological Diseases and Stroke, 1971.

———. *Journal of a Flight from Nobility in an Attempt, in Latin America and French East Africa, to Recover from the Aftermath of Notoriety, February 21, 1977 to April 28, 1977.* Bethesda, MD: National Institute of Neurological Disorders and Stroke, 1981.

———. "Kuru." *Transactions of the Royal Society of Tropical Medicine and Hygiene* 57 (1963): 151–69.

———. *Kuru Epidemiological Patrols from the New Guinea Highlands to Papua, 1957.* Bethesda, MD: National Institute of Neurological Diseases and Blindness, 1963.

———. "Kuru in New Guinea and the origin of the NINDB study of slow, latent and temperate virus infections of the nervous system of man." In *Slow, Latent, and Temperate Virus Infections,* edited by D. Carleton Gajdusek, Clarence J. Gibbs Jr., and Michael Alpers, 3–12. Washington, DC: U.S. Department of Health, Education, and Welfare, 1965.

———. *Melanesian Journal: Expedition to New Hebrides, Solomon Islands, Manus, New*

Britain, and New Guinea, January 23, 1965–April 7, 1965. Bethesda, MD: National Institute of Neurological Disorders and Stroke, 1989.

———. *Melanesian Journal, February 22, 1963–July 23, 1963.* Bethesda, MD: National Institute of Neurological Diseases and Stroke, 1973.

———. *Melanesian and Micronesian Journal: Return Expeditions to the New Hebrides, Caroline Islands, and New Guinea, July 29, 1965–December 20, 1965.* Bethesda, MD: National Institute of Neurological Disorders and Stroke, 1993.

———. *New Guinea Journal, June 10, 1959–August 15, 1959.* Bethesda, MD: National Institute of Neurological Diseases and Blindness, 1963.

———. *New Guinea Journal, Part I, October 2, 1961–March 5, 1962.* Bethesda, MD: National Institute of Neurological Diseases and Blindness, 1968.

———. *New Guinea Journal, Part II, March 6, 1962–August 4, 1962.* Bethesda, MD: National Institute of Neurological Diseases and Blindness, 1968.

———. *Solomon Islands, New Britain, and East New Guinea Journal, January 7, 1960–May 6, 1960.* Bethesda, MD: National Institute of Neurological Diseases and Blindness, 1964.

———. *South Pacific Expedition: To the New Hebrides and to the Fore, Kukukuku, and Genetei Peoples of New Guinea, January 26, 1967 to May 12, 1967.* Washington, DC: U.S. Department of Health and Human Services, 1967.

———. *Stumbling along the Tortuous Road to Unanticipated Nobility: Melanesian, Indonesian and Malaysian Expedition, November 16, 1975–December 31, 1976.* Bethesda, MD: National Institute of Neurological Disorders and Stroke, 1996.

———. "Subacute spongiform encephalopathies: Transmissible cerebral amyloidoses caused by unconventional viruses." In *Virology,* 2nd ed., edited by B. N. Fields, D. M. Knipe, and P. M. Howley, 2289–2324. New York: Raven Press, 1990.

———. "Subacute spongiform virus encephalopathies caused by unconventional viruses." In *Subviral Pathogens of Plants and Animals: Viroids and Prions,* edited by K. Maramorosh and J. J. McKelvey, 483–544. New York: Academic Press, 1985.

———. "Unconventional viruses and the origin and disappearance of kuru." In *Nobel Lectures, Physiology or Medicine, 1971–1980,* edited by Jan Lindsten, 167–216. Singapore: World Scientific Publishing, 1992.

———. *Viliuisk Encephalomyelitis: Journals of Unrequited Quests for Etiology, 1976, 1979, 1991, 1992, 1993.* Bethesda, MD: National Institute of Neurological Disorders and Stroke, 1996.

———. *A Year in the Middle East: Expeditions in Iran and Afghanistan with Travels in Europe and North America, 1954.* Bethesda, MD: National Institute of Neurological Disorders and Stroke, 1991.

———, ed. *Correspondence on the Discovery and Original Investigations of Kuru: Smadel-Gajdusek Correspondence, 1955–1958.* Bethesda, MD: National Institutes of Health, 1976.

Gajdusek, D. Carleton, Michael P. Alpers, and C. J. Gibbs Jr. "Kuru: Epidemiological and virological studies of a unique New Guinean disease with wide significance to general medicine." In *Essays on Kuru,* edited by R. W. Hornabrook, 125–45. Faringdon, England: E. W. Classey, 1976.

Gajdusek, D. Carleton, and Clarence J. Gibbs Jr. "Preface." In *Slow, Latent, and Temperate Virus Infections,* edited by D. Carleton Gajdusek, Clarence J. Gibbs Jr., and Michael Alpers, ix–x. Washington, DC: U.S. Department of Health, Education, and Welfare, 1965.

———. *Unconventional Viruses Causing Spongiform Encephalopathies: A Fruitless Search for the Coat and Core.* Washington, DC: U.S. Department of Health, Education, and Welfare, 1979.

Gajdusek, D. Carleton, Clarence J. Gibbs Jr., and Michael Alpers. "Experimental transmission of a kuru-like syndrome to chimpanzees." *Nature* 209 (1966): 794–96.

———. "Transmission and passage of experimental 'kuru' to chimpanzees." *Science* 155 (1967): 212–14.

Gajdusek, D. Carleton, Clarence J. Gibbs Jr., and Michael Alpers, eds. *Slow, Latent, and Temperate Virus Infections.* Bethesda, MD: U.S. Department of Health, Education, and Welfare, 1965.

Gajdusek, D. Carleton, Clarence J. Gibbs Jr., D. M. Asher, and E. David. "Transmission of experimental kuru to the spider monkey (*Ateles geoffreyi*)." *Science* 162 (1968): 693–94.

Gajdusek, D. Carleton, and Ian R. Mackay. "An autoimmune reaction against human tissue antigens in certain acute and chronic diseases: Clinical correlations." *Archives of Internal Medicine* 101 (1958): 30–46.

Gajdusek, D. Carleton, and N. G. Rogers. "Specific serum antibodies to infectious disease agents in Tarahumara Indian adolescents of northwestern Mexico." *Pediatrics* 16 (1955): 819–35.

Gajdusek, D. Carleton, and V. Zigas. "Degenerative disease of the central nervous system in New Guinea: The endemic occurrence of 'kuru' in the native population." *New England Journal of Medicine* 257 (1957): 974–78.

———. "Kuru: Clinical, pathological, and epidemiological study of an acute progressive degenerative disease of the central nervous system among natives of the eastern highlands of New Guinea." *American Journal of Medicine* 26 (1959): 442–69.

———. "Studies in kuru. I: The ethnologic setting of kuru." *American Journal of Tropical Medicine and Hygiene* 10 (1961): 80–91.

Gajdusek, D. Carleton, V. Zigas, and J. Baker. "Studies on kuru. III. Patterns of kuru incidence: Demographic and geographic epidemiological analysis." *American Journal of Tropical Medicine and Hygiene* 10 (1961): 599–627.

Garruto, Ralph M., M. A. Little, G. D. James, and D. E. Brown. "Natural experimental models: The global search for biomedical paradigms among traditional, modernizing, and modern populations." *Proceedings of the National Academy of Sciences* 96 (1999): 10536–43.

Gibbs Jr., Clarence J. "Spongiform encephalopathies—slow, latent, and temperate virus infections—in retrospect." In *Prion Diseases of Humans and Animals,* edited by Stanley Prusiner, John Collinge, John Powell, and Brian Anderton, 53–62. New York: Ellis Horwood, 1992.

Gibbs Jr., Clarence J., H. L. Amyx, A. Bacote, C. L. Masters, and D. Carleton Gajdusek. "Oral transmission of kuru, Creutzfeldt-Jakob disease, and scrapie to non-human primates." *Journal of Infectious Diseases* 142 (1980): 205–8.

Gibbs Jr., Clarence J., and D. Carleton Gajdusek. "Attempts to demonstrate a transmissible agent in kuru, amyotrophic lateral sclerosis, and other subacute and chronic progressive nervous system degenerations of man." In *Slow, Latent, and Temperate Virus Infections,* edited by D. Carleton Gajdusek, Clarence J. Gibbs Jr., and Michael Alpers, 39–48. Washington, DC: U.S. Department of Health, Education, and Welfare, 1965.

Gibbs Jr., Clarence J., D. Carleton Gajdusek, D. M. Asher, et al. "Creutzfeldt-Jakob disease (spongiform encephalopathy): Transmission to the chimpanzee." *Science* 161 (1968): 388–89.

Gibbs Jr., Clarence J., D. Carleton Gajdusek, and R. Latarget. "Unusual resistance to ionizing radiation of the viruses of kuru, Creutzfeldt-Jakob disease, and scrapie." *Proceedings of the National Academy of Science* 75 (1978): 6268–70.

Glasse, Robert M. "Cannibalism in the kuru region of New Guinea." *Transactions of the New York Academy of Sciences* 29 (1967): 748–54.

———. "Encounters with the Huli: Fieldwork at Tari in the 1950s." In *Ethnographic Presents: Pioneering Anthropologists in the Papua New Guinea Highlands,* edited by Terence E. Hays, 232–49. Berkeley: University of California Press, 1992.

———. "The Huli descent system: A preliminary account." *Oceania* 29 (1959): 171–84.

———. "Revenge and redress among the Huli: A preliminary account." *Mankind* 5 (1959): 273–89.

Glasse, Robert M., and Shirley Lindenbaum. "Fieldwork in the South Fore: The process of ethnographic inquiry." In *Prion Diseases of Humans and Animals,* edited by Stanley Prusiner, John Collinge, John Powell, and Brian Anderton, 77–91. New York: Ellis Horwood, 1992.

———. "The highlands of New Guinea: A review of ethnographic and related problems." In *Essays on Kuru,* edited by R. W. Hornabrook, 6–27. Faringdon, England: E. W. Classey, 1976.

———. "Kuru at Wanitabe." In *Essays on Kuru,* edited by R. W. Hornabrook, 38–52. Faringdon, England: E. W. Classey, 1976.

Glasse, Shirley. "The social effects of kuru." *Papua and New Guinea Medical Journal* 7 (1964): 36–47.

Glick, Leonard B. "Medicine as an ethnographic category: The Gimi of the New Guinea highlands." *Ethnology* 6 (1967): 31–56.

Goldfarb, Lev G., and D. Carleton Gajdusek. "Viliuisk encephalomyelitis in the Iakut people of Siberia." *Brain* 115 (1992): 961–78.

Greely, Henry T. "Human genome diversity: What about the other human genome project?" *Genetics* 2 (2001): 222–27.

Griffith, J. S. "Self-replication and scrapie." *Nature* 215 (2 September 1967): 1043–44.

Gunther, John. "Medical services, history." In *Encyclopaedia of Papua and New Guinea,* vol. 2., edited by Peter Ryan, 748–56. Carlton: Melbourne University Press, 1972.

———. "Post-war medical services in Papua New Guinea: A personal view." In *A History of Medicine in Papua New Guinea: Vignettes of an Earlier Period,* edited by Burton G. Burton-Bradley, 47–76. Sydney: Australasian Medical Publishing Co., 1990.

Hadlow, W. J. "Neuropathology and the scrapie-kuru connection." *Brain Pathology* 5 (1995): 27–31.

———. "Scrapie and kuru." *Lancet* ii (5 September 1959): 289–90.

———. "The scrapie-kuru connection: Recollections of how it came about." In *Prion Diseases of Humans and Animals,* edited by Stanley Prusiner, John Collinge, John Powell, and Brian Anderton, 40–46. London and New York: Ellis Horwood, 1992.

Hershey, A. D. "The injection of DNA into cells by phage." In *Phage and the Origins of Molecular Biology,* exp. ed., edited by John Cairns, Gunther S. Stent, and James D. Watson, 100–108. New York: Cold Spring Harbor Laboratory Press, 1992.

Hershey, A. D., and Martha Chase. "Independent functions of viral protein and nucleic acid in growth of bacteriophage." *Journal of General Physiology* 36 (1952): 39–56.

Hides, Jack G. *Papuan Wonderland*. Glasgow: Blackie and Son, 1936.

———. *Through Wildest Papua*. London: Blackie and Son, 1935.

Hill, A. F., M. Desbrulais, S. Joiner, et al. "The same prion strain causes vCJD and BSE." *Nature* 389 (1997): 448–50.

Hornabrook, R. W. "Kuru: The disease." In *Essays on Kuru*, edited by R. W. Hornabrook, 53–82. Faringdon, England: E. W. Classey, 1976.

———. "Kuru—some misconceptions and their explanation." *Papua New Guinea Medical Journal* 9 (1966): 11–15.

———. "Kuru—a subacute cerebellar degeneration. The natural history and clinical features." *Brain* 91 (1968): 53–74.

Hoskin, J. O., L. G. Kiloh, and J. E. Cawte. "Epilepsy and guria: The shaking syndromes of New Guinea." *Social Science and Medicine* 3 (1969): 39–48.

Hsiao, K. K., H. F. Baker, T. J. Crow, et al. "Linkage of the prion protein mis-sense variant to Gerstmann-Sträussler syndrome." *Nature* 338 (1989): 342–45.

Hsiao, K. K., and S. B. Prusiner. "Inherited human prion diseases." *Neurology* 40 (1990): 1820–27.

Hunter, Nora. "Natural transmission and genetic control of susceptibility of sheep to scrapie." *Current Topics in Microbiology and Immunology* 172 (1991): 165–80.

Jenkins, Carol. "Medical anthropology in the Western Schrader Range, Papua New Guinea." *National Geographic Research* 3 (1987): 412–30.

Kerr, Clark. *The Uses of the University*. 1963. Cambridge, MA: Harvard University Press, 2001.

Kimberlin, R. H. "Scrapie agent: Prions or virinos?" *Nature* 297 (13 May 1982): 107–8.

Kimberlin, R. H., C. A. Walker, and H. Fraser. "The genomic identity of different strains of mouse scrapie is expressed in hamsters and preserved on reisolation in mice." *Journal of General Virology* 70 (1989): 2017–25.

Klatzo, I., D. C. Gajdusek, and V. Zigas. "Pathology of kuru." *Laboratory Investigation* 8 (1959): 799–847.

Klitzman, Robert L. *The Trembling Mountain: A Personal Account of Kuru, Cannibals, and Mad Cow Disease*. Cambridge, MA: Perseus Publishing, 1998.

Klitzman, Robert L., Michael P. Alpers, and D. Carleton Gajdusek. "The natural incubation period of kuru and the episodes of transmission in three clusters of patients." *Neuroepidemiology* 3 (1984): 3–20.

Kurland, L. T., and D. W. Mulder. "Epidemiologic investigations of amyotrophic lateral sclerosis. 1. Preliminary report on geographic distribution, with special reference to the Mariana Islands, including clinical and pathological observations." *Neurology* 4 (1954): 355–78, 438–48.

Lawrence, Peter. "Social anthropology and the training of administrative officers at ASOPA." *Anthropological Forum* 1 (1964): 195–208.

Leahy, Michael J. *Explorations into Highland New Guinea, 1930–1935*, edited by Douglas E. Jones. Bathurst, New South Wales: Crawford House Press, 1994.

———. "Stone-age people of the Mt. Hagen area, mandated territory of New Guinea." *Man* 35 (1935): 185–86.

Leahy, Michael, and Maurice Crain. *The Land That Time Forgot: Adventures and Discoveries in New Guinea*. London: Hurst and Blackett, 1937.

Lee, H., N. Sambuughin, L. Cervenakova, et al. "Ancestral origins and worldwide distribution of the PRNP 200K mutation causing familial Creutzfeldt-Jakob disease." *American Journal of Human Genetics* 64 (1999): 1063–70.

Lévi-Strauss, Claude. *The Elementary Forms of Kinship and Marriage*, translated by J. H. Bell, J. R. von Stürmer, and Rodney Needham. Boston: Beacon Press, 1969.

Lewis, Sinclair. *Arrowsmith*. 1925. New York: New American Library, 1961.

Lindenbaum, Shirley. "Kuru, prions, and human affairs: Thinking about epidemics." *Annual Review of Anthropology* 30 (2001): 363–85.

———. "Kuru sorcery." In *Essays on Kuru*, edited by R. W. Hornabrook, 28–37. Faringdon, England: E. W. Classey, 1976.

———. *Kuru Sorcery: Disease and Danger in the New Guinea Highlands*. New York: McGraw-Hill, 1979.

———. "Sorcery and structure in Fore society." *Oceania* 41 (1971): 277–87.

———. "Thinking about cannibalism." *Annual Review of Anthropology* 33 (2004): 475–98.

———. "A wife is the hand of man." In *Man and Woman in the New Guinea Highlands*, edited by Paula Brown and Georgeda Buchbinder, 54–62. Washington, DC: American Anthropological Association, 1976.

Lindenbaum, Shirley, and Robert Glasse. "Fore age mates." *Oceania* 39 (1969): 165–73.

Lindenbaum, Shirley, and Margaret Lock, eds. *Knowledge, Power, and Practice: The Anthropology of Medicine and Everyday Life*. Berkeley: University of California Press, 1993.

Lwoff, André. "The concept of virus." *Journal of General Microbiology* 17 (1957): 239–53.

Mackay, Ian R., and F. Macfarlane Burnet. *Autoimmune Diseases: Pathogenesis, Chemistry, and Therapy*. Springfield, IL: Thomas, 1963.

Malinowski, Bronislaw. *Argonauts of the Western Pacific: An Account of Native Enterprise and Adventure in the Archipelagoes of Melanesian New Guinea*. 1922. New York: E. P. Dutton and Co., 1961.

———. "An ethnographic theory of the magical world." 1935. In *Coral Gardens and Their Magic*. Vol. 2: *The Language of Magic and Gardening*, 213–50. Bloomington: Indiana University Press, 1965.

———. "Magic, science, and religion." 1925. In *Magic, Science, and Religion and Other Essays*, 1–71. Boston: Beacon Press, 1948.

———. "Practical anthropology." *Africa* 2 (1929): 22–38.

Manuelidis, Laura. "The force of prions." *Lancet* 355 (2000): 2083.

Marsh, Richard F. "Transmissible mink encephalopathy." In *Prion Diseases of Humans and Animals*, edited by Stanley Prusiner, John Collinge, John Powell, and Brian Anderton, 300–307. New York: Ellis Horwood, 1992.

Marsh, Richard F., and R. H. Kimberlin. "Comparison of scrapie and transmissible mink encephalopathy in hamsters. II: Clinical signs, pathology and pathogenesis." *Journal of Infectious Diseases* 131 (1975): 104–10.

Masters, C. L., D. Carleton Gajdusek, and Clarence J. Gibbs Jr. "Creutzfeldt-Jakob disease virus isolations from the Gerstmann-Sträussler syndrome with an analysis of amyloid plaque deposition in the virus-induced spongiform encephalopathies." *Brain* 104 (1981): 559–88.

Mathews, John D. "The changing face of kuru." *Lancet* i (1965): 1138–41.

———. "Kuru as an epidemic disease." In *Essays on Kuru,* edited by R. W. Hornabrook, 83–104. Faringdon, England: E. W. Classey, 1976.

———. "Kuru: A puzzle in cultural and environmental medicine." M.D. thesis, University of Melbourne, 1971.

———. "Scientific method in epidemiology and research in community health problems." *Community Health Studies* 1 (1977): 26–30.

———. "A transmission model for kuru." *Lancet* i (1967): 821–25.

Mathews, John D., Robert Glasse, and Shirley Lindenbaum. "Kuru and cannibalism." *Lancet* ii (1968): 449–52.

McArthur, Norma. "The age incidence of kuru." *Annals of Human Genetics* 27 (1964): 341–52.

———. "Cross-currents: A demographic study." In *Essays on Kuru,* edited by R. W. Hornabrook, 107–16. Faringdon, England: E. W. Classey, 1976.

———. *Island Populations of the Pacific.* Canberra: Australian National University Press, 1967.

McCarthy, J. K. *Patrol into Yesterday: My New Guinea Years.* Melbourne: F. W. Cheshire, 1963.

McKinley, M. P., D. C. Bolton, and S. B. Prusiner. "A protease-resistant protein is a structural component of the scrapie prion." *Cell* 35 (1983): 57–62.

McKinley, M. P., M. B. Braunfield, C. G. Bellinger, and S. B. Prusiner. "Molecular characteristics of prion rods purified from scrapie-infected hamster brains." *Journal of Infectious Disease* 154 (1986): 110–20.

Mead, Margaret. *Blackberry Winter: My Earlier Years.* New York: William Morrow and Co., 1972.

Mead, Simon, M. P. H. Stumpf, Jerome Whitfield, et al. "Balancing selection at the prion protein gene consistent with prehistoric kurulike epidemics." *Science* 300 (2003): 640–43.

Medori, R., H. Tritschler, A. LeBlanc, et al. "Fatal familial insomnia, a prion disease with a mutation at codon 178 of the prion protein gene." *New England Journal of Medicine* 326 (1992): 444–49.

Melville, Herman. *Moby Dick, or, The Whale.* 1851. New York: Bantam Books, 1981.

Merz, Patricia A., R. A. Somerville, H. M. Wisniewski, and K. Iqbal. "Abnormal fibrils from scrapie-infected brain." *Acta Neuropathologica* 54 (1981): 63–74.

Mestel, Rosie. "Putting prions to the test." *Science* 273 (1996): 184–89.

Morris, J. A., and D. Carleton Gajdusek. "Encephalopathy of mice following inoculation of scrapie sheep brain." *Nature* 197 (1963): 1084–86.

Muller, H. J. "Variation due to change in an individual gene." *American Naturalist* 56 (1922): 48–49.

Nadel, S. F. *A Black Byzantium.* London: Oxford University Press, 1942.

Nietzsche, Friedrich. *The Genealogy of Morals: An Attack,* translated by Francis Golffing. 1887. Garden City, NY: Doubleday and Co., 1956.

Oesch, B., D. Westaway, M. Wälchli, et al. "A cellular gene encodes scrapie PrP 27–30 protein." *Cell* 40 (1985): 735–46.

O'Leary, Timothy J. *North and Aloft: A Personal Memoir of Service and Adventure with the Royal Flying Doctor Service in Far Northern Australia.* Brisbane: Amphion Press, 1988.

Palmer, M. S., A. J. Dryden, J. T. Hughes, and J. Collinge. "Homozygous prion protein genotype predisposes to sporadic Creutzfeldt-Jakob disease." *Nature* 352 (1991): 340–42.

Pattison, Iain H. "Fifty years with scrapie: A personal reminiscence." *Veterinary Record* 123 (1988): 661–66.

———. "Resistance of the scrapie agent to formalin." *Journal of Comparative Pathology* 75 (1965): 159–64.

———. "A sideways look at the scrapie saga: 1732–1991." In *Prion Diseases of Humans and Animals,* edited by Stanley Prusiner, John Collinge, John Powell, and Brian Anderton, 15–22. New York: Ellis Horwood, 1992.

Pattison, Iain H., and K. M. Jones. "The possible nature of the transmissible agent of scrapie." *Veterinary Record* 80 (1967): 2–9.

Prusiner, Stanley B. "An approach to the analysis of biological particles using sedimentation analysis." *Journal of Biological Chemistry* 253 (1978): 916–21.

———. "Novel proteinaceous infectious particles cause scrapie." *Science* 216 (1982): 136–44.

———. "The prion diseases." *Scientific American* 272 (1995): 48–56.

———. "The prion hypothesis." In *Prions: Novel Infectious Pathogens Causing Scrapie and Creuzfeldt-Jakob Disease,* edited by Stanley B. Prusiner and M. P. McKinley, 18–37. London: Academic Press, 1987.

———. "Research on scrapie." *Lancet* ii (1982): 494–95.

Prusiner, Stanley B., John Collinge, J. Powell, and B. Anderton, eds. *Prion Diseases of Humans and Animals.* London and New York: Ellis Horwood, 1992.

Prusiner, Stanley B., D. Carleton Gajdusek, and Michael P. Alpers. "Kuru with incubation periods exceeding two decades." *Annals of Neurology* 12 (1982): 1–9.

Prusiner, Stanley B., D. F. Groth, D. C. Bolton, S. B. Kent, and L. E. Hood. "Purification and structural studies of a major scrapie prion protein." *Cell* 38 (1984): 127–34.

Prusiner, Stanley B., and Maclyn McCarty. "Discovering DNA encodes heredity and prions are infectious proteins." *Annual Review of Genetics* 40 (2006): 25–45.

Radcliffe-Brown, A. R. "Applied anthropology." In *Report of the 20th Meeting of the Australian and New Zealand Association for the Advancement of Science, Brisbane 1930,* 267–80. Sydney: ANZAAS, 1931.

Read, Kenneth E. *The High Valley.* New York: Charles Scribner's Sons, 1965.

———. *Other Voices: The Style of a Male Homosexual Tavern.* Novato, CA: Chandler and Sharp, 1980.

———. *Return to the High Valley: Coming Full Circle.* Berkeley: University of California Press, 1986.

Rohwer, Robert. "The scrapie agent: A virus by any other name." *Current Topics in Microbiology and Immunology* 172 (1991): 195–232.

Schofield, F. D., and A. D. Parkinson. "Social medicine in New Guinea: Beliefs and practices affecting health among the Abelam and Wam peoples of the Sepik District." *Medical Journal of Australia* ii (1963): 1–8, 29–34.

Scott, M., D. Foster, D. F. Groth, S. J. DeArmond, and Stanley B. Prusiner. "Spontaneous neurodegeneration in transgenic mice with mutant prion protein." *Science* 250 (1990): 1587–90.

Scragg, R. F. R. "From medical tultul to doctor of medicine." In *A History of Medicine in Papua New Guinea: Vignettes of an Earlier Period,* edited by Burton G. Burton-Bradley, 15–46. Sydney: Australasian Medical Publishing Co., 1990.

Siakotos, A. N., D. C. Gajdusek, C. J. Gibbs Jr., R. D. Traub, and C. Bucana. "Partial purification of the scrapie agent from mouse brain by pressure disruption and zonal centrifugation in sucrose-sodium chloride gradients." *Virology* 70 (1976): 230–37.

Sigurdsson, Björn. "Observations on three slow infections of sheep." *British Veterinary Journal* 110 (1954): 255–70, 307–22, 341–54.

Simmons, R. T., D. C. Gajdusek, and L. C. Larkin. "A blood group genetical survey in New Britain." *American Journal of Physical Anthropology* 18 (1960): 101–8.

Simmons, R. T., J. J. Graydon, and D. C. Gajdusek. "A blood group genetical survey in Australian Aboriginal children of Cape York Peninsula." *American Journal of Physical Anthropology* 16 (1958): 59–77.

Simpson, D. A., H. Lander, and H. N. Robson. "Observations of kuru. II: Clinical features." *Australasian Annals of Medicine* 8 (1959): 8–15.

Sorenson, E. Richard. *The Edge of the Forest: Land, Childhood and Change in a New Guinea Proto-Agricultural Society.* Washington, DC: Smithsonian Institution Press, 1976.

———. *Journal Account of an Expedition to the Kuru Region.* Bethesda, MD: National Institute of Neurological Disease and Blindness, n.d. (c. 1965).

———. "Socio-cultural change among the Fore of New Guinea." *Current Anthropology* 13 (1972): 349–83.

———. "Where did the liminal flowers go? The study of child behavior and development in cultural isolates." *Anthropology of Consciousness* 7 (1996): 9–31.

Sorenson, E. Richard, and D. Carleton Gajdusek. "Investigation of nonrecurring phenomena: The research cinema film." *Nature* 200 (1963): 112–14.

———. "The study of child behavior and development in primitive cultures: A research archive for ethnopediatric film investigations of styles in the patterning of the nervous system." *Pediatrics* 37 (Supplement) 1 (1966): 149–243.

Steadman, Lyle B., and Charles F. Merbs. "Kuru and cannibalism?" *American Anthropologist* 84 (1982): 611–27.

Stent, Gunther S. *Molecular Genetics: An Introductory Narrative.* San Francisco: W. H. Freeman, 1971.

———. *Nazis, Women, and Molecular Biology: Memoirs of a Lucky Self-Hater.* Kensington, CA: Briones Books, 1998.

Taubes, Gary. "The game of the name is fame. But is it science?" *Discover* (December 1986): 28–52.

Telling, G. C., M. Scott, I. Mastrianni, et al. "Prion propagation in mice expressing human and chimeric PrP transgenes implicates the interaction of cellular PrP with another protein." *Cell* 83 (1995): 79–90.

Valentine, C. A. *Culture and Poverty.* Chicago: University of Chicago Press, 1968.

———. "An introduction to the history of changing ways of life in the island of New Britain." Ph.D. dissertation, University of Pennsylvania, 1958.

Vonnegut, Kurt. *Cat's Cradle.* 1963. New York: Dial Press, 2006.

Watson, James D. *The Double Helix: A Personal Account of the Discovery of the Structure of DNA.* New York: Atheneum, 1968.

———. "Growing up in the phage group." In *Phage and the Origins of Molecular Biology*, exp. ed., edited by John Cairns, Gunther S. Stent, and James D. Watson, 239–45. New York: Cold Spring Harbor Laboratory Press, 1992.

Weissmann, Charles. "Molecular biology of prion diseases." *Trends in Cell Biology* 4 (1994): 10–14.

——. "Some thoughts on the pursuit of success in science." *Biological Chemistry* 379 (1998): 233–34.

Weissmann, Charles, H. Bueler, M. Fischer, H. Bluethmann, and M. Aguet. "Susceptibility to scrapie in mice is dependent on PrPc." *Philosophical Transactions of the Royal Society of London, Series B* 343 (1994): 431–34.

Whitbread, Jane. "Stalking a new kind of killer." *Look* 33/10 (May 13, 1969): 22–28.

Wiener, Norbert. "A rebellious scientist after two years." *Bulletin of the Atomic Scientists* 4 (1948): 338–39.

Wilesmith, J. W., J. B. M. Ryan, and M. J. Atkinson. "Bovine spongiform encephalopathy—epidemiological studies on the origin." *Veterinary Record* 128 (1991): 199–203.

Will, R. G., S. N. Cousens, C. P. Farrington, et al. "Deaths from variant Creutzfeldt-Jakob disease." *Lancet* 353 (1999): 979.

Will, R. G., J. W. Ironside, M. Zeidler, et al. "A new variant of Creutzfeldt-Jakob disease in the United Kingdom." *Lancet* 347 (1996): 921–25.

Williams, E. S., and S. Young. "Chronic wasting disease of captive mule deer: A spongiform encephalopathy." *Journal of Wildlife Diseases* 16 (1980): 89–98.

Williams, G. R., Ann Fischer, J. L. Fischer, and L. T. Kurland. "Evaluation of the kuru genetic hypothesis." *Journal Génétique Humaine* 13 (1964): 11–21.

Wilson, David R., R. D. Anderson, and W. Smith. "Studies in scrapie." *Journal of Comparative Pathology* 60 (1950): 267–82.

Wyatt, J. M., G. R. Pearson, T. N. Smerdon, et al. "Naturally occurring scrapie-like spongiform encephalopathy in five domestic cats." *Veterinary Record* 129 (1991): 233–36.

Zigas, Vincent. *Laughing Death: The Untold Story of Kuru.* Clifton, NJ: Humana Press, 1990.

Zigas, Vincent, and D. Carleton Gajdusek. "Kuru: Clinical, pathological and epidemiological study of a recently discovered acute progressive degenerative disease of the central nervous system reaching 'epidemic' proportions among natives of the eastern highlands of New Guinea." *Papua and New Guinea Medical Journal* 3 (1959): 1–24.

——. "Kuru: Clinical study of a new syndrome resembling paralysis agitans in natives of the Eastern Highlands of Australian New Guinea." *Medical Journal of Australia* ii (1957): 745–54.

SECONDARY SOURCES

Abraham, Itty. "The contradictory spaces of postcolonial technoscience." *Economic and Political Weekly* (January 21, 2006): 210–17.

Anderson, Warwick. "Frank Macfarlane Burnet." In *Dictionary of Medical Biography*, vol. 1, edited by W. F. Bynum and Helen Power Bynum, 284–85. Westport, CT: Greenwood Press, 2007.

——. "Natural histories of infectious disease: Ecological vision in twentieth-century biomedical science." *Osiris* 19 (2004): 39–61.

——. "The possession of kuru: Medical science and biocolonial exchange." *Comparative Studies in Society and History* 42 (2000): 713–44.

——. "Postcolonial technoscience." *Social Studies of Science* 32 (2002): 643–58.

Anderson, Warwick, and Vincanne Adams. "Pramoedya's chickens: Postcolonial studies of technoscience." In *The Handbook of Science and Technology Studies,* 3rd ed., edited by Edward J. Hackett, Olga Amsterdamska, Michael Lynch, and Judy Wajcman, 181–204. Cambridge, MA: MIT Press, 2007.

Andrews, Lori, and Dorothy Nelkin. *Body Bazaar: The Market for Human Tissue in the Biotechnology Age.* New York: Crown, 2001.

Appadurai, Arjun. "Introduction: Commodities and the politics of value." In *The Social Life of Things: Commodities in Cultural Perspective,* edited by Arjun Appadurai, 3–63. Cambridge: Cambridge University Press, 1986.

Appel, Toby A. *Shaping Biology: The National Science Foundation and American Biological Research, 1945–1975.* Baltimore: Johns Hopkins University Press, 2000.

Arens, William. *The Man-Eating Myth: Anthropology and Anthropophagy.* New York: Oxford University Press, 1979.

———. "Rethinking anthropophagy." In *Cannibalism and the Colonial World,* edited by Francis Barker, Peter Hulme, and Margaret Iverson, 39–62. Cambridge: Cambridge University Press, 1998.

Bamford, Sandra. *Biology Unmoored: Melanesian Reflections on Life and Biotechnology.* Berkeley: University of California Press, 2007.

Bartolovich, Crystal. "Consumerism, or the cultural logic of late capitalism." In *Cannibalism and the Colonial World,* edited by Francis Barker, Peter Hulme, and Margaret Iverson, 204–37. Cambridge: Cambridge University Press, 1998.

Bashkow, Ira. *The Meaning of Whitemen: Race and Modernity in the Orokaiva Cultural World.* Chicago: University of Chicago Press, 2006.

Beasley, Annette. "Frontier science: The early investigation of kuru in Papua New Guinea." In *Challenging Science: Issues for New Zealand Society in the 21st Century,* edited by K. Dew and R. Fitzgerald, 146–66. Palmerston North: Dunmore Press, 2004.

———. "Kuru truths: Obtaining Fore narratives." *Field Methods* 18 (2006): 21–42.

———. "The promised medicine: Fore reflections on the scientific investigation of kuru." *Oceania* 76 (2006): 186–202.

Benison, Saul, ed. *Tom Rivers: Reflections on a Life in Medicine.* Cambridge, MA: MIT Press, 1969.

Benjamin, Walter. "The work of art in the age of mechanical reproduction." In *Illuminations,* edited by Hannah Arendt and translated by Harry Zohn, 117–252. New York: Schocken Books, 1968.

Biagioli, Mario. *Galileo Courtier: The Practice of Science in the Culture of Absolutism.* Chicago: University of Chicago Press, 1993.

———. "Rights or Rewards? Changing frameworks of scientific authorship." In *Scientific Authorship: Credit and Intellectual Property in Science,* edited by Mario Biagioli and Peter Galison, 253–80. New York: Routledge, 2003.

Bloch, Maurice, and Jonathan Parry. "Introduction: Death and the regeneration of life." In *Death and the Regeneration of Life,* edited by M. Bloch and J. Parry, 1–44. Cambridge: Cambridge University Press, 1982.

Blumberg, Baruch S. *Hepatitis B: The Hunt for a Killer Virus.* Princeton, NJ: Princeton University Press, 2002.

Bourdieu, Pierre. *Outline of a Theory of Practice.* Cambridge: Cambridge University Press, 1977.

Bowker, Geoffrey C., and Susan Leigh Star. *Sorting Things Out: Classification and Its Consequences.* Cambridge, MA: MIT Press, 1999.

Boyd, David. "A tale of 'first contact': The Hagahai of Papua New Guinea." *Research in Melanesia* 20 (1996): 103–40.

Brown, Paula, and Donald Tuzin, eds. *The Ethnography of Cannibalism.* Washington, DC: Society for Psychological Anthropology, 1983.

Campbell, I. C. "Anthropology and the professionalisation of colonial administration in Papua and New Guinea." *Journal of Pacific History* 33 (1998): 69–90.

Carrier, James G., ed. *History and Tradition in Melanesian Anthropology.* Berkeley: University of California Press, 1992.

Cassel, John. "Social science theory as a source of hypotheses in epidemiological research." *American Journal of Public Health* 54 (1964): 1488–94.

Chen, Kwang-Ming, ed. *The Mysterious Diseases of Guam.* Guam: Richard F. Taitano Micronesian Area Research Center, University of Guam, 2004.

Clarke, Adele E. "Research materials and reproductive science in the United States." In *Ecologies of Knowledge: Work and Politics in Science and Technology,* edited by Susan Leigh Star, 183–225. Albany: State University of New York Press, 1995.

Collins, K. J., and J. S. Weiner. *Human Adaptability: A History and Compendium of Research in the International Biological Programme.* London: Taylor and Francis, 1977.

Comfort, Nathaniel C. *The Tangled Field: Barbara McClintock's Search for the Patterns of Genetic Control.* Cambridge, MA: Harvard University Press, 2001.

Comstock, George W. "Cohort analysis: W. H. Frost's contributions to the epidemiology of tuberculosis and chronic disease." *Sozial und Präventiv Medizin* 46 (2000): 7–12.

Connolly, Bob, and Robin Anderson. *First Contact: New Guinea's Highlanders Encounter the Outside World.* New York: Viking, 1987.

Cooke, Jennifer. *Cannibals, Cows, and the CJD Catastrophe.* Milsons Point, New South Wales: Random House Australia, 1998.

Coombe, Rosemary J. *The Cultural Life of Intellectual Properties: Authorship, Appropriation, and the Law.* Durham, NC: Duke University Press, 1998.

Creager, Angela N. H. *The Life of a Virus: Tobacco Mosaic Virus as an Experimental Model, 1930–1965.* Chicago: University of Chicago Press, 2002.

Cunningham, Hilary. "Colonial encounters in postcolonial contexts: Patenting indigenous DNA and the human genome diversity project." *Critique of Anthropology* 18 (1998): 205–33.

De Certeau, Michel. *The Possession at Loudun.* 1970. Chicago: University of Chicago Press, 2000.

De Chadarevian, Soraya. "Sequences, conformation, information: Biochemists and molecular biologists in the 1950s." *Journal of the History of Biology* 29 (1996): 361–86.

De Chadarevian, Soraya, and Harmke Kamminga, eds. *Molecularizing Biology and Medicine: New Practices and Alliances, 1910s to 1970s.* Amsterdam: Harwood Academic Publishers, 1998.

Dening, Greg. *History's Anthropology: The Death of William Gooch.* Lanham, MD: University Press of America, 1988.

———. *Islands and Beaches: Discourse on a Silent Land: Marquesas 1774–1880*. Honolulu: University of Hawaii Press, 1980.

Denoon, Donald, with Kathleen Dugan and Leslie Marshall. *Public Health in Papua New Guinea: Medical Possibility and Social Constraint, 1884–1984*. Cambridge: Cambridge University Press, 1989.

Downs, Ian F. *The Australian Trusteeship: Papua New Guinea 1945–75*. Canberra: Australian Government Publishing Service, 1980.

Draper, Robert. "The genius who loved boys." *GQ* (November 1999): 313–30.

Dror, Otniel E. "'Voodoo death': Fantasy, excitement, and the untenable boundaries of biomedical science." In *The Politics of Healing: Histories of Alternative Medicine in Twentieth-Century North America*, edited by Robert D. Johnston, 71–81. New York: Routledge, 2004.

Faris, James C. "Pax Britannica and the Sudan: S. F. Nadel." In *Anthropology and the Colonial Encounter*, edited by Talal Asad, 153–70. Atlantic Highlands, NJ: Humanities Press, 1973.

Farreras, Ingrid G., Caroline Hannaway, and Victoria A. Harden, eds. *Mind, Brain, Body and Behavior: The Foundations of Neuroscience and Behavioral Research at the National Institutes of Health*. Amsterdam: IOS Press, 2004.

Findlen, Paula. "The economy of scientific exchange in early modern Italy." In *Patronage and Institutions*, edited by Bruce Moran, 5–24. Rochester, NY: Boydell, 1991.

Fleck, Ludwik. *Genesis and Development of a Scientific Fact*, edited by Thaddeus J. Trenn and Robert K. Merton, translated by Fred Bradley and Thaddeus J. Trenn. 1934. Chicago: University of Chicago Press, 1979.

Fleming, Donald. "Emigré physicists and the biological revolution." *Perspectives in American History* 2 (1968): 152–89.

Foster, Robert J. "Dangerous circulation and revelatory display: Exchange practices in a New Ireland society." In *Exchanging Products: Producing Exchange*, edited by Jane Fajans, 15–32. Sydney: University of Sydney Press, 1993.

Fox, Daniel M. "The politics of the NIH extramural program, 1937–1950." *Journal of History of Medicine and Allied Sciences* 42 (1987): 447–66.

Fox, Renée. "The evolution of American bioethics: A sociological perspective." In *Social Science Perspectives on Medical Ethics*, edited by George Weisz, 210–17. Dordrecht: Kluwer, 1990.

Fox, Renée, and Judith P. Swazey. "Examining American bioethics: Its problems and prospects." *Cambridge Quarterly of Healthcare Ethics* 14 (2005): 361–73.

———. "Medical morality is not bioethics: Medical ethics in China and the United States." *Perspectives in Biology and Medicine* 27 (1984): 336–60.

Freud, Sigmund. *Totem and Taboo*. 1919. Harmondsworth: Penguin, 1938.

Frow, John. "Invidious distinctions: Waste, difference, and classy stuff." In *Culture and Waste: The Creation and Destruction of Value*, edited by G. Hawkins and S. Muecke, 25–38. Lanham, MD: Rowman and Littlefield, 2003.

Fujimura, Joan. *Crafting Science: A Sociohistory of the Quest for the Genetics of Cancer*. Cambridge, MA: Harvard University Press, 1996.

Galison, Peter. "Computer simulations and the trading zone." In *The Disunity of Science: Bound-

aries, Contexts, and Power, edited by Peter Galison and David J. Stump, 118–57. Stanford, CA: Stanford University Press, 1996.

——. *Image and Logic: A Material Culture of Microphysics.* Chicago: University of Chicago Press, 1997.

——. "Material culture, theoretical culture and delocalization." In *Companion to Science in the Twentieth Century,* edited by John Krige and Dominique Pestre, 669–82. London: Routledge, 2003.

Gamage, Bill. "Police and power in the pre-war Papua New Guinea Highlands." *Journal of Pacific History* 31 (1996): 162–77.

——. *The Sky Travellers: Journeys in New Guinea.* Carlton South: Melbourne University Press, 1998.

Gaudillière, Jean-Paul, and Hans-Jorg Rheinberger, eds. *From Molecular Genetics to Genomics: Mapping the Cultures of Twentieth-Century Genetics.* London: Routledge, 2004.

Geary, Patrick. "Sacred commodities: The circulation of medieval relics." In *The Social Life of Things: Commodities in Cultural Perspective,* edited by Arjun Appadurai, 169–91. Cambridge: Cambridge University Press, 1986.

Geertz, Clifford. "Found in translation: On the social history of the moral imagination." In *Local Knowledge: Further Essays in Interpretive Anthropology,* 36–54. New York: Basic Books, 1983.

George, Kenneth M. "Head hunting, history, and exchange in upland Sulawesi." *Journal of Asian Studies* 50 (1991): 536–64.

Gieryn, Thomas F. "Boundary-work and the demarcation of science from non-science: Strains and interests in professional ideologies of scientists." *American Sociological Review* 48 (1983): 781–95.

Godelier, Maurice. *The Enigma of the Gift,* translated by Nora Scott. Chicago: University of Chicago Press, 1999.

Goffman, Erving. *Relations in Public: Microstudies of the Public Order.* New York: Basic Books, 1971.

Goodenough, Ward H. "Moral outrage: Territoriality in human guise." *Zygon* 32 (1997): 5–27.

Gray, Geoffrey. "'Being honest to my science': Reo Fortune and J. H. P. Murray, 1927–30." *Australian Journal of Anthropology* 10 (1999): 56–76.

——. "There are many difficult problems: Ernest William Pearson Chinnery—government anthropologist." *Journal of Pacific History* 38 (2003): 313–30.

——. "'You are . . . my anthropological children': A. P. Elkin, Ronald Berndt, and Catherine Berndt, 1940–1956." *Aboriginal History* 29 (2005): 77–106.

——, ed. *Before It's Too Late: Anthropological Reflections, 1950–1970.* Sydney: Oceania Publications, 2001.

Gregory, Christopher A. *Gifts and Commodities.* London: Academic Press, 1982.

Griffin, James T. "John Gunther and medicine in Papua New Guinea." In *Health and Healing in Tropical Australia and Papua New Guinea,* edited by Roy MacLeod and Donald Denoon, 88–102. Townsville, Queensland: James Cook University Press, 1991.

Griffin, James T., Hank Nelson, and Stewart Firth. *Papua New Guinea: A Political History.* Richmond, Vic.: Heinemann, 1979.

Hagstrom, Warren O. *The Scientific Community.* New York: Basic Books, 1965.

Harden, Victoria A. *Inventing the NIH: Federal Biomedical Research Policy, 1887–1937.* Baltimore: Johns Hopkins University Press, 1986.

———. *Rocky Mountain Spotted Fever: History of a Twentieth-Century Disease.* Baltimore: Johns Hopkins University Press, 1990.

Hasluck, Paul. *A Time for Building: Australian Administration in Papua New Guinea, 1951–1963.* Carlton: Melbourne University Press, 1976.

Hayden, Cori. *When Nature Goes Public: The Making and Unmaking of Bioprospecting in Mexico.* Princeton, NJ: Princeton University Press, 2002.

Hays, Terence E. "A historical background to anthropology in the New Guinea highlands." In *Ethnographic Presents: Pioneering Anthropologists in the Papua New Guinea Highlands,* edited by Terence E. Hays, 1–36. Berkeley: University of California Press, 1992.

Hirsch, Eric, and Marilyn Strathern, eds. *Transactions and Creations: Property Debates and the Stimulus of Melanesia.* Oxford: Berghahn, 2004.

Hoskins, Janet. "Introduction: Head hunting as practice and trope." In *Head Hunting and the Social Imagination in Southeast Asia,* edited by Janet Hoskins, 1–49. Stanford, CA: Stanford University Press, 1996.

Hoyningen-Huene, Paul, and Howard Sankey, eds. *Incommensurability and Related Matters.* Dordrecht: Kluwer, 2001.

Hughes, Sally Smith. "Making dollars out of DNA: The first major patent in biotechnology and the commercialization of molecular biology, 1974–1980." *Isis* 92 (2001): 541–75.

Humphrey, Caroline, and Stephen Hugh-Jones. "Introduction: Barter, exchange, and value." In *Barter, Exchange, and Value: An Anthropological Approach,* 1–20. Cambridge: Cambridge University Press, 1992.

Jaarsma, Sjoerd R. "Conceiving New Guinea: Ethnography as a phenomenon of contact." In *In Colonial New Guinea: Anthropological Perspectives,* edited by Naomi M. McPherson, 27–44. Pittsburgh, PA: University of Pittsburgh Press, 2001.

James, Craig R., Ron Stall, and Sandra M. Gifford, eds. *Anthropology and Epidemiology: Interdisciplinary Approaches to the Study of Health and Disease.* Dordrecht: D. Reidel, 1986.

Jebens, Holger. "'Vali did that too': On Western and Indigenous cargo discourses in West New Britain (Papua New Guinea)." *Anthropological Forum* 14 (2004): 117–39.

Jong, Simcha. "How organizational structures in science shape spin-off firms: The biochemistry departments of Berkeley, Stanford, and UCSF and the birth of the biotech industry." *Industrial and Corporate Change* 15 (2006): 251–83.

Judson, Horace Freeland. *The Eighth Day of Creation: Makers of the Revolution in Biology.* New York: Simon and Schuster, 1979.

Kay, Lily E. *The Molecular Vision of Life: Caltech, the Rockefeller Foundation, and the Rise of the New Biology.* New York: Oxford University Press, 1993.

Keller, Evelyn Fox. *The Century of the Gene.* Cambridge, MA: Harvard University Press, 2000.

———. *A Feeling for the Organism: The Life and Work of Barbara McClintock.* San Francisco: W. H. Freeman, 1983.

———. "Physics and the emergence of molecular biology: A history of cognitive and political synergy." *Journal of the History of Biology* 23 (1990): 389–409.

Kevles, Daniel J. "Cold war and hot physics: Science, security, and the American state, 1945–56." *Historical Studies in the Physical and Biological Sciences* 20 (1990): 239–64.

———. "Principles and politics in federal R&D policy, 1945–1990: An appreciation of the Bush report." Introduction to *Science, the Endless Frontier: A Report to the President*, by Vannevar Bush, ix–xxxiii. Washington, DC: National Science Foundation, 1990.

———. "Renato Delbecco and the new animal virology: Medicine, methods, and molecules." *Journal of the History of Biology* 26 (1993): 409–42.

Kevles, Daniel J., and Gerald L. Geison. "The experimental life sciences in the twentieth century." *Osiris* 10 (1995): 97–121.

Keyes, Martha E. "The prion challenge to the 'central dogma' of molecular biology, 1965–1991. Part II: The problem with prions." *Studies in History and Philosophy of Biological and Biomedical Sciences* 30 (1999): 181–218.

Kidd, J. S. "Scholarly excess and journalistic restraint in the popular treatment of cannibalism." *Social Studies of Science* 18 (1988): 749–54.

Kim, Kiheung. *The Social Construction of Disease: From Scrapie to Prion.* London: Routledge, 2007.

Kirsh, Stuart. "Property effects: Social networks and compensation claims in Melanesia." *Social Anthropology* 9 (2001): 147–63.

Kleinman, Arthur. "Moral experience and ethical reflection: Can ethnography reconcile them? A quandary for 'the new bioethics.'" *Daedalus* 128 (1999): 69–98.

Kleinman, Daniel L. *Impure Cultures: University Biology and the World of Commerce.* Madison, WI: University of Wisconsin Press, 2003.

———. *Politics on the Endless Frontier: Postwar Research Policy in the United States.* Durham, NC: Duke University Press, 1995.

Knorr-Cetina, Karin D. "The couch, the cathedral, and the laboratory: On the relationship between experiment and laboratory in science." In *Science as Practice and Culture*, edited by Andrew Pickering, 113–38. Chicago: University of Chicago Press, 1992.

———. *Epistemic Cultures: How the Sciences Make Knowledge.* Cambridge, MA: Harvard University Press, 1999.

———. "The ethnographic study of scientific work: Towards a constructivist interpretation of science." In *Science Observed: Perspectives on the Social Study of Science*, edited by Karin D. Knorr-Cetina and Michael Mulkay, 115–40. London: Sage, 1983.

Knowles, Chantal. "Reverse trajectories: Beatrice Blackwood as collector and anthropologist." In *Hunting the Gatherers: Ethnographic Collectors, Agents, and Agency in Melanesia, 1870s–1930s*, edited by Michael O'Hanlon and Robert Welsch, 251–71. New York: Berghahn, 2000.

Kohler, Robert E. *Landscapes and Labscapes: Exploring the Lab-Field Border in Biology.* Chicago: University of Chicago Press, 2002.

———. *Lords of the Fly: Drosophila Genetics and the Experimental Life.* Chicago: University of Chicago Press, 1994.

Krimsky, Sheldon. "The profit of scientific discovery and its normative implications." *Chicago-Kent Law Review* 75 (1999): 15–39.

Kuhn, Thomas S. *The Structure of Scientific Revolutions.* Chicago: University of Chicago Press, 1962.

Kuklick, Henrika. *The Savage Within: The Social History of British Anthropology, 1885–1945.* Cambridge: Cambridge University Press, 1991.

Kuper, Adam. *Anthropology and Anthropologists: The Modern British School.* London: Routledge, 1996.

Kwa, Chung-lin. "Representations of nature mediating between ecology and science policy: The case of the International Biological Programme." *Social Studies of Science* 17 (1987): 413–42.

Latour, Bruno. "Give me a laboratory and I will raise the world." In *Science Observed: Perspectives on the Social Study of Science,* edited by Karin D. Knorr-Cetina and Michael Mulkay, 141–70. London: Sage, 1983.

———. *The Pasteurization of France,* translated by Alan Sheridan and John Law. Cambridge, MA: Harvard University Press, 1988.

Latour, Bruno, and Steven Woolgar. *Laboratory Life: The Construction of Scientific Facts.* Princeton, NJ: Princeton University Press, 1986.

Latukefu, Sione, ed. *Papua New Guinea: A Century of Colonial Conflict.* Port Moresby: National Research Institute and University of Papua New Guinea, 1989.

Lederberg, Joshua. "Infectious history." *Science* 288 (2000): 287–93.

———. "The ontogeny of the clonal selection theory of antibody formation: Reflections on Darwin and Ehrlich." *Annals of the New York Academy of Sciences* 546 (1988): 175–87.

Lederman, Rena. *What Gifts Engender: Social Relations and Politics in Mendi, Highland Papua New Guinea.* Cambridge: Cambridge University Press, 1986.

Leslie, Stuart W. *The Cold War and American Science: The Military-Industrial-Academic Complex at MIT and Stanford.* New York: Columbia University Press, 1993.

Lindee, M. Susan. *Moments of Truth in Genetic Medicine.* Baltimore: Johns Hopkins University Press, 2005.

———. *Suffering Made Real: American Science and the Survivors at Hiroshima.* Chicago: University of Chicago Press, 1994.

Lock, Margaret. "Interrogating the human genome diversity project." *Social Science and Medicine* 39 (1994): 603–6.

Lynch, Michael E. "Sacrifice and the transformation of the animal body into a scientific object: Laboratory culture and ritual practice in the neurosciences." *Social Studies of Science* 18 (1988): 265–89.

MacLeod, Roy. "Introduction." In *Nature and Empire: Science and the Colonial Enterprise,* edited by Roy MacLeod. *Osiris* 15 (2000): 1–13.

Marks, Jonathan. *What It Means to Be 98% Chimpanzee: Apes, People and Their Genes.* Berkeley: University of California Press, 2002.

Mauss, Marcel. *The Gift: Forms and Functions of Exchange in Archaic Societies,* translated by Ian Cunnison. London: Cohen and West, 1970.

Max, D. T. *The Family That Couldn't Sleep: A Medical Mystery.* New York: Random House, 2006.

McCalman, Iain, with Benjamin Penny and Misty Cook, eds. *Mad Cows and Modernity: Cross-Disciplinary Reflections on the Crisis of Creutzfeldt-Jakob Disease.* Canberra: Humanities Research Center, Australian National University, 1998.

McKaughan, Daniel J. "The influence of Niels Bohr on Max Delbrück: Revisiting the hopes inspired by 'Light and Life.'" *Isis* 96 (2005): 507–29.

McLaren, Ann. "In the footprints of Reo Fortune." In *Ethnographic Presents: Pioneering Anthropologists in the Papua New Guinea Highlands,* edited by Terence E. Hays, 37–67. Berkeley: University of California Press, 1992.

McNeil, Maureen. "Postcolonial technoscience." *Science as Culture* 14 (2005): 105–12.

McPherson, Naomi M. "'Wanted: Young man, must like adventure': Ian McCallum Mack, patrol officer." In *In Colonial New Guinea: Anthropological Perspectives,* edited by Naomi M. McPherson, 82–110. Pittsburgh, PA: University of Pittsburgh Press, 2001.

McSherry, Corinne. *Who Owns Academic Work? Battling for Intellectual Property.* Cambridge, MA: Harvard University Press, 2001.

Meskell, Lynn, and Peter Pels, eds. *Embedding Ethics.* Oxford: Berg, 2005.

Mirowski, Philip, and Esther-Mirjam Sent, eds. *Science Bought and Sold: Essays in the Economics of Science.* Chicago: University of Chicago Press, 2002.

Moore, Clive. *New Guinea: Crossing Boundaries and History.* Honolulu: University of Hawaii Press, 2003.

Morange, Michel. *A History of Molecular Biology,* translated by M. Cobb. Cambridge, MA: Harvard University Press, 1998.

Mullins, Nicholas C. "The development of a scientific specialty: The phage group and the origins of molecular biology." *Minerva* 10 (1972): 51–82.

Munn, Nancy D. *The Fame of Gawa: A Symbolic Study of Value Transformation in a Massim (Papua New Guinea) Society.* Durham, NC: Duke University Press, 1986.

Neel, James V. *Physician to the Gene Pool: Genetic Lessons and Other Stories.* New York: John Wiley and Sons, 1994.

Nelson, Hank. *Black, White and Gold: Gold Mining in Papua New Guinea, 1878–1930.* Canberra: Australian National University Press, 1976.

———. "Kuru: The pursuit of the prize and the cure." *Journal of Pacific History* 31 (1996): 178–201.

Olby, Robert C. *The Path to the Double Helix.* Seattle: University of Washington Press, 1974.

Oudshoorn, Nellie. "On the making of sex hormones: Research materials and the production of knowledge." *Social Studies of Science* 20 (1990): 5–33.

Pannell, Sandra. "Travelling in other worlds: Narratives of head hunting, appropriation and the other in the 'Eastern Archipelago.'" *Oceania* 62 (1992): 162–78.

Park, Buhm Soon. "The development of the intramural research program at the National Institutes of Health after World War II." *Perspectives in Biology and Medicine* 46 (2003): 383–402.

Parry, Bronwyn. *Trading the Genome: Investigating the Commodification of Bio-Information.* New York: Columbia University Press, 2004.

Parry, Jonathan. "On the moral perils of exchange." In *Money and the Morality of Exchange,* edited by Jonathan Parry and Maurice Bloch, 64–93. Cambridge: Cambridge University Press, 1989.

Pennington, T. Hugh. *When Food Kills: BSE, E. Coli, and Disaster Science.* Oxford: Oxford University Press, 2003.

Pickstone, John V. "Objects and objectives: Notes on the material cultures of medicine." In *Technologies of Medicine,* edited by Ghislaine Lawrence, 13–24. London: Science Museum, 1994.

———. *Ways of Knowing: A New History of Science, Technology and Medicine.* Manchester: Manchester University Press, 2000.

Pietz, William. "The death of the deodand, accursed objects and the monetary value of human life." *Res* 31 (1997): 93–108.

———. "The problem of the fetish I." *Res* 9 (1985): 5–17.

———. "The problem of the fetish IIIa." *Res* 16 (1988): 105–23.

———. "The spirit of civilization: Blood sacrifice and monetary debt." *Res* 28 (1995): 23–38.

Polier, Nicole. "Commoditization, cash, and kinship in postcolonial Papua New Guinea." In *Commodities and Globalization: Anthropological Perspectives,* edited by Angelique Huagerud, M. Priscilla Stone, and Peter D. Little, 197–217. Lanham, MD: Rowman and Littlefield, 2000.

Povinelli, Elizabeth A. "Radical worlds: The anthropology of incommensurability and inconceivability." *Annual Review of Anthropology* 30 (2001): 319–34.

Pratt, Mary Louise. *Imperial Eyes: Travel Writing and Transculturation.* London and New York: Routledge, 1992.

Price, Derek J. de Solla. *Little Science, Big Science.* New York: Oxford University Press, 1963.

Provine, William B. "Introduction (to Section 11: England)." In *The Evolutionary Synthesis: Perspectives on the Unification of Biology,* edited by Ernst Mayr and William B. Provine, 329–33. Cambridge, MA: Harvard University Press, 1980.

Rabinow, Paul. *Making PCR: A Story of Biotechnology.* Chicago: University of Chicago Press, 1996.

Radford, Robin. *Highlanders and Foreigners in the Upper Ramu: The Kainantu Area, 1919–1942.* Carlton: Melbourne University Press, 1987.

Rajan, Kaushik Sunder. *Biocapital: The Constitution of Post-Genomic Life.* Durham, NC: Duke University Press, 2006.

Rasmussen, Nicolas. "The mid-century biophysics bubble: Hiroshima and the biological revolution in America, revisited." *History of Science* 35 (1997): 245–93.

Ratzan, Scott C., ed. *The Mad Cow Crisis.* New York: New York University Press, 1998.

Reardon, Jenny. "The human genome diversity project: A case study in co-production." *Social Studies of Science* 31 (2001): 357–88.

Reingold, Nathan. "Vannevar Bush's new deal for research: Or, the triumph of the old order." *Historical Studies in the Physical and Biological Sciences* 17 (1987): 299–344.

Rheinberger, Hans-Jörg. *Toward a History of Epistemic Things: Synthesizing Proteins in the Test Tube.* Stanford, CA: Stanford University Press, 1997.

Rhodes, Richard. *Deadly Feasts: Tracking the Secrets of a Terrifying New Plague.* New York: Simon and Schuster, 1997.

Ridley, R., and H. Baker. *Fatal Protein: The Story of CJD, BSE, and Other Prion Diseases.* Oxford: Oxford University Press, 1998.

Ritvo, Harriet. "The roast beef of old England." In *Mad Cows and Modernity: Cross-Disciplinary Reflections on the Crisis of Creutzfeldt-Jakob Disease,* edited by Iain McCalman, with Benjamin Penny and Misty Cook, 97–123. Canberra: Humanities Research Center, Australian National University, 1998.

Roll-Hansen, Nils. "The application of complementarity to biology: From Niels Bohr to Max Delbrück." *Historical Studies in the Physical and Biological Sciences* 30 (2000): 417–42.

Rosaldo, Michelle Z. "Skulls and causality." *Man* 12 (1977): 168–70.

Roscoe, Paul. "Margaret Mead, Reo Fortune, and Mountain Arapesh warfare." *American Anthropologist* 105 (2003): 581–91.

Rosenberg, Charles E. "Cholera in nineteenth-century Europe: A tool for social and economic analysis." *Comparative Studies in Society and History* 8 (1966): 452–63.

———. *The Cholera Years: The United States in 1832, 1849, and 1866.* Chicago: University of Chicago Press, 1962.

———. "Explaining epidemics." In *Explaining Epidemics and Other Studies in the History of Medicine,* 293–304. Cambridge: Cambridge University Press, 1992.

———. "Framing disease: Illness, society, and history." In *Framing Disease: Studies in Cultural History,* edited by Charles E. Rosenberg and Janet Golden, xiii–xxxvi. New Brunswick, NJ: Rutgers University Press, 1991.

———. "Meanings, policies, and medicine: On the bioethical enterprise and history." *Daedalus* 128 (1999): 27–46.

———. "What is an epidemic? AIDS in historical perspective." *Daedalus* 118 (1989): 1–17.

Rothman, David J. *Strangers at the Bedside: A History of How Law and Bioethics Transformed Medical Decision Making.* New York: Basic Books, 1991.

Rowland, Lewis P. *NINDS at 50: An Incomplete History Celebrating the Fiftieth Anniversary of the National Institute of Neurological Disorders and Stroke.* Bethesda, MD: National Institutes of Health, 2001.

Rumsey, Alan. "The white man as cannibal in the New Guinea Highlands." In *The Anthropology of Cannibalism,* edited by L. Goldman, 105–21. Westport, CT: Bergin and Garvey, 1999.

Ryan, D'Arcy. "Gift Exchange in the Mendi Valley: An Examination of the Socio-Political Implications of the Ceremonial Exchange of Wealth among the People of the Mendi Valley, Southern Highlands District, Papua." Ph.D. thesis, University of Sydney, 1962.

Sagan, Eli. *Cannibalism: Human Aggression and Cultural Form.* New York: Harper and Row, 1974.

Sahlins, Marshall D. "Cannibalism: An exchange." *New York Review of Books* 26 (March 22, 1979): 45–53.

———. "Poor man, rich man, big-man, chief: Political types in Melanesia and Polynesia." *Comparative Studies in Society and History* 5 (1963): 285–303.

———. *Stone Age Economics.* Chicago: Aldine-Atherton, 1972.

Sanday, Peggy Reeves. *Divine Hunger: Cannibalism as a Cultural System.* Cambridge: Cambridge University Press, 1986.

Scheper-Hughes, Nancy. "Organ trade: The new cannibalism." *New Internationalist* (April 1998): 14–17.

Schieffelin, Edward L., and Robert Crittenden, eds. *Like People You See in a Dream: First Contact in Six Papuan Societies.* Stanford, CA: Stanford University Press, 1991.

Schwartz, Maxime. *How the Cows Turned Mad,* translated by Edward Schneider. Berkeley: University of California Press, 2003.

Scragg, R. F. R. "Sir John Gunther." In *Encyclopaedia of Papua and New Guinea,* vol. 2, edited by Peter Ryan, 317–31. Carlton: Melbourne University Press, 1972.

Seguin, Eve. "The BSE saga: A cannibalistic tale." *Science as Culture* 12 (2003): 3–22.

———. "The UK BSE crisis: Strengths and weaknesses of existing conceptual approaches." *Science and Public Policy* 27 (2000): 293–301.

———, ed. *Infectious Processes: Knowledge, Discourse and the Politics of Prions.* Basingstoke, Hampshire, U.K.: Palgrave Macmillan, 2004.

Sexton, Christopher. *The Seeds of Time: The Life of Sir Macfarlane Burnet.* Oxford: Oxford University Press, 1991.

Shapin, Steven. "Trust, honesty, and the authority of science." In *Society's Choices: Social and Ethical Decision Making in Biomedicine,* edited by Ruth E. Bulger, Elizabeth M. Bobby, and Harvey V. Fineberg, 388–408. Washington, DC: National Academy Press, 1995.

Sharp, Lesley A. *Bodies, Commodities and Biotechnologies: Death, Mourning, and Scientific Desire in the Realm of Human Organ Transfer.* New York: Columbia University Press, 2007.

———. "The commodification of the body and its parts." *Annual Review of Anthropology* 29 (2000): 287–328.

Simmel, Georg. *The Philosophy of Money,* 3rd enlarged ed., edited by David Frisby and translated by Tom Bottomore and David Frisby. 1907. London: Routledge, 2004.

Sinclair, James. *Kiap: Australia's Patrol Officers in Papua New Guinea.* Bathurst, New South Wales: Robert Brown and Associates, 1984.

Soo, Mary, and Cathryn Carson. "Managing the research university: Clark Kerr and the University of California." *Minerva* 42 (2004): 215–36.

Souter, Gavin. *New Guinea: The Last Unknown.* Sydney: Angus and Robertson, 1963.

Spark, Ceridwen. "Learning from the locals: Gajdusek, kuru, and cross-cultural interaction in Papua New Guinea." *Health and History* 7 (2005): 213–19.

Stanton, Jenny. "Blood brotherhood: Techniques, expertise, and sharing in hepatitis B research in the 1970s." In *Technologies of Medicine,* edited by Ghislaine Lawrence, 120–33. London: Science Museum, 1994.

Star, Susan Leigh, and James R. Griesemer. "Institutional ecology, 'translations,' and boundary objects: Amateurs and professionals in Berkeley's Museum of Comparative Zoology, 1907–39." *Social Studies of Science* 19 (1989): 387–420.

Stetten, DeWitt Jr., and W. T. Carrigan, eds. *NIH: An Account of Research in Its Laboratories and Clinics.* New York: Academic Press, 1984.

Stocking, George W. *After Tylor: British Social Anthropology, 1888–1951.* Madison: University of Wisconsin Press, 1995.

———. "Gatekeeper to the field: E. W. P. Chinnery and the ethnography of the New Guinea mandate." *History of Anthropology Newsletter* 9 (1982): 3–12.

Stone, Linda, and Paul F. Lurquin. *A Genetic and Cultural Odyssey: The Life and Work of L. Luca Cavalli-Sforza.* New York: Columbia University Press, 2005.

Strasser, Bruno J. "Collecting and experimenting: The moral economy of biological research, 1960s–1980s." In *History and Epistemology of Molecular Biology and Beyond: Problems and Perspectives,* edited by Soraya de Chadarevian and Hans-Jörg Rheinberger, 105–22. Berlin: Max-Planck-Institut für Wissenschaftsgeschichte, 2005.

Strathern, Andrew J. "The kula in comparative perspective." In *The Kula: New Perspectives on Massim Exchange,* edited by Jerry W. Leach and Edmund Leach, 73–88. Cambridge: Cambridge University Press, 1983.

———. "Looking backward and forward." In *Ethnographic Presents: Pioneering Anthropologists in the Papua New Guinea Highlands,* edited by Terence E. Hays, 250–70. Berkeley: University of California Press, 1992.

———. *The Rope of Moka: Big-men and Ceremonial Exchange in Mt. Hagen, New Guinea.* Cambridge: Cambridge University Press, 1971.

Strathern, Andrew, and Pamela J. Stewart. *Arrow Talk: Transaction, Transition, and Contradiction in New Guinea Highlands History.* Kent, OH: Kent State University Press, 2000.

Strathern, Marilyn. "Cutting the network." *Journal of the Royal Anthropological Institute, New Series* (1996): 517–35.

———. "The decomposition of an event." *Cultural Anthropology* 7 (1992): 244–54.

———. "Divisions of interest and languages of ownership." In *Property, Substance and Effect: Anthropological Essays on Persons and Things,* 136–58. London: Athlone Press, 1999.

———. *The Gender of the Gift: Problems with Women and Problems with Society in Melanesia.* Berkeley: University of California Press, 1988.

———. "Potential property: Intellectual rights and property in persons." In *Property, Substance and Effect: Anthropological Essays on Persons and Things,* 161–78. London: Athlone Press, 1999.

———. "Qualified value: The perspective of gift exchange." In *Barter, Exchange and Value: An Anthropological Approach,* edited by Caroline Humphrey and Stephen Hugh-Jones, 169–91. Cambridge: Cambridge University Press, 1992.

Strickland, Stephen P. *Politics, Science and Dread Disease: A History of United States Medical Research Policy.* Cambridge, MA: Harvard University Press, 1972.

Sturdy, Steve. "Reflections: Molecularization, standardization, and the history of science." In *Molecularizing Biology and Medicine: New Practices and Alliances, 1910s–1970s,* edited by Soraya de Chadarevian and Harmke Kamminga, 273–92. Amsterdam: Harwood Academic, 1998.

Tambiah, Stanley Jeyaraja. *Magic, Science, Religion, and the Scope of Rationality.* Cambridge: Cambridge University Press, 1990.

Taussig, Michael. *The Devil and Commodity Fetishism.* Chapel Hill, NC: University of North Carolina Press, 1980.

Thomas, Lewis. *Late Night Thoughts on Listening to Mahler's Ninth Symphony.* New York: Penguin, 1983.

Thomas, Nicholas. *Entangled Objects: Exchange, Material Culture and Colonization in the Pacific.* Cambridge, MA: Harvard University Press, 1991.

Tierney, Patrick. *Darkness in El Dorado: How Scientists and Journalists Devastated the Amazon.* New York: W. W. Norton, 2000.

Trouillot, Michel-Rolph. "Anthropology and the savage slot: The poetics and politics of Otherness." In *Recapturing Anthropology,* edited by Richard G. Fox, 17–44. Santa Fe: School of American Research Press, 1991.

Van Helvoort, Ton. "The controversy between John H. Northrop and Max Delbrück on the formation of bacteriophage: Bacterial synthesis or autonomous multiplication?" *Annals of Science* 49 (1992): 545–75.

———. "What is a virus? The case of tobacco mosaic disease." *Studies in the History and Philosophy of Science* 22 (1991): 557–88.

Ventura Santos, Ricardo. "Indigenous peoples, postcolonial contexts, and genomic research in the late twentieth century: A view from Amazonia (1960–2000)." *Critique of Anthropology* 22 (2002): 81–104.

Verrips, Jojada. "Dr. Jekyll and Mr. Hyde: Modern medicine between magic and science." In *Magic and Modernity: Interfaces of Revelation and Concealment,* edited by Birgit Meyer and Peter Pels, 223–40. Stanford, CA: Stanford University Press, 2003.

Wald, Priscilla. "What's in a cell? John Moore's spleen and the language of bioslavery." *New Literary History* 36 (2005): 205–25.

Waldby, Catherine, and Robert Mitchell. *Tissue Economies: Blood, Organs and Cell Lines in Late Capitalism.* Durham, NC: Duke University Press, 2006.

Weber, Max. *The Protestant Ethic and the Spirit of Capitalism,* translated by Talcott Parsons. 1904. New York: Charles Scribner's Sons, 1930.

Weiner, Annette B. *Inalienable Possessions: The Paradox of Keeping-While-Giving.* Berkeley: University of California Press, 1992.

———. "Reproduction: A replacement for reciprocity." *American Ethnologist* 7 (1980): 71–85.

———. "Sexuality among the anthropologists, reproduction among the informants." *Social Analysis* 12 (1982): 52–65.

Westermark, George. "Anthropology and administration: Colonial ethnography in the Papua New Guinea Eastern Highlands." In *In Colonial New Guinea: Anthropological Perspectives,* edited by Naomi M. McPherson, 45–63. Pittsburgh, PA: University of Pittsburgh Press, 2001.

Williams, Raymond. *Marxism and Literature.* Oxford: Oxford University Press, 1977.

Winkler, Claudia. "Ignoble Nobelman: The saga of Carleton Gajdusek, a brilliant scientist—and an accused pedophile." *Weekly Standard* (October 7, 1996): 16–25.

Wise, Tigger. *The Self-Made Anthropologist: A Life of A. P. Elkin.* Sydney: George Allen and Unwin, 1985.

Wright, Susan. *Molecular Politics: Developing American and British Regulatory Policy for Genetic Engineering, 1972–1982.* Chicago: University of Chicago Press, 1994.

Yam, Philip. *The Pathological Protein: Mad Cow, Chronic Wasting, and Other Deadly Prion Diseases.* New York: Copernicus Books, 2003.

Yoxen, Edward J. "Where does Schrödinger's *What is Life?* belong in the history of molecular biology?" *History of Science* 17 (1979): 17–52.

Zilsel, Paul R. "The mass production of knowledge." *Bulletin of the Atomic Scientists* 20 (1964): 28–29.

INDEX

Berndt, Ronald *(continued)*
241n57; Fore research of, 12, 20–23, 241n5;
on kuru, 3, 81, 161; relations with Fore
and Papuan highlanders, 12, 20–22,
240n50; on sorcery, 23–25, 27, 81
biocolonialism, 219–20
bleeding: as biomedical practice, 70, 74, 95–
96, 154, 182; as Fore cultural practice, 70,
96; as Fore medical practice, 92, 98
blood: collection of, 70, 74–76, 92, 95–96,
99, 154; Fore beliefs about, 92, 96, 98;
gender of, 96; as marketable commodity,
99–100
Blumberg, Baruch, 159, 186, 273n5
Bohr, Neils, 39, 244n16
Bolton, David, 195
Boone, Ken, 211, 278n97
Boyer, Herbert, 192
brains. *See* autopsies; cannibalism; exchange
relations
Brightwell, Mert, 152–54, 163, 172, 266n77,
266n80, 266n83, 270n47
Brown, Paul: and Alpers, 152, 174; and autopsies, 142, 150–52; and Brightwell, 153,
266n83; career of, 265n65; and exchange
relations with Fore, 151–52; and Gajdusek,
185, 217, 221; on prion hypothesis, 197;
and Prusiner, 192–93
BSE (bovine spongiform encephalopathy),
190, 199–203, 205–6, 210, 276n57
Burkitt, Denis, 159
Burnet, F. Macfarlane: aspirations for medical research in New Guinea, 56–58, 60,
85, 125, 131, 182, 248n99, 248–49n2; early
career of, 39, 42, 47–49, 244n14; and Gajdusek, 42, 49–50, 54, 60, 62, 68, 75–76,
83–86, 88, 118, 126, 129–31, 177, 252n82;
Gunther correspondence with, 56, 60,
82–83, 85–86, 118; influence on Gajdusek's research philosophy, 48; on kuru
and cannibalism, 168, 269n23; on kuru
and genetics, 119, 159, 174, 177, 182; and
kuru research, 30–31, 267n105
Burnet, Ian, 59

California Institute of Technology (Caltech),
38–42, 189, 195
Calman, Kenneth, 201
cannibalism: anxieties about, 169, 209–10;
anthropological understandings of, 24–
25, 166–69, 241n66, 256n46, 256n49,
269n26, 270n44; Berndts' understanding
of, 24–25, 241n66; colonial and popular
reports of, 15–16, 201, 241n70; fantasies of
medical, 106–8; Gajdusek's understanding of, 52, 61, 67, 93, 106–8, 169, 185,
269n28, 273n2; Glasses' understanding
of, 166–69; and kuru, 166–177, 181, 185;
as metaphor, 169; ritual, 1–4, 31, 166–69,
201
cargo: Berndts' understanding of, 20–21;
and exchange relations, 92, 94–95, 99,
102, 214; Fore interest in, 12, 21, 23, 69,
72, 163; Gajdusek's understanding of, 53,
94–95, 102; movements of, 16, 25–26, 52,
69, 96, 108, 257n62; scientific, 86, 179.
See also exchange relations
Cavalli-Sforza, L. Luca, 182–84, 273n83
Cilento, R. W., 52
Cleland, Donald, 122
clonal selection theory, 49, 246n58
Collinge, John, 3, 205–6, 208–11, 216, 223,
277n84
Colman, John, 14, 16, 27, 30, 243n95
colonial science: and contact zone, 6, 112–
14; economy of, 220; material cultures of,
6, 112, 237n7; nature of, 88; and trading
zone, 113
commodities, 100–101, 199, 220, 227. *See
also* exchange relations: barter; gifts and
gift economies in; and specimens
Commonwealth Serum Laboratories (CSL),
51, 135
complement fixation test, 54, 169, 247n71
creolization, 113–14
Creutzfeldt-Jakob disease (CJD), 80–81;
BSE compared with, 200; incidence of, in
Britain, 190, 200; injected into nonhuman
primates, 139; kuru connection with, 201,

immunology, 42, 49, 248n101
Inamba, 99, 104, 165, 180, 182, 193–94
incommensurability, limited, 114–15, 257n65
Inglis, Ken, 163, 268n9
Institute for Research on Animal Diseases, 188
Institute of Medical Research (IMR; Goroka), 182–83, 219
Institut Pasteur (Tehran), 45–47
International Biological Programme (IBP), 182

Jenkins, Carol, 219–20, 278n6
Julius, Charles, 61

Kainantu: Berndts at, 18; Fore at, 112, 165; Fortune at, 11; Gajdusek at, 61, 74, 121–22, 129, 223; hospital at, 26, 33, 64, 129, 144–45; during 1930s, 10; rough track to, 12–13; Zigas at, 67, 120, 123
Kelly, W. J., 14, 26
Kerr, Clark, 191
Kimberlin, Richard, 195
Kirkwood, John, 39–40, 244n13
Klatzo, Igor, 80, 135
Kukukuku (highland Anga): on adoption of boys by Gajdusek, 112; Colman on patrol among, 27, 30; and exchange, interest in, 11, 73, 238n10; and Gajdusek, 70–73, 89, 139, 144, 161, 176, 273n85
Kurland, Leonard, 135, 260n65
kuru: and cannibalism, 167–70, 175, 181, 185; and Fore mourning practices, 1, 3, 107, 171; genetic explanation for, 63, 80, 87, 119, 124–27, 131–32, 159, 162, 167, 169–70, 174; infectious explanation for, 60, 62, 159–60, 181; as infectious protein, 190, 198; as "laughing sickness," 81–82, 201; nutritional explanation for, 170; popular press descriptions of, 80–81, 201, 204; psychosomatic explanation for, 3, 28, 81; sorcery explanation for, 1, 3, 4, 14–15, 24–25, 27–28, 30, 59, 61, 81, 92, 167, 170–71, 216; stories of, 164, 166–67; symptoms of,

79–81; transmissibility of, 141. *See also* prions; slow viruses; sorcery
Kutne, Alphonse, 181

laboratory, the, 109–10; commercialization of, 198–99; as "local condensation point," 76; as node between regimes of value, 76; as site of conflict, 143; as site of modern magic, 109
Lander, Harry, 251n59, 253n107
Larkin, Lois (Lois Larkin Baker), 54–55, 78–80, 136, 255n29, 262–63n16
Lévi-Strauss, Claude, 226, 228
Lindenbaum, Shirley: collecting Fore stories, 164–67; and Fore, relations with, 163–66, 268n14; on kuru and ritual cannibalism, 168, 170, 173, 177, 181; and study of social consequences of kuru and sorcery, 170–72; at Wanitabe, 163, 268n10, 268n16
Linsley, Gordon T., 12–14

Mackay, Ian R., 54–55, 84, 177, 179, 271n54
"mad cow" disease. *See* BSE
Masasa, 32–33, 72–73, 93, 104, 112, 213–14, 223
Masiarz, Frank, 194
Masland, Richard, 129, 135
Mathews, John D., 176–81, 271n59, 271n63, 272n70
Mbaginta'o, Ivan (Gajdusek), 111, 223–26, 279n17
McArthur, Norma, 173–74, 270n41
McCarthy, J. K., 10, 86, 224–25, 238n10
Mead, Margaret, 11, 226, 238n12
Moke. *See* Okapa
Moredun Research Institute, 190
Morison, Robert S., 42, 56–57, 85–86, 121, 245n30, 248n99, 258n18
Morris, J. Anthony, 139

Nadel, S. F., 17, 240n34
National Institute of Neurological Diseases and Blindness (NINDB), 129–30, 134–35, 141–42

viruses *(continued)*
    Hall Institute, 47, 52, 60; tobacco mosaic, 40, 189. *See also* bacteriophage; slow viruses

Walsh, R. J., 180, 259n46
Walter and Eliza Hall Institute. *See* Hall Institute
Walter Reed Army Medical Center, 42, 47, 137
Wanevi, 72, 223
Wanitabe, 31–32, 151, 163–68, 170–72, 209, 268n10
Watson, James D., 39–41, 244n18
Wegstein, Joe, 224, 279n23
Weissmann, Charles, 195–99
Wellcome Medical Museum, 132
Wellcome Trust, 205, 206
West, Harry, 27, 32
White, H. N., 123, 260–61n65

Whitfield, Jerome, 206–11, 234
Wilson, Edwin B., 38, 243n8

Youden, William J., 35, 230

Zigas, Vin (Vaclovas), 3, 29, 31, 58, 60, 77, 107, 145; career of, after kuru research, 260n62; and collaboration with Gajdusek, 61–64, 67, 80, 85, 107, 117, 119, 128, 141, 176; Fore perceptions of, 92; and kuru, initial investigations of, 27–28, 30, 64; memoir by, 29, 242n85
Zigas, Vin (Vaclovas), relations with: Alpers, 264n50; Baker, 67–68, 118; Fisher, 122; Fortune, 161; Gajdusek, 61, 117–18, 120, 123; Gray, 122–23; Gunther, 30, 86–87, 118, 120, 251n70, 253n105; Hornabrook, 270n47; Robson, 117, 120; Scragg, 120, 123; Simpson, 259n33